T0329838

Dis-positions: Troubling Methods and Theory in STS

Series Editors: **Mike Michael**, University of Exeter and **Alex Wilkie**, Goldsmiths, University of London

---

Turning the mirror on science and technology studies, this pioneering book series explores the pivotal changes in the discipline. It occupies a unique position in the field as a platform for adventurous projects that redraw the disciplinary boundaries of STS.

Forthcoming in the series

Adventures in Aesthetics:
Rethinking Aesthetics beyond the Bifurcation of Nature
Edited by **Melanie Sehgal** and **Alex Wilkie**

1000 Platforms
Adrian Mackenzie

Find out more at

bristoluniversitypress.co.uk/dis-positions

ECOLOGICAL REPARATION

Repair, Remediation and Resurgence in Social and Environmental Conflict

Edited by
Dimitris Papadopoulos, Maria Puig de la Bellacasa
and Maddalena Tacchetti

First published in Great Britain in 2023 by

Bristol University Press
University of Bristol
1–9 Old Park Hill
Bristol
BS2 8BB
UK
t: +44 (0)117 374 6645
e: bup-info@bristol.ac.uk

Details of international sales and distribution partners are available at bristoluniversitypress.co.uk

British Library Cataloguing in Publication Data
A catalogue record for this book is available from the British Library

ISBN 978-1-5292-3954-6 hardcover
ISBN 978-1-5292-3955-3 paperback
ISBN 978-1-5292-3957-7 ePub
ISBN 978-1-5292-3956-0 ePdf

Cover design: Andrew Corbett
Front cover image: Photograph by Dimitris Papadopoulos, 2021, from unEcology in the Heart of Rural England
Bristol University Press uses environmentally responsible print partners.
Printed and bound in Great Britain by CPI Group (UK) Ltd, Croydon, CR0 4YY

Contents

List of Figures

Notes on Contributors

Abisola Balogun-Katung, Newcastle University, UK

Cristobal Bonelli, University of Amsterdam, The Netherlands

Patrick Bresnihan, Maynooth University, Ireland

Patrick Brodie, McGill University, Montréal, Canada

Linda Brothwell, artist, Bristol, UK

Steven D. Brown, Nottingham Trent University, UK

Juan Camilo Cajigas, Pontificia Universidad Javeriana, Bogotá, Colombia

Nerea Calvillo, University of Warwick, UK

Ruth Catlow, artist, Furtherfield, UK

Marco Checchi, Northumbria University, UK

Timothy Choy, University of California, Davis, USA

Donna Ciarlo, London South Bank University, UK

Leila Dawney, University of Exeter, UK

Marisol de la Cadena, University of California, Davis, USA

Jérôme Denis, MINES ParisTech, France

Vanessa Farr, University of Cape Town, SA

Andrea Ghelfi, Università di Firenze, Italy

Lesley Green, University of Cape Town, SA

Eleanor Hadley Kershaw, University of Exeter, UK

Sara Heitlinger, City, University of London, UK

Lara Houston, Anglia Ruskin University, UK

Steven J. Jackson, Cornell University, New York, USA

Santiago Martínez-Medina, Universidad Nacional de Colombia, Bogotá, Colombia

Naomi Millner, University of Bristol, UK

Fredy Mora-Gámez, University of Vienna, Austria

Atsuro Morita, Osaka University, Japan

Alejandra Osejo, Rice University, Houston, USA

Dimitris Papadopoulos, University of Nottingham, UK

Constantin Petcou, atelier d'architecture autogérée (AAA), Paris, France

Doina Petrescu, University of Sheffield, UK

David Pontille, MINES ParisTech, France

Maria Puig de la Bellacasa, University of Warwick, UK

Paula Reavey, London South Bank University, UK

Eliana Sánchez-Aldana, Universidad de los Andes, Bogotá, Colombia

Martin Savransky, Goldsmiths, University of London, UK

Sam Siva, Land in Our Names, UK

Maddalena Tacchetti, University of the West of Scotland, UK

Alex Taylor, City, University of London, UK

Manuel Tironi, Pontifica Universidad Católica de Chile, Santiago, Chile

Kazutoshi Tsuda, Kyoto Institute of Technology, Japan

Aristotle Tympas, National and Kapodistrian University of Athens, Greece

Claire Waterton, Lancaster University, UK

Acknowledgements

During the preparation of *Ecological Reparation* we captured some of the ongoing conversations with this volume's contributors in a series of video interviews that discuss their work and its relevance for remediating and repairing as well as claiming reparations for damaged more than human ecologies.

These videos are accessible on the YouTube channel *Ecological Reparation* www.youtube.com/c/EcologicalReparation.

The production of the channel would not be possible without the videographic work of Gerard Ortín Castellví and Sangbum Ahn who created the interview capsules. We are deeply grateful to them for their inspiring ideas and their invaluable contribution in shaping the aesthetic and intellectual orientation of this project. Heartfelt thanks go to Gerard Ortín Castellví for designing the logo of the channel. We are also thankful to Juliana Mainard-Sardon and Giulia Champion for supporting the production and dissemination of the channel. A big thanks goes to Hanna Rullmann, Oriol Campi and Tom Fischer for support with sound recording and video editing.

Special thanks got to Paul Stevens, our editor at Bristol University Press, for generously supporting this complex book project from its very beginning, to Georgina Bolwell and Emma Cook for their creative editorial support, and to our series editors Mike Michael and Alex Wilkie.

Dimitris Papadopoulos and Maddalena Tacchetti gratefully acknowledge the support of the Arts and Humanities Research Council, UK (grant number AH/R013640/1). Dimitris Papadopoulos gratefully acknowledges the support of The Leverhulme Trust, UK (grant number RF-2018–338\4) and Maria Puig de la Bellacasa the support of the Arts and Humanities Research Council, UK (grant number AH/T00665X/1). We are grateful to the Institute for Science and Society and the School of Sociology and Social Policy, University of Nottingham, UK and the Centre for Interdisciplinary Methodologies, University of Warwick, UK for their support.

Note on the Figures

We wish to acknowledge that, though many of the figures included in this book are of a low resolution, they are integral to the authors' content. The decision to publish these figures also accepts that these are the best versions available. For those reading in digital formats, it will be possible to zoom in on any image to achieve greater clarity.

Dis-Positions Series Preface

The aim of the Dis-Positions book series is to bring together recent work in the emerging intersections of such disciplines as sociology, anthropology, history, geography, design and philosophy, not least as they bear on the broad field of science and technology studies (STS). Conversely, if STS is undergoing major shifts in how it engages with 'the social' and the question of 'societies', it raises vital matters of concern for these various disciplines and their inter-connections. Dis-Positions thus provides a platform on which varieties of generative mutualities across these areas of scholarship can be presented. In this respect, Dis-Positions is undergirded by a desire to promote novel fields of inquiry, adventurous theoretical and empirical projects and inventive methodological practices. It seeks to encourage authors to address live debates, while drawing on and interrogating developments across academic areas, in the process disturbing and repatterning STS.

In pursuing this ethos, Dis-Positions comprises a consolidated, rigorous and proactive space through which creative and critical new perspectives in STS and beyond can find a voice. Under this rubric fall: discussions of the posthuman, post-colonial, affective and aesthetic; methodological inventions that incorporate speculative, engaged, entangled, and sociomaterial practices; empirical novelty that ranges from emergent technoscientific innovations to reformulations of the ordinary; and conceptually creative and critical developments that capture processual and pluralistic thought, extensions of assemblage and practice theories, and the turns to affect and post-performativity.

We are immensely proud to have the superb edited volume by Dimitris Papadopoulos, Maria Puig de la Bellacasa and Maddalena Tacchetti, *Ecological Reparation: Repair, Remediation and Resurgence in Social and Environmental Conflict*, as the inaugural volume of the Dis-Positions book series. *Ecological Reparation* is concerned with how to rethink social-environmental degradation so that we are better equipped to respond to – remediate and recover – the damaged ecologies that comprise a deteriorating global context. However, the volume is not a typical example of academic practice: rather, it is also immersed in the endeavours of social and environmental justice movements and policy. Indeed, together these latter serve as a necessary backdrop of, and an intended constituency for, *Ecological Reparation*. It is no easy task to marry these analytic, practical and political objectives but Papadopoulos, Puig de la Bellacasa and Tacchetti have done a superb job of pursuing these aims within and across the various chapters that make up *Ecological Reparation*.

In several ways, this collection encapsulates the principles underpinning Dis-Positions. The editors have to be congratulated for so successfully collecting contributions that span such an impressive range of conceptual, empirical, practical and disciplinary approaches. These contributions – innovatively and provocatively arranged around a series of concepts-in-tension – each focus on ecological reparation, but in ways that open it up through diverse ethnographic, artistic, analytic, political and practitioner perspectives. Drawing expansively on a variety of textual and pictorial forms, and on a wealth of data, materials and resources, the volume cumulatively exposes the mundane, if precarious, processes and practices whereby ecological reparation in its multiplicity, and through its many relationalities, comes to take shape and, at least potentially, comes to be actualized in specific settings. Insofar as this collection attends to these everyday moments of ecological reparation, it can inspire 'new imaginaries of possible liveable futures' and 'a renewed sense of hope' about human and nonhuman mutual ecological embeddedness. In its complex multi-disciplinarity, in its diverse empirical, methodological, theoretical, practical and generic provocations, and in its formal and substantive audacity, *Ecological Reparation* embodies the ideals of the Dis-Positions book series.

Mike Michael,
University of Exeter
Alex Wilkie,
Goldsmiths, University of London

Introduction: No Justice, No Ecological Peace: The Groundings of Ecological Reparation

Dimitris Papadopoulos, Maria Puig de la Bellacasa and Maddalena Tacchetti

Ecologies, repair, justice

There is an urgency to face the precipice of ecological devastation, take a leap into the fractures of a broken Earth, and experiment with more sustainable and just ways of living. And yet, in spite of incessant calls to attend to multiple fronts of emergency, major obstacles remain. One of them, as many analyses of the hope-shattering failures of COP26 and previous climate agreements have shown (Kutney, 2014; Rosen, 2015; Spash, 2016; Arora and Mishra, 2021), is the recognition of the inescapable entanglements of ecological care with care for people. No justice, no *ecological* peace. Attempting a modest contribution to efforts addressing this situation, *Ecological Reparation* engages with social-environmental degradation by trying to rethink concepts and practices that may be needed to repair and to remediate both damaged ecologies and persistent inequities in ways that support resurgence against more than human injustice. This book takes up this task from a diversity of theoretical and political fronts, unpacking some of the workings at stake in the conceptual coupling of the *ecological* with *reparation*.

The intellectual and political grounds upon which this book has grown are nurtured by a surge of thinking about new multifaceted and generative meanings for *ecology*. Encompassing traditional scientific meanings, fostering 'the ecological' has come to signify an embodied understanding of the highly specific and worldly environmental, socio-political and mental connections that situate different beings, human and non-human (Guattari, 1995) – or, perhaps, in more abstract conceptual undertakings, connections between multiple forms of matter and experience within interdependent environments (Hoerl and Burton, 2017; see also Kingsland, 2005; Herzogenrath, 2009; Morton, 2010; Wall, 2014; Papadopoulos, 2022). Exploring a sense of general ecology has promoted the decentring of the human and confronted human exceptionalism, yet increasingly too the revival of ecological thought has been challenged by perspectives that insist that all more than human affairs exist precariously and are marked by situated histories of social, gendered, racial, colonial and spatial conflicts (Nightingale, 2011; Wright, 2018). In other words, there is no such thing as a neutral ecology for people and nonhumans; ecological interventions pose issues of justice at their core, even when these are difficult to apprehend from where one stands (Plumwood, 2008).

Socio-political conflict and the need to restore justice are in turn inherent requirements to the notion of *reparation*. Most commonly, however, reparation has remained within a perspective of social wrongs, typically referring to measures of reconciliation – such as monetary compensations, ceremonial commemoration, or top-down policy interventions – in the context of war and human rights abuses (Walker, 2010; White, 2017). Critics have focused on how these are often bypassing strong claims for reparative and restorative justice (Mora-Gámez, 2016; Macleod et al, 2017; Macleod, 2019). Moreover, both state- or community-based reparation models start and end in a socio-humanist concept of the individual or collective subjects of reparation. As a result, reparation is most often regarded as making amends to right prior wrongs against human groups, and while reparations may include or have effects in ecological relations that constitute a human 'environment', even eco-social reparations such as land reparations are mostly conceived as responding to damage perceived first and foremost from the perspective of human collectives.

Coupling ecology with reparation brings attention to the interconnectedness between ecological and reparative obligations. Such claims and actions of, and for, ecological reparation do exist already in different ways. Addressing environmental degradation and species extinction through ecological restoration measures, for example, aims at assisting the recovery of ecosystems and biodiversity that have been degraded, damaged or destroyed. This is a form of ecological reparation done in the spirit of also restoring justice by benefiting people who live from and care for such ecosystems, and whose lives have been the most affected by persistent ecological loss and damage. It is most prominently Indigenous peoples around the world who are involved in these struggles and in articulating these interdependent relationships, while continuing to uphold the memories and battling the effects of interlinked ecocides and genocides. Standing for justice that does not separate human from non-humans, Indigenous thought and struggles are defending alliances of people and ecologies, and the inseparability of Earth/place belongings with cultural belongings (Gilio-Whitaker, 2020; Whyte, 2022). Ecological reparation involves a recognition that harm and recovery of environments and human societies are intrinsically linked. As Leila Darwish (2013: 4) says, the lands and relations that need repairing and remediating are the same that have been devastated through acts of colonization and environmental racism against Indigenous, Black and People of Colour; and against those uprooted by means of displacement, dispossession and land grabbing; and against those living in areas targeted as sites for military bases, waste dumps or heavy industrialization. Ecological reparation in this way requires solidarity as well as recognizing the lead of those who have remained guardians of ecological relations with non-humans at the peril of their own lives and the safety of their communities.

This book attempts to participate in the collective conversations engaged in fostering liveable connections between people and ecologies from where we stand, as three editors in the Global North with specific intellectual and political inheritances. A fundamental premise in this book is that we are facing conditions in which restoring lost value is simply impossible, inasmuch as socio-ecological devastation has reached a critical point for which no reparative measures could possibly make up for the losses and damages of the environment and human societies (Cairns, 2003; Jackson and Hobbs, 2009; Hale et al, 2014; Perez Murcia, 2014). And so, while 'romanticised reparations' (Cadieux et al, 2019: 649), may mean longing for the return to an idyllic before, this collection acknowledges that there is no ecologically neutral state or innocence to begin with: we carry the marks of colonial (Ferdinand, 2019), racist (Cole and Foster, 2001), and sexist (Di Chiro, 2008) injustice and violence, deeply intertwined with resource extractivism and environmental destruction (Cairns, 2003). Our second fundamental premise is the interest we share in investigating the hopes and possibilities in non-innocent ways for ecological reparation from the perspective of the fragile everyday materialities that involve humans and non-humans. Here notions of care and repair become central because appreciating the intrinsic fragility of worlds we are inhabiting allows us to illuminate another dimension of reparation: the maintenance of life through subtle and often 'minor acts' (Deleuze and Guattari, 1987) and processes of repair, healing, care and reinventions by humans and non-humans, despite the ongoing, and possibly inevitable, experiences of breakdown (Caney, 2006; Jackson, 2014; Cadieux et al, 2019; Denis and Pontille, 2019; Puig de la Bellacasa, 2022).

A focus on minor acts of care and repair also turns attention towards alternatives to grand-scale projects both of environmental remediation *and* state reparation that aim at redressing ecosystem damages and human rights violations by focusing on one dimension only – lands, environment, humans, nonhumans, *or* their communities – potentially affecting close and distant ecological relations in unpredictable ways. The coupling of ecological relations with reparations could then also indicate a move beyond top down, individualized and abstracted interventions, towards processes of repair of material relations in their complex embeddedness in communities. We hope that opening fields of inquiry into ecological reparation may point at the ability of ecologies both to recover from socio-environmental destruction and for ecological thinking to nurture alternative ways of restoration and maintenance of the cohabiting conditions for humans and non-humans. This is not only an ethical imperative to right prior wrongs, but a theoretical, practical and everyday move, necessary for continuing living on a damaged planet (Gan et al, 2017). All the various contributions to *Ecological Reparation* provide further insights into alternative ways of understanding both ecological relations and their reparation, conceived as *repair and healing* of what has been harmed by different forms of conflict, as well as the reclaiming of (rights and life in) such ecologies, by bringing to light voices, knowledges and practices of a diversity of peoples and non-humans.

Ecological reparation as transformative movement

This book brings social movement actions and practices into discussion with committed academic research, in particular by new forms of social and environmental justice mobilizations that are moving away from simply lobbying or protesting against recognized institutions and governance. Instead, the focus turns to practice-based community projects aimed at directly modifying the socio-ecological conditions of life in intersected ways beyond a designated issue or affected group (Papadopoulos, 2018). These 'more-than-social-movements' (Ghelfi and Papadopoulos, 2022) enact ecological reparation in its twofold meaning of both repairing and reclaiming the battered ecologies they inhabit, and by materially taking back spaces and practices that they have been denied by years of socioeconomic injustice. They make reparation a matter of justice for the socio-ecological damages against humans, more-than-humans and environments. We are thinking of mundane processes and practices of care and solidarity that recover and craft alternative 'ecologies of existence' (Tacchetti et al, 2022), actioning reparation as direct repair of human-nonhuman relations in which they are involved beyond, or alongside, more visible institutionalized forms of reconciliation and state reparation.

More-than-social-movements are not only the background and topic of the contributions to this edited volume but also the horizon. Along with a novel theoretical/conceptual framework, *Ecological Reparation* engages with socio-ecological mobilizations and critical policy developments as sites of concrete interventions and transformations. As such, the volume generates a dialogue between academic and social and environmental justice movements by bringing together scholars investigating practices of repairing damaged ecologies through various societal, environmental, material involvements and across different locations and geographies. It redirects attention from more visible players in processes of institutional social and ecological reparation towards all those who are typically absent or silenced in constituted political settings as well as in mainstream media and academic debates and who are quotidianly engaging precarious practices of repairing their and our worlds: rural labourers, Indigenous communities, urban activists, grassroots infrastructure maintainers, practitioners of ecological transitions in the Global North and South, but also inorganic actors, natural elements, animals, the dead, spirits, objects and landscapes. While the volume addresses socio-ecological struggles, it also wants to celebrate the inspiration of insurgent and inventive modes of conservation, mending, care and empowerment of more than human ecologies.

Life continues in the interstices and cracks of broken ecologies. Resurgence is fuelled and maintained by the humans and non-humans that remain despite having been exploited and marginalized along with the ecologies they inhabit. Ecological degradation is always a local phenomenon but the forces that sustain it crisscross different locations and regions. And yet, it affects us all in deeply intersected ways as, increasingly, most regions of the planet are facing the threat of the ecological crisis. As the various interventions in the book show, ecological reparation is enacted and endured in differential ways in different places. Everywhere we witness continuous experimentations with matter, processes and practices, as well as temporalities and rhythms offering alternatives to dominant, quick techno-scientific fixes that accelerate large-scale transformation processes of enclosure, accumulation, extraction and ultimately destruction (Papadopoulos, 2018). By paying attention to and engaging with neglected experiences, *Ecological Reparation* illuminates the ethics of care and responsibility underlying repair not only conceived as a way of patching up what has been damaged, but also as creating and sustaining resurgence through practices that encourage novel alliances, sensitivities, affectivities, intimacies and material relations.

We did not seek with this book to achieve a single definition of ecological reparation, but to incite a collective exploration around the notion by opening it to a range of perceptions and practical-ontological approaches. There was no imposed brief, but an invitation to engage with the potentially multidimensional meanings of ecological reparation. Some contributors explore ecological relations as an entry term into more than human relationalities; others address ecology as expressing a specific form of interconnected materiality; for some reparation is mainly about justice; and others explore it as one of the meanings of repair. The authors have diverse academic backgrounds – such as critical science and technology studies, human, environmental and cultural geography, environmental humanities, architecture, art, design and crafts, political economy, social theory, anthropology. Topics vary from practices of maintenance and repair to the making of the material commons, from alternative ways of farming, such as agroecological and permaculture practices, to socio-anthropological research on the meaning of the 'ecological' in different settings, from mental health institutions to the streets as sites of protest.

Each one of the contributions to *Ecological Reparation* engages with a distinct topic in novel and sometimes experimental ways. The volume contains pieces of different formats and styles: research papers, shorter creative provocations, a graphic novella and a photo-essay. Our contributors reflect existing conversations intersecting academic and para-academic work in East Asia, Europe, Latin America, North America, and South Africa. As a result, the book is a collection of visions that offers a kaleidoscopic conversation; an expressive research approach that illuminates what the unexplored field of ecological reparation may look like when allowed to emerge as a distinct area of research and practical inquiry from a multiplicity of diverse theoretical, empirical and practical perspectives. In structuring the book in nine distinct parts, we have in some way imposed our own situated path of connections and understanding but we hope readers will also explore their own ways of connecting. The chapters are organized around different threads, which open lines of inquiry to define and redefine the meaning of ecological reparation by combining contributions from different fields of inquiry in nonlinear ways. The threads – as detailed next – are composed as a series of conceptual pairs in tension, denoting countervailing forces, processes and movements of damaging and repairing, which co-exist in non-innocent troubled ecologies. The various sections tackle different aspects of their interplay through multiple contributions, which allow us to navigate the meaning of ecological reparation from different perspectives and in different contexts.

Overview

Part I Depletion<>Resurgence explores ecological resurgence in contexts where soil and land depletions are intimately connected with social injustice. In the first contribution, Naomi Millner draws on her activist and research work with agricultural *campesinos* (peasants) in El Salvador who experiment with alternative agricultural practices, combining permaculture, agroecology and recovered Indigenous practices, to regenerate degraded soils. Her piece brings to the fore how in these processes of repairing ecologies it is not only the future of food production that is at stake but also collective healing by sharing experiences of the wider eco-social trauma of colonial histories and genocidal violence. Soils here emerge as archives of material, human and ecological memory of past injustice. Millner emphasizes how experimentation with biofertilizer techniques based on composting to restore and replenish soil capacities becomes an opportunity for *experience* to be reconsidered as *expertise*, for ways of knowing that have been marginalized to be recognized, in order to promote new ways of commoning that foster a reparative ecology around an ethos of decolonial healing.

In the second contribution, Manuel Tironi speculatively pauses on the hesitation that arises when different narrations of what ecology means and how reparation is done collide. He offers three equivocal theses/meanings in practices of ecological reparation by exploring the role of spectres, (unthinkable) politics and abundance restitution. Engaging with diverse ethnographic stories of Indigenous communities inhabiting ecosystems damaged by extractivism, industrialism and settler colonialism in Chile, from the Tubul-Raqui wetlands to the high plains of the Atacama salt flat, Tironi invites new ways of approaching how divergent instantiations of the labour of reparation and its decolonization might encounter and empower possibilities for more than human interdependencies, that may be experimentalized by alternative Indigenous communities and non-anthropocentric sustainable transitions.

In the last contribution of Part I, Lesley Green and Vanessa Farr turn to material flows – circulating water and sewage – as a mode of attention that exacerbates the awareness of socio-environmental spatial divides for humans and non-humans in Cape Town. We learn about a Vlei (lake), near a site designated as a wetland of international importance, where migrating birds make their stops and Cape Town's only population of hippopotamuses live. The lake is about a kilometre from the major solid waste dump of southern Cape Town, and also in the middle of the Cape Flats where about a million people live mostly on or below the breadline. Zooming out from a 'Google map eye view' to a ground view – 'from pixels to soil' – the authors interrogate the toxicities affecting the flow of the water from the mountain into the aquifer, and via the two local rivers that feed it (both of which are highly polluted urban rivers). Green and Farr identify different conservationist approaches involved in managing these flows, constrained by the city's neoliberal politics – including efforts to cover up freshwater quality data. They highlight the need to shift away from financialized calculations of ecosystem services in urban environments and the creation of sacrifice conservation zones, as they explore the potential of expanding a 'critical zone' approach to human systems that may expose the necropolicies at play in the management of waste flows across water – in mountains, aquifers, rivers and lakes.

Part II Deskilling<>Experimenting investigates experimental practices of repair that embrace and cultivate different skills and practices that have been gradually lost in the dominant individualizing skill-based working cultures of Global North societies. All interventions highlight the importance of transversal experimentations with a multiplicity of techniques, materials, knowledges and practices as a key ingredient of ecological reparation. The first contribution focuses on ongoing mending in the significant area/acreage of grass pastures in the UK which are used to feed beef and dairy cattle. Drawing on interviews with a group of UK farmers who call themselves 'Pasture for Life' farmers, Claire Waterton explores how they are experimenting with novel ways of feeding and moving their herds around, in daily or half-daily rhythms, around the pasture of their farm holdings, and how these practices and everyday actions transform relationships within the farmer-cattle-ecology of the farm in new creative and sustainable ways.

Atsuro Morita and Kazutoshi Tsuda explore attempts to reconcile nature and infrastructures in Japan through figures and practices of ecological circularity. As a country that has suffered from economic and population decline in the past decades, sustainability in Japan concerns not only fighting climate change but also the very possibility of the survival of rural communities and provincial cities. Exploring the Fab Lab movement's vision and global efforts to scale grassroots techno-activism up to an urban scale, this contribution focuses on the close relationship between a circular imagination about material flows that reduce energy use and waste and the technical capabilities of citizens. With a specific approach to the circular flow, the Fab Lab movement sees nurturing technical capabilities of citizens as essential for this purpose.

In the last contribution to Part II, Martin Savransky asks how to think about 'repair' when what may need reparation or regeneration is the very fabric of reality, and the arts of living, the more-than-human modes of evaluation by which lives become worth living and deaths become worth living for. Such a question, he suggests, demands perhaps what Nietzsche called 'philosophising with a hammer': experimenting into another mode of thinking. He tests this approach by thinking with practices that emerged in the wake of the 2011 Great East Japan Earthquake and Tsunami disaster to care for those afflicted by loss and grief and who were confronted with the emergence of ghosts. By learning how to tell their stories otherwise in order to regenerate their collective capacities to think, feel and imagine, the survivors recompose not only an ecology, but a whole cosmos on a devastated Earth. Savransky elaborates a concept of 'Cosmological Reparation' that emerges in the incommensurable space opened up between destruction and redemption, between loss and compensation.

Part III Contaminating<>Cohabiting examines cohabiting with non-human others as a form of repair in contaminated ecosystems. Eleanor Hadley Kershaw examines two approaches to mending damaged ecologies with microbes. First, therapeutic efforts to 'rewild' human microbiomes through the restoration of mutualistic microbes (bacteria and helminths) to treat diseases of microbial absence. Second, synthetic biology endeavours to engineer microbes to assimilate greenhouse gases and produce value-added chemicals in industrial ecosystems. Turning to the analytical and methodological lens of 'ecologies of participation' Kershaw considers ways of accounting – and making space – for these diverse modes of engagement with microbial mending.

Nerea Calvillo explores alternative approaches to air pollution that instead of considering air as a resource 'out there' that can be isolated, abstracted and fixed, attend to it as part of complex situated involvements of interdependent entities. In this logic, even remediation projects focused on the repair of one element or function can be damaging in other ways. Calvillo develops a notion of 'involvement' as analytical tool to think about human/non-human entanglements by telling the story of the troubles encountered during the setting up of an experimental installation dedicated to make air pollution visible ('Yellow Dust', for the 2017 Seoul Biennale of Architecture). Her piece engages with the air's materiality, and the possibility of its possible repair, as part of ecological re-compositions that intertwine a range of materials, spaces, technologies and humans, making visible the ethical, political and social implications of always messy and uncertain processes of repair.

Tim Choy literally *draws* our attention into a museum of breathers, maintained by an autonomous robot who gathers anything that fits the 'curatorial cut' of the 'breathing condition'. The museum's curatorial effect is a prompt to inhabit the simultaneous sharedness and irreducible differences of co-breathing as an ongoing condition of complicity and conspiracy for Earth-bound living. As the robot curator rescues a new 'specimen' – a lab rat – to add to the museum and convinces her to join, he discovers the breathers have escaped, into a forest, evading the discipline of the Museum of Breathers to become a *MoB*. The background of this piece is Choy's ongoing interrogation of what it means to share an unequal atmosphere as a conspirator or co-breather with others (Choy and Zee, 2015).

Part IV Enclosing<>Reclaiming Land scrutinizes practices of effective access to, use of and control over land for community food growing as well as practices of land reparation that drive the emergence of new peasantries and new rural forms of life in contexts damaged by land grabbing and commons enclosure. Sam Siva, from the British Black and People of Colour collective *Land in Our Names* (LION), presents LION's work and unpacks the centrality of land reparations to ecological reparation. Supporting Black and People of Colour in reconnecting with land through food growing and claiming access to nature, LION puts access to land at the centre of ecological justice. Siva affirms the interconnectedness between ecological repair, healing and concrete reparations for damages done to people and non-humans through colonialism and extractivism while emphasising the centrality of racism to these enterprises and their persistent inheritances in the present. Together, this text and a follow-up interview with the author connect the personal and the political, struggles and hopes, to demonstrate the deep connection of exclusionary racial dynamics to contemporary concerns of justice at the heart of ecological reparation – such as climate change, the often exclusionary whiteness of the countryside, the rise of malnutrition and poor mental health among marginalised social groups.

Patrick Brenishan and Patrick Brodie examine how landscapes and communities in the midlands of Ireland that have been ruined by industrial extraction engage with these legacies materially and imaginatively. Drawing on empirical work with communities, where industrial peat extraction has been taking place since the 1940s, leaving vast areas of drained, cutaway bog inhospitable to most plants and animals, the authors explore how remnants of past ecologies and practices might be salvaged and repurposed for the making of life after fossil fuels. The contribution ultimately traces out some forms of bog life – bog commons, bog ecologies and bog modernity – that unsettle colonial and post-colonial logics and temporalities of industrial extraction and green storage infrastructure of carbon emissions, disrupting linear accounts of both progress and ruination.

Andrea Ghelfi offers an analysis of the practices and semiotics of *Genuino Clandestino*, a network based in Italy that was created in 2010 to defend and reinvent a plurality of forms of rural existence. Ghelfi presents three key dimensions of becoming a peasant in this context: farming as ecological practice, food communities that link the countryside and urban spaces, and the civic and collective use of land. Through an analysis of practices and semiotics proposed by the new peasantry movement, this text offers a novel look at agroecology as a science, a practice, a form of life and a movement simultaneously.

Conversely, in the final contribution to Part IV Juan Camilo Cajigas describes how small farmers and activists of agroecological collectives in the Tequendama region of Colombia contest received notions of development, place and environmental activism. Cajigas argues that exploring the meanings of these practices as forms of 'obey life'-politics that require thinking outside the liberal frameworks of nature conservation. Generating material conditions for practices of reparative ecology, these agroecological responses to the ongoing rearrangement of Earth's material forces reconfigure modern understandings of environmental citizenship. Rather than taking grounds on individuality, universal thinking, self-interest and antagonism among political actors, here obligation is defined as a relation established by the unfolding dynamic of place.

Part V Loss<>Recollecting examines mending through remembering practices, with a particular focus on the strategies of relating to and being affected by memorially significant artefacts. Steve Brown, Paula Reavey, Donna Ciarlo and Abisola Balogun-Katung examine memorial work around mental health as repair work grounded in specific ecological material settings. Based on a study on practices of recollection among mental health services users detained in a secure forensic psychiatric unit in the UK, the authors explore how memory sustained by an ecology of artefacts and objects comes to matter to patients who live in conditions that operate as a 'system of forgetting' and damage the connections between past and present that make life meaningful. When remembering is traumatic or conflictual in some way, it has the potential to become an act of repair.

The chapter that follows, by Leila Dawney and Linda Brothwell, is the result of a conversation between the authors connecting research on the spatiality of austerity and deindustrialization with work of repairing damaged objects in public spaces. They show how the slow, spatial violence that has eroded material aspects of public life in the UK (schools, streets, parks, libraries) has deteriorated the sense of a shared life and of being cared for, or being held by, public objects and spaces. Focusing on Linda Brothwell's public bench repairs artistic work, the authors argue that small acts of care for objects can be considered as rituals of mourning for a lost public, as hopeful minor acts that offer consolation in the face of ongoing structural and democratic destruction and can demonstrate alternative approaches to the shared life, responsibility and belonging.

Finally, by drawing on ethnographic data describing interconnected material practices of commemoration and crafting, Fredy Mora Gámez explores a novel line of inquiry, at the intersection between social movements studies and science and technology studies, around the notions of infrastructural repair and everyday experience. The piece describes alternative forms of reparation and solidarity in the streets of Bogotá (Colombia), such as street memorials by social movements and the artisanal work of Venezuelan migrant communities using de-valued Bolivars (Venezuelan bank notes) to make crafted objects. Mora Gámez reads these worlds as exceeding governmental reparation and asylum solidarity, engaging bodies in the material politics of alternative self-organised forms of reparation and solidarity.

Part VI Representing<>Self-governing presents different examples of self-sustaining and self-governing ecologies that exist in precarious urban contexts. Doina Petrescu and Constantin Petcou explore how the strategic design project R-Urban enables mending ecologies in deprived suburban neighbourhoods. R-Urban seeks to create possibilities for a commons-based mending ecology through a network of civic hubs in which inhabitants can develop resilience practices and create locally closed ecological circuits. The authors give a critical account of the successes and difficulties encountered by the five R-Urban hubs built since 2011 in Paris and London, trying to demonstrate how creating commons grounded on relations of collective care, regeneration and resilience can be a workable pathway for repairing ecologies in deprived urban contexts.

Marco Checchi, on the other hand, discusses various forms of repairing at place in Ri.Maflow, a workers' recuperated factory in the north of Italy. After the bankruptcy of the Maflow Factory, workers occupied the site and founded a cooperative inspired by the principles of mutualism and solidarity. For Checchi, the occupation of the factory with new productive and sustainable activities is an example of how workers manage to repair an ecology damaged by economic, social and environmental crises. His account also shows how occupying the factory can be seen as a way of directly taking back compensation for the damage suffered by both this eco-social territory and its inhabitants.

Finally, in the last contribution to Part VI, inspired by recent protests in Chile, Cristobal Bonelli and Marisol de la Cadena seek to generate ethnographic concepts capable of thinking with, and expressing, the eventfulness of the streets – literally so. Relaying the immanent stories told by Chilean streets during the protests that erupted in 2019, the authors experiment with images, texts and figures that irrupted in walls and their surroundings. They read street images as emerging from wounds inflicted by colonialism-capitalism on bodies and territories, as presences that haunt politics as usual and that configure an open-source archive against the grain of history. This is an archive that cannot be contained by taxonomic categories and thus enacts an ongoing demand for reparation.

The three contributions in Part VII Isolating<>Embodying explore mending ecologies as a socio-ecological craft which is realized in bodily and affective interconnections with human and non-human others. Eliana Sánchez-Aldana investigates textile crafts performances that accompany grief by understanding them as practices through which one makes-unmakes the self. With a craft-text piece, that weaves multiple times, voices and theoretical threads, she tells us a story of ecologies and attachments emerging within the encounters between textile objects and the human: between her manual loom, the threads, the weaver, and her process of mourning as she remakes the fabric and herself. What emerges is an understanding of a hand-made ecology weaving textile and human together, in which craft is both verb and noun.

Jérôme Denis and David Pontille highlight the attention to fragility at work in two situations of urban maintenance: graffiti removal and water network maintenance. Maintainers' attention is deployed through sensorial explorations that are both routine-based and open to surprises specific to emergent material ecologies. During their inquiries, maintainers compose with a material ecology in which the various substances they deal with, the tools they use, and their own bodies interact with each other. Articulated through theoretical, regulatory and contractual documents, this attention to fragility is both a matter of perception and politics. Paying attention to maintainers' continuous attention, Denis and Pointille approach reparation as a process that cannot be 'settled' once and for all, emphasizing the infra-ordinary fragility of things that exist beyond experiences of breakdown and disaster.

Finally, Aristotle Tympas discusses the contrasts between wind energy production based on small-scale and local storage and, on the other hand, wind farming which relies on large-scale generation, long-distance transmission, and profiting through consumption. Drawing on historical observations of evolving wind structures and technologies, he makes explicit a difference between open and closed wind energy structures. The latter are most present in today's industrial wind farm turbines, that he identifies as a 'black box' because they display only those technical parts that we associate with clean energy (the wind) while encasing and making invisible their material connections to the fossil fuel technologies they depend on. Ecological reparation in this context thus requires 'opening up black boxes' and transitioning towards local open wind technologies that explicitly expose their dependence on physical human and non-human labours/agencies.

Last, Part VIII Growth<>Flourishing reconceptualizes ecological reparation as a practice of reimagining and making alternative worlds of existence. Rather than following a paradigm of linear growth, these contributions are cultivating imaginations and affects of hopeful flourishing within damaged environments and worlds. Lara Houston, Sara Heitlinger, Ruth Catlow and Alex Taylor explore blockchain as an infrastructure that presents both productive imaginaries but also challenges towards adopting restorative and sustainable work. The authors explore this possibility by exposing the results of a series of workshops with community growers, organizers, artists and technologists, acting out fictional scenarios set in 2025, when all of London became a city farm, following 'The Great Food Emergency' of 2020. For organizations and participants, reparation meant to bring into being more-than-human-value systems that radically decentred human knowledge and experience.

Santiago Martínez-Medina and Alejandra Osejo discuss the implications for ecological repair in projects and politics of biodiversity conservation of the Colombian *páramos* by exploring differences and tensions between relations and knowledge of natural scientists, social scientists and campesinos (peasants). Environmental institutions working on nature conservation have developed scientific maps to define *páramos* boundaries. But despite their scientific accuracy, these ways of mapping can contribute to socio-ecological conflict by disregarding the affective relations between *páramos*, peasants and other inhabitants. To better understand the latter, the authors describe how they used participatory approaches to work with *páramo* peasants and allow them to communicate their own understanding of intimate relations with landscape through drawings. Reflecting on the contrast between these forms of mapping the *páramos*, the authors propose that biodiversity scientists may learn to be affected by these 'inexact materials'.

In the last contribution to Part VIII and to the whole volume, Steven Jackson expands on his thinking on care and repair towards a worldly and ordinary form of hope. Reclaiming the concept of hope from salvationist roots in (some) readings of Judeo-Christian theology and critical theory, the author offers instead a deflationary account in which hope functions as a regular and ordinary feature of being in the world – and one which deepens rather than reduces our entanglements with the nature cultures around us. There could not be a better way to end the book than a call for ordinary hope. Ultimately, by navigating through and provoking new imaginaries of possible liveable futures, *Ecological Reparation* stands in the space opened by deepened awareness of the inheritances that make everyday liveability impossible for so many, but with renewed attention to the capacity for resurgence by claiming reparations and continuously relearning ways to care, feel and practice ecological embeddedness and interconnectedness.

References

Arora, N.K. and Mishra, I. (2021) 'COP26: more challenges than achievements', *Environmental Sustainability*, 4(4): 585–8.

Cadieux, K.V., Carpenter, S., Liebman, A., Blumberg, R. and Upadhyay, B. (2019) 'Reparation ecologies: Regimes of repair in populist agroecology', *Annals of the American Association of Geographers*, 109(2): 644–60.

Cairns, J. (2003) 'Reparations for environmental degradation and species extinction: A moral and ethical imperative for human society', *Ethics in Science and Environmental Politics*, 3: 25–32.

Caney, S. (2006) 'Environmental degradation, reparations, and the moral significance of history', *Journal of Social Philosophy*, 37(3): 464–82.

Choy, T. and Zee, J. (2015) 'Condition – Suspension', *Cultural Anthropology*, 30(2): 210–23.

Cole, L.W. and Foster, S.R. (2001) *From the Ground Up: Environmental Racism and the Rise of the Environmental Justice Movement*, New York: New York University Press.

Darwish, L. (2013) *Earth Repair: A Grassroots Guide to Healing Toxic and Damaged Landscapes*, Gabriola, BC: New Society Publishers.

Deleuze, G. and Guattari, F. (1987) *A Thousand Plateaus. Capitalism and Schizophrenia*, Minneapolis, MN: University of Minnesota Press.

Denis, D. J. (2019) 'Why do maintenance and repair matter?' in B. Anders, F. Ignacio, and C. Roberts (eds) *The Routledge Companion to Actor-Network Theory*, New York: Routledge, pp 283–93.

Di Chiro, G. (2008) 'Living environmentalisms: Coalition politics, social reproduction, and environmental justice', *Environmental Politics*, 17(2): 276–98.

Ferdinand, M. (2019) *Pour une ecologie decoloniale. Penser l'ecologie depuis le monde caribeen*, Paris: Editions du Seuil.

Ghelfi, A. and Papadopoulos, D. (2021) 'Ungovernable Earth: Resurgence, translocal infrastructures, and more-than-social movements', Environmental Values, 30(6).

Gilio-Whitaker, D. (2020) *As Long as Grass Grows: The Indigenous Fight for Environmental Justice, from Colonization to Standing Rock*, Boston, MA: Beacon Press.

Guattari, F. (1995) *Chaosmosis. An Ethicoaesthetic Paradigm*, Bloomington, IN: Indiana University Press.

Hale, B., Lee, A. and Hermans, A. (2014) 'Clowning around with conservation: Adaptation, reparation and the new substitution problem', *Environmental Values*, 23(2): 181–98.

Herzogenrath, B. (2009) *Deleuze/Guattari & Ecology*, Basingstoke: Palgrave Macmillan.

Hoerl, E. and Burton, J. (2017) *General Ecology: The New Ecological Paradigm*, London: Bloomsbury.

Jackson, S.J. (2014) 'Rethinking repair', in T. Gillespie, P. Boczkowski, and K. Foot (eds) *Media Technologies: Essays on Communication, Materiality and Society*, Cambridge, MA: MIT.

Jackson, S.T. and Hobbs, R.J. (2009) 'Ecological restoration in the light of ecological history', *Science*, 325(5940): 567–9.

Kingsland, S.E. (2005) *The Evolution of American Ecology, 1890–2000*, Baltimore, MD: Johns Hopkins University Press.

Kutney, G. (2014) *Carbon Politics and the Failure of the Kyoto Protocol*, London: Routledge.

Macleod, C.I. (2019) 'Expanding reproductive justice through a supportability reparative justice framework: The case of abortion in South Africa', *Cult Health Sex*, 21(1): 46–62.

Macleod, C.I., Beynon-Jones, S., and Toerien, M. (2017) 'Articulating reproductive justice through reparative justice: Case studies of abortion in Great Britain and South Africa', *Cult Health Sex* 19(5): 601–15.

Mora-Gámez, F. (2016) *Reparation beyond Statehood. Assembling Rights Restitution in Post-conflict Colombia*, University of Leicester: PhD Dissertation.

Morton, T. (2010) *The Ecological Thought*, Cambridge, MA: Harvard University Press.

Nightingale, A.J. (2011) 'Bounding difference: Intersectionality and the material production of gender, caste, class and environment in Nepal', *Geoforum*, 42(2): 153–62.

Papadopoulos, D. (2018) *Experimental Practice. Technoscience, Alterontologies, and More-Than-Social Movements*, Durham, NC: Duke University Press.

Papadopoulos, D. (2022) 'Implicated by scale: Anthropochemicals and the experience of ecology', *Sociological Review*, 70(2): 330–51.

Perez Murcia, L.E. (2014) 'Social policy or reparative justice? Challenges for reparations in contexts of massive displacement and related serious human rights violations', *Journal of Refugee Studies*, 27(2): 191–206.
Plumwood, V. (2008) 'Shadow places and the politics of dwelling', *Australian Humanities Review*, 44(March): 139–50.
Puig de la Bellacasa, M. (2022) 'Embracing breakdown – Soil ecopoethics and the ambivalences of remediation', in D. Papadopoulos, M. Puig de la Bellacasa, and N. Myers (eds) *Reactivating Elements. Chemistry, Ecology, Practice*, Durham, NC: Duke University Press, pp 195–230.
Rosen, A.M. (2015) 'The wrong solution at the right time: The failure of the Kyoto protocol on climate change', *Politics & Policy*, 43(1): 30–58.
Spash, C.L. (2016) 'This changes nothing: The Paris Agreement to ignore reality', *Globalizations*, 13(6): 928–33.
Tacchetti, M., Toro, N.Q., Papadopoulos, D. and Puig de la Bellacasa, M. (2022) 'Crafting ecologies of existence: More than human community making in Colombian textile craftivism', *Environment and Planning E: Nature and Space*, 5(3): 1383–404.
Tsing, A., Swanson, H., Gan, E., Bubandt, N. (eds) (2017) *Arts of Living on a Damaged Planet Ghosts and Monsters of the Anthropocene*, Minneapolis, MN: University of Minnesota.
Walker, M.U. (2010) 'What is reparative justice?', The Aquinas Lecture, Milwaukee, WI: Marquette University Press.
Wall, D. (2014) *The Commons in History: Culture, Conflict, and Ecology*, Cambridge, MA: MIT.
White, R. (2017) 'Reparative justice, environmental crime and penalties for the powerful', *Crime, Law and Social Change*, 67(2): 117–32.
Whyte, K. (2022) 'Settler colonialism, ecology, and environmental injustice', in J. Dhillon (ed) *Indigenous Resurgence: Decolonization and Movements for Environmental Justice*, New York: Bergahn Books, pp 127–46.
Wright, W.J. (2021) 'As above, so below: Anti-Black violence as environmental racism', *Antipode*, 53(3): 791–809.

PART I

Depletion<>Resurgence

1

Experiments *in Situ*: Soil Repair Practices as Part of Place-Based Action for Change in El Salvador

Naomi Millner

Introduction

The soil has never been flat, empty or inert. The solid phase of minerals, gases and organic matter that make up what is known as the 'pedosphere' teems with microbial communities in perpetual interaction; with the remains of once-living things in re-composition; with disintegrating rock and crystalline minerals; with slow-growing root-systems and, between them, burrowing macro-organisms. We would not have food without these inter-relating elemental processes, or without the millennia of experimental practices that have developed for understanding and working with them.

The major centres of plant domestication that emerged with modern agriculture depended on these vernacular practices, although local soil knowledge was progressively marginalized as generalizable scientific principles were extracted (Barrera-Bassols and Zinck, 2003). In the contemporary world, peasant agriculture is often considered primitive or 'under-developed', even although it has survived precisely via innovation and adaption (van der Ploeg, 2014). The process of marginalization has been accelerated through the rise of economic vocabularies of profit and production through the course of the twentieth century, which have worked to redefine the soil as 'exploitable and potentially modifiable genetic material' (Krzywoszynska and Marchesi, 2020: 195). This seems paradoxical, however, because many contemporary agricultural and medical plant-based products were originally sourced from local agricultural knowledge, usually without recompense (Hayden, 2020). It is also deeply concerning, because an 'extractivist' approach to agriculture cannot be applied endlessly without exhausting the soil's vitality (Rist and Dahdouh-Guebas, 2006; Puig de la Bellacasa, 2015; Gray and Sheikh, 2018).

This relegation of everyday agricultural expertise has not, however, prevented the elaboration of vibrant agricultural alternatives. In Mexico, while Green Revolution technologies were being tested during the 1960s, experimental ecological practices were also being consolidated, based on an extensive cataloguing of existing traditional and indigenous approaches (Astier et al, 2017).[1] Building on objections to precursors to Green Revolution projects, voiced as early as the 1940s, networks of agronomists, geographers and anthropologists worked to document the merits of traditional and small-scale farming practices, and consolidate more holistic understandings of soil, seeds and farming systems (Yapa, 1993). Since the 1960s, these networks have become associated under the term 'agroecology', denoting agricultural approaches that value ecological factors over other metrics. Agroecological networks have long prioritized the experience and expertise of small-scale farmers and disseminate principles for *in situ* experimentation to enable the effectiveness of traditional techniques to be systematically evaluated. In this chapter I suggest that this focus on experience and experimentation is vital within broader ecologies of 'mending' environmental damage, and to the ideas of ecological reparation put forward by this book. Specifically, I suggest that the loop of collective reflection on experience afforded by the examples I document allow for a supplementation of official histories with small stories; with purposefully censored collective memory; and with everyday expertise. Together, these fuller archives allow for a wider array of possible solutions to soil damage, while also allowing contestation to take place over stories told over land, ownership, lives and communities. The loop of experimentation ensures that this widened ground of the past sets the conditions for more ample possible futures – where, for example, formerly discarded traditional methods of terracing soil or managing water are carefully evaluated (as in agroecology), or entirely new ideas can be tried out, if they fit a shared ethos of practice.

This focus on experimentation does not, then, just mean 'hit-and-miss'. In his book *Experimental Practice*, Dimitris Papadopoulos frames collective experimentation with more-than-human natures and material processes as central to the discovery of modes of living beyond dominant ideas of accumulation, profit and extraction. When it comes to reparation, such practices are interesting because they are materially exploratory at the same time as they look to alter social and political horizons of the possible. One way they do this is to creatively show up what has been 'invisibilized' or rendered imperceptible in the current political situation (Papadopoulos, 2018: 17). Papadopoulos, with others, calls the process that ensues 'commoning'. I read commoning as a verb for continuously re-making the world with more room for sharing, more space for alternative visions of the future and a priority on care for human and nonhuman ecologies of relation. Inspired by the histories of agroecology and associated realms of practice such as permaculture, in this chapter I seek to explore how far experimental practices of soil repair can be seen as examples of commoning in these terms. In doing so, I will highlight the importance of the *experience* of everyday actors, and the process of *experimentation* with material aspects of these worlds, as two ingredients that are critical to the emergence of 'solutions' that are not only about making soil work for humans, but making humans work in relation to wider, more-than-human ecologies.

The case presented in this chapter is engaged in experimental soil repair through composting practices and the creation of 'biofertilizers' – recipes that use local organic ingredients to boost soil health. Soil repair is necessary because the world's soils are highly degraded – about a third of all soils are classified this way in a recent report to the United Nations, while the world's topsoil could be fully unproductive within 60 years (Maximillian et al, 2019). In many cases, this degradation can be closely linked with the rapid uptake of Green Revolution technologies from the 1940s onwards. For example, El Salvador, where the Permaculture Institute of El Salvador is based, was a key site where techniques developed in Mexico were expanded during the 1960s. Like other chapters in this volume exploring the restoration of *campesino* (peasant) expertise, I show how agroecological approaches to soil repair place *campesino* expertise at the heart of agricultural innovation and ecological reparation. By revaluing pre-existing and constantly evolving knowledge practices, agroecology also functions to repair wrongs brought about by centuries of oppression in Central America and elsewhere. In concluding the chapter, I put forward these principles as tactics for wider agendas of ecological reparation.

Permaculture as ecological repair

Permaculture and agroecology are terms sometimes used interchangeably, yet they delimit quite distinct communities of practice. Where agroecology focuses on fostering ecological, small-scale models for agriculture, permaculture is a system of forest gardening that aims to establish 'permanent ecological cultures': a web of interacting social and biophysical systems that constantly regenerate with minimum external inputs. Often associated with urban contexts in the Global North, permaculture evolved from tree ecology[2] at the turn of the twentieth century but was consolidated into a set of 'design principles'[3] in Australia during the 1960s (for example, Mollison and Holmgren, 1978). Practitioners create new designs by first mapping out a given site and its features, and also how ecological systems there already interact to: prevent water loss; act as barriers to pests; restore lost nutrients, and so on. While it is possible to think of permaculture as a colonial import from the Global North to solve socio-environmental issues cultivated by colonial institutions, its emphasis on popular education means that permaculture in each place varies constantly according to local practices and concerns. Like agroecology, permaculture trains participants in field observation skills and experimentation, allowing all learners to test their own designs, and to teach others. This observational training also encourages inquiry into local and ancestral practices of growing (and knowing), with the understanding that these have been developed over many centuries of trial-and-error agriculture (Altieri and Toledo, 2011). Acknowledging indigenous and traditional knowledge as a repository of pragmatically useful practices avoids placing indigenous and peasant voices in the past, without reifying localism as the only possible solution (Millner, 2017).

In both agroecology and permaculture, small-scale farmers or *campesinos* are experts of their own experience, and protagonists in social and environmental repair. This is critical to the model of social transformation in play, because 'solutions' arise directly out of experiences of oppression and marginalization, rather than from external actors or institutions. Such is also the emphasis in the long history of Latin American popular education, such as Freire's (1972) experiments with literacy in 1960s Brazil. In a context marked by authoritarian suppression, Freire's programmes focused on teaching adults without formal education to read, as part of processes of learning to 'read' and 'name' oppressive power relations. What this model assumes is that systems of oppression can't be addressed without engaging with the power relations of knowledge and expertise directly, which is to say, without dismantling the rules that dictate who gets to count as an expert, and who does not ('the poor'). Popular education turns these rules on their head, claiming that all people are experts of their own experience. The principle is that when these experiences are shared and investigated systematically, they will give rise to ways of naming and transforming oppression, and new social collectives.

It is important to grasp that agroecology and permaculture, alongside other models of popular education and methodologies of inquiry like action research, all build on related understandings about experience and experimentation, devised through histories of resistance praxis. In Brazil and Latin America of the 1960s, another vital influence on this 'praxis' – or action-oriented change practice – was liberation theology, a social movement that emerged from within the Roman Catholic church as a critique of its sedimented authoritarianism. The resulting calls to transformative collective action were central to revolutionary uprisings in El Salvador and across Latin America from the 1970s onwards (Hammond, 1999), which tended to focus on extreme inequalities in land ownership.

Such examples remind us that, as much as European critical thought may have informed vocabularies for understanding how social change takes place, the energies for such movements in practice come from contexts of anti-oppression struggle and contexts where reworking unequal power relations is necessary for survival. As popular education spread in subsequent decades, the forms of naming and codifying social change derived from European critical theories – also important to Freire – were subjected to considerable critique on this basis, leading to important feminist, anti-racist and post-colonial rearticulations of the central ideas (for example, Ellsworth, 1989; Zemblyas, 2018). Such movements also entered dialogue with the kinds of praxis also being fermented within anti-racist movements and indigenous resistance to neo-colonialism, leading to pedagogies that emphasized emotion as much as rational understanding (Motta, 2014).

Permaculture and agroecological approaches are thus part of the same genealogy (family history) of popular education practices and share with them a concern to break apart hierarchies that dictate who can make knowledge, and who cannot. Methodologies in this family also share an emphasis on working on and through particular action contexts rather than creating knowledge blueprints that are meant to be applied everywhere. As in decolonial approaches to knowledge, which insist that colonial relations persist in the ways that institutions function, and expertise is organized, emphasis falls, here, on the power of collective work to create new vocabularies that authorize marginalized ways of knowing out of shared experiences of oppression. As the work of scholars such as Maria Puig de la Bellacasa (2017, 2015, 2010) and others have emphasized in relation to matters of soil, when such collective praxis is brought to bear on material horizons, we may see emerge quite distinct kinds of temporality and ethos to those animating the dominant forms of agricultural production. The idea here is that not only forms of knowledge but modes of enacting care and even time (for example, slowness) have been marginalized by the progress-tempo of the Green Revolution, and can expected to be 'repaired' as part of new kinds of knowledge politics.

In the next two sections I build on this discussion to zoom into soil repair practices contextualized by permaculture collectives and agroecology movements in El Salvador and Nepal. Focusing on regenerative composting practices, I underline the aspects of experience and experimentation in play in both sites to show what animates these "more-than-social" movements.

Compost as experimental food security practice in El Salvador

Julio, pictured in Figure 1.1, takes a fistful of the composted material, showing me how it holds together. It has been turned daily for ten days, allowing it to heat up and then cool. At the beginning it was a layer-cake: a carefully constructed mount of chicken waste, coffee grounds, soil extracted from termite mounds, ash, soil, and rotted-down compost from under the banana

Figure 1.1: Julio makes bokashi

Figure 1.2: The layer cake

tree (see Figure 1.2). Mountain micro-organisms were added; molasses, and then yeast. Then it was left alone, for the 'magic' to happen. Now it is ready. Julio holds up a fistful of the stuff: "This is how we know it is ready. It makes a *puño* [fist], like this."

Julio was one of 20 people taking a six-month permaculture course in Suchitoto, El Salvador for three days per fortnight, which I observed as part of ethnographic fieldwork in 2012–15 (see Millner, 2017). The majority were from *campesino* families, aged from 16 to 60, and either, like Julio, finishing education or combining this training with part-time jobs. Most were attending with a view to improving the productivity of family plots; for those under 40, motivations additionally included desires to create sustainable agricultural alternatives, and career opportunities beyond farming. In one training session, Oscar, town councillor for Suchitoto, explained that the fermented compost known as 'bokashi' should be seen as the essence of regional food security. While the Green Revolution had rendered farmers reliant on chemical fertilizers and one-season seeds, he argued, you could make bokashi for free from waste products. The resulting mix would 'reboot' and 'heal' even exhausted soils.

There are several words used for compost on a Salvadoran farm: *abono* is decomposed vegetable matter, while *fertilizante* denotes chemical fertilizers. Bokashi, on the other hand, is active, fermented organic matter. The term entered the lexicon via Costa Rica in agroecological exchanges, where farmers learned to mix animal and vegetable wastes with other substrates (for example, eggshells, ash, soil and coffee grounds); a blend of microorganisms sourced from mountainous areas, and active ingredients such as molasses and yeast to help break down these materials. The fermented compost that results is important to *campesinos* involved in agroecology and permaculture, because it can be made for free from local ingredients and has an observable impact on yield and crop quality.

When introduced to bokashi I naively assumed that it was a traditional technique, like the terracing of soils, long-practiced in Central America. I soon discovered, however, that the concept was developed experimentally in Japan. The process is credited to Professor Higa, a Japanese horticulturalist and agricultural scientist who sought alternatives to the accelerating use of chemical fertilizers during the 1970s, when the Green Revolution reached Japan (Higa and Parr, 1994). Higa's studies focused initially on individual species of microorganisms, but, after combining the 'rejects' of his experiments, he accidentally discovered the effectiveness of certain families of microorganisms (Prisa, 2019). The mixture taken up in agroecology movements from the 1980s onwards carried the name Effective Microorganisms (EM), while the popular term 'bokashi', literally meaning 'all in', consists of inoculation of the liquid EM mixture into selected organic material (Higa and Parr, 1994). In the Salvadoran process this recipe had been adapted to involve collecting naturally-occurring 'mountain microorganisms', to activate the process.

Bokashi is not an 'indigenous' concept, then, but part of a repertoire of lay scientific practices developed experimentally in the context of agroecological networks – and which today has expanded as far as urban gardens and kitchens in the forms of 'bucket' bokashi (Kinnunen, 2017). Bokashi arrived in Costa

Rica via the *campesino-a-campesino* (CaC) (farmer-to-farmer) movement that moved agroecology practices across Central America during the 1980s and 1990s (Holt-Giménez, 2006).[4] However, as agroecological farmers have adapted and shared bokashi, so its recipes have multiplied, reflecting local availability but also the variability of traditional practices. Usually, the process involves arranging, in a covered area, the selected solid materials into a pile, together with a source of easily-assimilated carbon (for example, molasses, honey or sugar) and a mixture of microorganisms dissolved in water. The pile is then covered with plastic or sacks and is turned several times a day for a period of 7–21 days (the recommendation varies) (Quiroz and Céspedes, 2019).

After building a collaboration with one of the founding members of IPES, I participated in the design of an action research process that aimed to explore how issues of national food security could be addressed through the principles of permaculture. Through the inquiry process, it became clear that participants' desire for a fresh expression of food security was grounded in violence/repair narratives centred on the Green Revolution. Where the state was focused on participating in international schemes to subsidize export-oriented commodities and further rounds of industrialization, communities wanted more autonomy to choose what and how to grow (often termed 'food sovereignty'). Ideas of reparation participants focused on bringing together the impacts wrought on soil with those on local economies and knowledge cultures – in turn reflected in centuries of earlier colonial violence.

The chemical impacts of the Green Revolution were particularly palpable in El Salvador, as one of the countries 'modernized' following the disputed successes of the Mexican Agricultural Program (MAP)[5] – when national governments were persuaded by a heavily anti-communist rhetoric that if they did not accept a 'green' revolution, they would see a 'red' one (Bebbington and Carney, 1990). *Campesinos* in El Salvador had already suffered huge losses to traditional agricultural autonomies earlier in the twentieth century, when the economy was progressively liberalized in the interests of large land-owning families investing in coffee, who allied with the military and progressively eroded rural rights (Paige, 1998). By the 1960s, elections were tightly controlled, and in some places, sterilization of Indigenous women and the poor was practiced as an imposed form of birth control (Roseberry, 1991). The Green Revolution then left farmers even more economically vulnerable, despite promising progress for all. The 'miracle' high-yielding varieties of seeds, distributed freely at first, needed to be purchased anew every year, as well as requiring costly fertilizers and pesticides. Although agricultural concerns were sidelined by the central focus on land inequality once war erupted in 1979, they returned to centre-ground once the peace process mandated by the United Nations was underway in the early 1990s.

Within the counter-narratives animating soil repair practices in agroecology and permaculture, the soil becomes – perhaps unsurprisingly – a focus for rethinking food security. The violence associated with the civil war and genocidal warfare on Indigenous groups and *campesinos* took place in such fields, which also stand as a material record of ecological degradation. Earlier rounds of violence in colonization are also remembered and talked about in everyday events – for example, when thinking about ancestral uses of medicinal plants. Soil is spoken of as a kind of 'archive' of more-than-human memory, which contains the elements of future agriculture and survival. In agroecological and permaculture movements, the erosion of traditional and indigenous farming practices is also associated both with the impoverishment of soils (implying the loss of food security) and the destruction of livelihoods and associated cultural knowhow.

It is understandable, then, that soil repair practices are often imagined within such practices as a kind of decolonial healing process (see Millner, 2017). The idea here is that repairing the damage incurred on the fabric of subsistence agricultural systems (soil) can take place through the rehabilitation of marginalized traditional and indigenous knowledge systems on the one hand, and the rebuilding of economic autonomies on the other. In the process, new innovations are celebrated. Soil repair, here, is not a question of restoring traditional practices *per se*, but developing a form of lay experimental science that rejects the notion of the field as a flat space without histories, or stage for modern geopolitics. Or, as Puig de la Bellacasa (2010) suggests, also in relation to permaculture, it is a biological principle based on ethos and doings that sets new horizons for care for species life.

The creative energy of such praxis manifests, for example, in the performance of sociodramas, which farmers use to communicate agroecological practices from one community to another. In the permaculture course just described, groups created their own interpretations of the Green Revolution, based on oral histories and new learning. In one performance, one of two visiting film students – the only non-*campesino* course participants – played the part of 'the soil', while others were representatives of the Monsanto biotechnology company, scientists or local farmers. The group used heroin-pushing as a metaphor to capture the way that hybrid corn was first given for free and only later cost farmers – who had traditionally saved their seed year on year – what they could not afford (see Figure 1.3). Another character, a hybrid corn plant, initially thrives, but later wilts, sending the soil into convulsions and requiring continuous injections of fertilizers from the worried *campesino*. More seed then has to be bought from Monsanto, because hybrid corn does not regenerate, while new pesticides are needed when plagues of insects invade the new monocrops, sending the *campesino* into abject poverty. Finally, the soil cannot take any more convulsions or injections, and dies. Concluding the play, the soil sits up on the empty stage, and announces "for the sake of

Figure 1.3: Playing the part of the soil

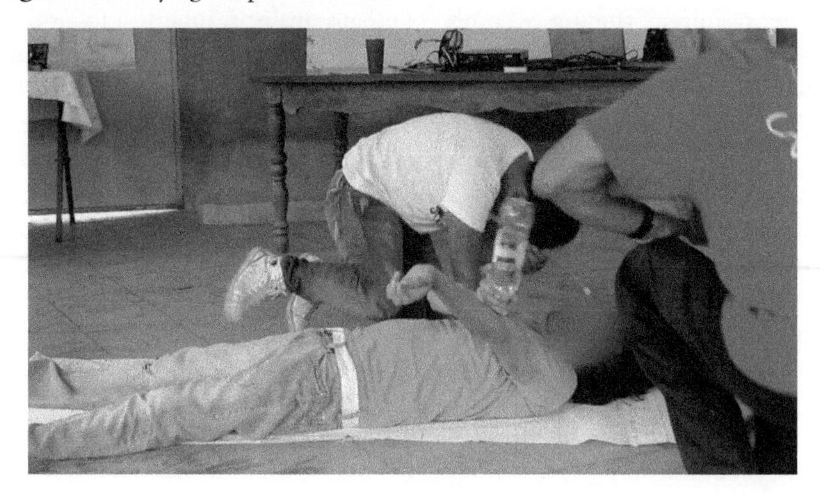

Mother Earth [*la Madre Tierra*], don't abandon me to this!" It is through such practices that that Madre Tierra, a figure of the traumatized earth as well as of non-modern forms of environmental knowledge, promises a horizon of healing that can be applied to colonial violence from beyond it.

While experimentation in action-oriented pedagogies of agroecology and permaculture is never about uncritically restoring past models of practice, indigenous techniques and motifs are often revalorized in this process. In particular, the notion of the primordial connection between 'madre' (mother), maize, and soil resonates across Central America, especially in the light of shared experiences of colonization. In Mayan portrayals of soil life, both what has been lost (land, cultural practices, soil fertility) and what is to be restored are joined together. The Yucatec Maya term for soil is *Lu'um*, which also means land, terrain, landscape and nature – as well as life-supporter, home, territory, womb, and graveyard (Barrera-Bassols and Toledo, 2005). *Lu'um* is a comprehensive relational domain of life, while *Santo Lu'um*, the 'spirit of the land' is one of the most important deities.[6] Soil is also associated with the capacity to heal and restore. This offers quite a stark contrast with Euro-western, 'modern' accounts of soil, which tend to frame it as dead, nonhuman matter.

It is important to note that the return to 'natural' or traditional solutions can be accompanied by other ideals, such as those associated with extreme nationalism. There is certainly a risk that other moral principles are translated in association with permaculture practices such that doing one implies the other. Into the idea of Madre Tierra, for example, may pass all kinds of other moral judgements about what is 'natural', including ideas about gender roles, technology or sexuality. However, what I hope to have shown within this section is that the experimental soil practices in play in the Salvadoran permaculture movement are exciting because they do not propose to work on *either* ecological damage *or* socio-cultural dispossession, but on the fabric that is the relationship between them. In the process, they trouble the idea that what is 'natural' and what is 'cultural' can be ever neatly divided. In the process, they also muddle what counts as 'tradition', including experimental practices devised in Japan and innovations developed in the field. From this observation we may derive a fresh definition for the idea of commoning I have associated with ecological repair: commoning rejects the idea of a set of inert 'natural resources' to be managed by a set of 'users', as if they can be neatly divided. The process of commoning – or expanding the possibilities for sharing among communities – reveals interdependency, interconnection and liveliness, focusing repair not on the restoration of the past, but on the cultivation of ecologies of practice that are regenerative, and can heal the ground of the future.

Commoning, experiment and experience in practice

Such reflections raise a query from within experimental soil repair practices: how and when does 'repair' risk being paired with agendas that aim to restore other kinds of pasts and imagined utopias, including those that are socially exclusionary? For Papadopoulos (2018) and others, it is essential that experimental practices authorize collectives to determine their own agendas, giving them the freedom to break with charismatic leaders or come to different conclusions. However, the resultant collectives are not necessarily free of nationalist or other attachments. In India, for example, agroecology in the form of the Zero Budget Natural Farming movement combines practices of fermentation with notions of bodily purification and guru worship that – in some regions – cultivate cult-like Hindu nationalism (Münster, forthcoming). The fact that the soil preparations *work* in practice and channel frustration with colonial administrations can keep participants from questioning organizational power structures. This raises the importance of thinking about the 'pedagogies' in play within agroecological networks, because popular education processes in theory rely on cultivating the capacities to address power relations in practice. This was important in the case I examined: at various points in the history of the permaculture organization new collectives formed by breaking away from, and objecting to, the consolidation of leadership around a small number of dominant – often male – voices, and the constitution of new, more expansive kinds of organizing.

We could call this process of collective-forming and disagreement itself a part of commoning, for commoning is not only a matter of creating a common basis for survival, but a matter of debating over the givens.

Experimentation is part of the process of socio-ecological reparation in that it involves intimate and iterative relationships with matter, but it also concerns negotiating over who and what gets to be part of the 'commons' in question, and where the limits are between those 'inside' and those 'outside' the collective. Meanwhile, experience is part of socio-ecological reparation in that it informs its processes with everyday expertise, sometimes long excluded from consideration, but it also is part of an ontological negotiation of what is considered relevant *as* experience or expertise; who or what is present in a given site or field or practice; and what gets to count as part of narratives of the past (and future).

We can also think about these processes of commoning on a micro level by zooming into the everyday processes of permaculture and thinking about the ways that participation scaffolds in experiment and experience in careful ways. As several young people in the Salvadoran permaculture context noted, when designing participatory processes, big open questions tend to be intimidating, meaning that those already used to leading the group tend to take the initiative, and already tried-and-tested paths tend to be taken. For a more collaborative 'texture', small group work is often required with tasks that feel relatively straightforward, for example, mapping one's own farms or the resources in the surrounding area; making a timeline of transformations to the region; or (especially for young people) interviewing elders to document traditional practice and oral memory of local transformations. As in focus group methodologies, where a researcher consults a group to understand an issue, this small group work often works best when people are asked to work together with whom they share interests (for example, similar age groups, from the same community, etc.). While organizing groups by socialized gender or occupation can obscure differences and tensions within these groups, time to talk with others in a 'low-risk' way is essential if disagreements are to be raised with others who hold less in common (Nyumba et al, 2018). Such tasks also build individual and group confidence in naming parts of experience that are difficult to articulate, enabling dialogue to proceed across differences, and building capacity ('scaffolding') in collective decision-making.

This constant process of listening-in, facilitating dialogue and framing-reframing is a vital part of the practical process of commoning in both permaculture and social research processes. Rather than asking a range of stakeholders to 'buy in' to a process created by someone else, especially if this person is viewed as an outsider in some way, the sense of who 'we' are is continuously expanding and shifting. In the domains of agroecology and permaculture this of course incorporates nonhuman and more-than-human elements, for farmers are not just individuals but communities of practice, including soils, fields, animals, microbial allies and insects.

Commoning also engages communities beyond the research collective, because commoning is about continuously expanding the resources and possibilities for sharing and co-existing. Thus, restoring degraded soils is a form of commoning because it works on the future ground of food production, benefiting nonhumans, those who eat the food, and future generations. However, producing evidence about the damage of chemical fertilizers, or the effectiveness of organic alternatives, is also part of commoning because it expands the options for soil repair (and experimental practices) in other sites. It is for these reasons that, in our own action research process, we opted to include quantitative measures of the effects of biofertilizers on soils in our inquiry: if a wider circle of stakeholders, such as regional policymakers, are to be persuaded to alter the frameworks of regulation that affect agricultural subsidies and advice, it must also be possible to evidence that what works for one farmer can be replicated to similar effect. This is the distinctive importance of experimental practice to processes such as this, for it combines immersion in material processes with the production of convincing evidence that make it possible to repair degraded soils *and* alter the broader geographies of expertise that have rendered the experience of farmers marginal or irrelevant.

Conclusion

Permaculturists care deeply about soil and can talk about its texture, 'feel', colour and properties. Yet, in working on its repair they are not only working on the basis of future food production but contributing the creation of shared narratives about social histories and the damage incurred by colonial and extractive systems. In the case I have presented, soil repair is about ecological remediation *and* a broader kind of socio-ecological reparation which has to do with historical and endemic geographies of expertise. Specifically, the modes of inquiry central to permaculture and agroecology disrupt the notion that rural farmers or *campesino* communities cannot be innovators or experts. By modelling training in systematic observation of ecological and social systems, these modes empower collectives to conduct their own inquiries and propose their own solutions. Because agroecology focuses on the production of repeatable, verifiable evidence, these processes can also be used to influence other actors, including those involved in regulation and policymaking.

Thinking this way asks us to look again at what we call 'the environment'. Historically, environmentalism and rural agriculture have tended to be thought about using entirely separate paradigms (Agrawal and Sivaramakrishnan, 2000), with the former framing Global North-led international organizations as experts over tracts of 'pristine' forest or land to the exclusion of those who live there (Braun, 2002), and the latter empowering industrial models for enhanced production among the supposedly ignorant poor. Both have systematically sidelined *campesinos*, rural people, Indigenous groups and women, and both have worked to make issues of human livelihood appear separate from questions of environmental degradation. Yet, as the work of postcolonial thinkers such as Franz Fanon and Aimé Césaire, postcolonial agronomist Amilcar Cabrál, and postcolonial poets Suzanne Césaire and Edouard Glissant remind us, matters of soil are crucially connected to the dignity and labour struggles of oppressed peoples (DeLoughrey et al, 2011).[7] Postcolonial literature and agroecological movements seek, through different registers, to reunite ecological and human justice issues, rendering soil a vital site to practice reparation of diverse kinds (Bresnihan and Millner, forthcoming).

Thinking this way, and through the examples in this chapter, ecological repair practices are those that not only treat social ills *or* environmental damage but those that break down the imaginary disconnect that makes them seem separate. Ecological repair practices are partly about enmeshing humans in the species worlds, soil and water bubbles and communities of fungi, microbes and roots that help sustain planetary life. On the other hand, they crucially involve processes of dialogue and exchange that allow claims to be made that reflect the experience of multiple communities of actors and expand the conditions for mutually beneficial living. Following this argument, I have suggested that what allows bokashi-making and other soil repair practices to participate in the kind of commoning that Papadopoulos imagines is the way that experimentation with soil remains, in permaculture, connected to the creation of systems that multiply the possibilities for convivial more-than-social relationships. This concerns the relations of humans with their wider ecologies but also the construction of feedback loops that promote ongoing mutual flourishing and possibilities for interspecies co-habitation on long-term scales.

Through this chapter I have also drawn attention to the way that experiment and experience have been central to reparative ecological practices, both in my case study and in the broader genealogies of social movements and resistance praxes that situate them. Although I explored composting as a domain of practice created at the intersection of agroecology and permaculture, I want to propose it as a useful metaphor for 'reparative' experimental practices in this sense. As a model for action in other domains, composting does not mean mixing everything together, throwing things in arbitrarily, but methodologically sifting the old and new to create something that can remediate, can act as a nourishing bed for future life (Abrahamsson and Bertoni, 2014; Jones, 2019). Bokashi, for example works on the past in the present, heals soil and prepares and actualizes different kinds of virtual futures, free of toxic deposits or the curse of non-fertility. Composting thus marks one appropriate response to being confronted by ruins, to devastated environments and to the wider trauma witnessed on and by soils, including colonial histories, violence associated with the Green Revolution and genocidal violence.

Notes

1. Green Revolution technologies set out to intensify agricultural production to tackle global hunger, partly through the genetic engineering of high-yielding, disease-resistant seeds. While dramatically increasing crop yields, these new seeds quickly reduced the diversity of genetic varieties under widespread cultivation and required ever higher inputs of chemical pesticides and fertilizers (Patel, 2013). Perhaps the greatest damage, however, was incurred on the microbial life of the soil, as the diverse communities of fungi, soil algae, and microorganisms we now know to be essential to plant capacities for absorbing nutrients were significantly compromised through widespread use of new chemicals.
2. Permaculture leans on the principles of tree ecology set out by Joseph Russell Smith (1929) in his book *Tree Crops: A Permanent Agriculture*, as well as developing concepts of agroecology (the application of ecology to the design and management of sustainable agroecosystems) and agroforestry.
3. See Permacultureprinciples.com.
4. The first agroecology initiative associated with this network was a small programme in Guatemala in 1972, which aimed to empower a group of Indigenous Maya-Kaqchikel to teach ecological techniques that were embedded in their traditional culture. These were so successful that further exchanges were organized between Indigenous farmers and *campesinos*, first in Guatemala, and later in Nicaragua. The model grew rapidly during the late 1980s when heavy flooding exposed the difference between traditionally terraced farms, and modern farms, which were stripped of topsoil.
5. The MAP, which preceded the larger-scale Green Revolution projects, are seen to have been of limited success despite huge investment, but the major 'Plan Puebla', beginning in 1967, was considered successful enough to form a prototype for other regions across Latin America (Edelman, 1980: 33). Such projects were contextualized by a deliberate policy of directing U.S. assistance primarily to countries featuring in their geopolitical strategies (Edelman, 1980, Paré, 1990), linked in this historical moment with defusing the threat of communism.
6. More than 80 descriptive terms referring to soil life and properties are architectured on this symbology, which correlate closely with conventional scientific classifications WinklerPrins and Barrera-Bassols (2004). The interdisciplinary study of ethnopedology also tells us that the people of Mesoamerica had a profound knowledge of soils; understood how to create it, manage and sustain it for productive agriculture – as is evidenced for example, in glyphs from the Aztec Códices, which date back to 1000–1400 AD (WinklerPrins and Barrera-Bassols, 2004, Barrera-Bassols and Toledo, 2005).
7. For Cabrál (2016: 261), soil is 'the inscribed body and erosion is the scar left by historical violence', while for Fanon, land should be understood as 'a primary site of postcolonial recuperation, sustainability and dignity' (1961, cited in DeLoughrey et al, 2011: 3).

References

Abrahamsson, S. and Bertoni, F. (2014) 'Compost politics: Experimenting with togetherness in vermicomposting', *Environmental Humanities*, 4: 125–48.

Agrawal, A. and Sivaramakrishnan, K. (2000) *Agrarian Environments: Resources, Representations, and Rule in India*, Durham, NC: Duke University Press.

Altieri, M. and Toledo, V. (2011) 'The agroecological revolution in Latin America: Rescuing nature, ensuring food sovereignty and empowering peasants', *Journal of Peasant Studies*, 38(3): 587–612.

Astier, M., Argueta, J., Orozco-Ramírez, Q., González, M., Morales, J., Gerritsen, P., Escalona, M. et al (2017) 'Back to the roots: Understanding current agroecological movement, science, and practice in Mexico', *Agroecology and Sustainable Food Systems*, 41(3–4): 329–48.

Barrera-Bassols, N. and Toledo, V.M. (2005) 'Ethnoecology of the Yucatec Maya: Symbolism, knowledge and management of natural resources', *Journal of Latin American Geography*, 4(1): 9–41.

Barrera-Bassols, N. and Zinck, J.A. (2003) 'Ethnopedology: A worldwide view on the soil knowledge of local people', *Geoderma*, 111(3–4): 171–95.

Bebbington, A. and Carney, J. (1990) 'Geography in the international agricultural research centers: Theoretical and practical concerns', *Annals of the Association of American Geographers*, 80(1): 34–48.

Braun, B. (2002) *The Intemperate Rainforest: Nature, Culture, and Power on Canada's West Coast*, Minneapolis, MN: University of Minnesota Press.

Bresnihan, P. and Millner, N. (forthcoming) *All We Want Is the Earth*, Bristol: Bristol University Press.

Cabral, A. (2016) *Resistance and Decolonization*, London: Rowman & Littlefield.

DeLoughrey, E.M., DeLoughrey, E., and Handley, G.B. (eds) (2011) *Postcolonial Ecologies: Literatures of the Environment*, Oxford: Oxford University Press.

Edelman, M. (1980) 'Agricultural modernization in smallholding areas of Mexico: A case study in the Sierra Norte de Puebla', *Latin American Perspectives*, 7(4): 29–49.

Ellsworth, E. (1989) 'Why doesn't this feel empowering? Working through the repressive myths of critical pedagogy', *Harvard Educational Review*, 59(3): 297–325.

Freire, P. (1972 [1968]) *Pedagogy of the Oppressed*, Trans. M. Bergman Ramos, New York: Herder.

Gliessman, S. (2013) 'Agroecology: Growing the roots of resistance', *Agroecology and Sustainable Food Systems*, 37(1): 19–31.

Gray, R. and Sheikh, S. (2018) 'The wretched earth: Botanical conflicts and artistic interventions, introduction', *Third Text*, 32(2–3): 163–75.

Hammond, J.L. (1999) 'Popular education as community organizing in El Salvador', *Latin American Perspectives*, 26(4): 69–94.

Hayden, C. (2020). *When Nature Goes Public: The Making and Unmaking of Bioprospecting in Mexico* (Vol. 1), Princeton: Princeton University Press.

Higa, T. and Parr, J.F. (1994) *Beneficial and Effective Microorganisms for a Sustainable Agriculture and Environment* (Vol. 1), Atami: International Nature Farming Research Center.

Holt-Giménez, E. (2006). *Campesino a Campesino: Voices from Latin America's Farmer-to-Farmer Movement for Sustainable Agriculture*, Oakland: OUP.

Jones, B.M. (2019) '(Com)Post-capitalism: Cultivating a more-than-human economy in the Appalachian Anthropocene', *Environmental Humanities*, 11(1).

Kinnunen, V. (2017) 'Bokashi composting as a matrixal borderspace', in V. Kinnunen and A. Valtonen (eds) *Living Ethics in a More-than-human World*, Rovaniemi: University of Lapland Press.

Krzywoszynska, A. and Marchesi, G. (2020) 'Toward a relational materiality of soils: Introduction', *Environmental Humanities*, 12(1): 190–204.

Maximillian, J., Brusseau, M.L., Glenn, E.P., and Matthias, A.D. (2019) 'Pollution and environmental perturbations in the global system', in M. Brusseau, I Pepper and C. Gerba (eds) *Environmental and Pollution Science*, London: Academic Press, pp 457–76.

Millner, N. (2017) '"The right to food is nature too": food justice and everyday environmental expertise in the Salvadoran permaculture movement', *Local Environment*, 22(6): 764–83.

Mollison, B. and Holmgren, D. (1978). *Permaculture*, Melbourne: Lesmurdie Progress Association.

Motta, S.C. (2014) 'Decolonization in Praxis: Critical educators, student movements, and feminist pedagogies in Colombia', in M. Cole and S. Motta (eds) *Constructing Twenty-First Century Socialism in Latin America*, New York: Palgrave Macmillan, pp 143–71.

Münster (forthcoming) 'The nectar of life fermentation and the ambivalences of natural farming in India', Working paper received by personal communication.

Nyumba, T.O., Wilson, K., Derrick, C.J., and Mukherjee, N. (2018) 'The use of focus group discussion methodology: Insights from two decades of application in conservation', *Methods in Ecology and Evolution*, 9(1): 20–32.

Paige, J.M. (1998) *Coffee and Power: Revolution and the Rise of Democracy in Central America*, Harvard: Harvard University Press.

Papadopoulos, D. (2018). *Experimental Practice: Technoscience, Alterontologies, and More-than-Social Movements*, Durham, NC: Duke University Press.

Paré, L. (1990) 'The challenges of rural democratisation in Mexico', *Journal of Development Studies*, 26(4): 79–96.

Patel, R. (2013) 'The long Green Revolution', *Journal of Peasant Studies*, 40(1): 1–63.

Prisa, D. (2019) 'Effective microorganisms for germination and root growth in Kalanchoe daigremontiana', *World Journal of Advanced Research and Reviews*, 3(3): 47–53.

Puig de la Bellacasa, M. (2010) 'Ethical doings in naturecultures', *Ethics, Place and Environment*, 13(2): 151–69.

Puig de La Bellacasa, M. (2015) 'Making time for soil: Technoscientific futurity and the pace of care', *Social Studies of Science*, 45(5): 691–716.

Puig de la Bellacasa, M. (2017) *Matters of Care: Speculative Ethics in More Than Human Worlds* (Vol. 41). Minnesota: University of Minnesota Press.

Quiroz, M. and Céspedes, C. (2019) 'Bokashi as an amendment and source of nitrogen in sustainable agricultural systems: A review', *Journal of Soil Science and Plant Nutrition*, 19(1): 237–48.

Rist, S. and Dahdouh-Guebas, F. (2006) 'Ethnosciences: A step towards the integration of scientific and indigenous forms of knowledge in the management of natural resources for the future', *Environment, Development and Sustainability*, 8(4): 467–93.

Roseberry, W. (1991) 'La falta de brazos', *Theory and Society*, 20(3): 351–82.

Russell Smith, J. (1929) *Tree Crops: A Permanent Agriculture*, New York: Harcourt, Brace & Co.

Van der Ploeg, J.D. (2014) 'Peasant-driven agricultural growth and food sovereignty', *Journal of Peasant Studies*, 41(6): 999–1030.

WinklerPrins, A.M. and Barrera-Bassols, N. (2004) 'Latin American ethnopedology: A vision of its past, present, and future', *Agriculture and Human Values*, 21(2–3): 139–56.

Yapa, L. (1993) 'What are improved seeds? An epistemology of the Green Revolution', *Economic Geography*, 69(3): 254–73.

Zembylas, M. (2018) 'Affect, race, and white discomfort in schooling: Decolonial strategies for "pedagogies of discomfort"', *Ethics and Education*, 13(1): 86–104.

2

Hesitant: Three Theses on Ecological Reparation (Otherwise)

Manuel Tironi

The *bude* can only exist if *winkas* change their way of thinking.
Juan Roa Antileo, Werken of the
Lafkenche community of Chilcoco

Act of stammering. To be irresolute. To be undecided. In times of emergency and crisis, when the world crumbles and life is pushed to the brink of the impossible, to hesitate is second to irresponsibility. Perplexity was an exercise that we could indulge ourselves into back when the catastrophe was a future-to-come, somewhere, eventually. Today, when ecological breakdown is reorganizing the distribution of life and death, hesitation is ethically inappropriate and politically banal.

And yet, here we are, hesitating. I am not thinking generically. My 'we' is a very specific one. It refers to the sociocultural aggregate that for the lack of a better category it could be called 'white people'. Or maybe it is 'Western scientists'. Or both. In any case, those that have self-categorized themselves as the other of the Other, the other of the 'Indigenous'. So, what I'm trying to situate is the moment in which, in times of heightened ecological sensitivity, 'we' stumble. And not because we find ourselves allowing climate deniers to perforate our principles and convictions – but because we meet the alterity against which we assert ourselves, with our convictions and their enemies; we meet other ways of defining what ecology means and how reparation is done, other ways of delimiting life from death, other modes of practicing time and grief. An ecological reparation otherwise.

Hesitation, for me, has unravelled partaking in the efforts of Likanantai and Mapuche communities to heal lands, waters and atmospheres damaged by settler-colonialism. In the presence of their narrations about and practices to repair wetlands, forests, oceans, and soils damaged by extractivism, I have stammered. I have stepped back. I have mumbled, at odd, confronting modes of organizing biotic and abiotic life in ways that defy my imagination. Other forms of existence, other chronologies and teleologies that mess up with (my) well-defined questions and answers. I like the image of hesitation because it moves away from the eventfulness and determination of the conflict. What I have experienced doing ecological reparation in the Salar de Atacama or Tubul-Raqui, rather than a clash of convictions, is a doubt. Close to Viveiro de Castro's equivocation (2004), a doubt is a mismatch about the 'real' plausibility of an empirical description, as we learn from González-Gálvez (2016: 52). But I take doubts as messier and stickier than equivocations, perhaps more affective and atmospheric. Doubts are attached to bodies and life projects. Doubts are minor equivocations, subtle suspicions that often remain as intimate ruminations and only ever fully escalate to public disagreement – perhaps because doubts are, in themselves, doubtful.

I hesitated, for example, when Juan Roa Antileo, *werken* (spokesperson) of the Lafkenche community of Chilcoco interpellated me saying that in order to allow for the existence of the *bude* (wetland) in Tubul-Raqui, we *winkas* (Chileans) needed to think differently. What does it mean to think differently? How can we think otherwise? Which onto-epistemic operations will we have to hesitate about to allow the *bude* – as a different being than a 'wetland' – to exist? What I have learned from these interpellations is that the decolonization of ecological reparation begins by rendering thinkable the unthinkable: to accept, even if doubtfully, that ecological reparation is not mainly about recognizing non-human life and rights, stopping anthropogenic damage, or even affirming Indigenous sovereignty – but about 'thinking differently', about cultivating a hesitancy about what life, death, grieving, mourning, damage, healing, otherness, past, future, (non) human, and sovereignty are and do.

I will not articulate a theoretical programme here, nor will I give a thorough account of the multiple hesitations that have punctuated my work in Likanantai and Mapuche territories. I prefer to address three concrete situations that cracked open doubts I have elevated as minor yet potentially fruitful analytics to approach ecological reparation and its decolonization. Three situations that function, then, both as ethnographic descriptions of ecological reparation as hesitation, *and* as methods or analytical strategies to think such vacillations. I gloss them as *theses*, which might sound too grandiloquent for the ethnographic and rather speculative nature of my reflections. But as a proposition that needs to be tested, a thesis gestures to a tentativeness that fits well my – doubtful – attempts. My theses are: ecological reparation in the context of Indigenous worlds entails working in the time, ethical imperative and material texture of *spectres*; it involves the articulation of a *politics of the unthinkable*; and it is practically and spiritually oriented towards the restitution of *abundance*. These three propositions counter or mess with well-established assumptions about ecological reparation – namely that ecological reparation deals with species and landscapes, involves doing regulatory or activist work, and is about restoring ecosystemic functionality. By troubling this canon, the three theses I present here account for the hesitancy I have learned to stay with.

Thesis #1: Ecological reparation is to account for spectres

On 27 February 2010, after an 8.8 magnitude earthquake, a 12-metre tsunami ravaged Tubul-Raqui. Located 600 km south of Santiago (Figure 2.1), Tubul-Raqui is one of the major coastal wetlands of the Western South American coast and home to diverse ecosystems. The tsunami inundated more than 430 m and propagated 3 km upstream, carrying large amounts of sand that deposited over marshes and meadows. The salt and freshwater interaction was dramatically reduced, reporting the total drying of hard bottom habitats and the total loss of aquatic fauna. This included the mortality of several aquatic birds, and importantly, the disappearance of banks of economically important bivalves, such as the *huepo* (Ensis macha) and the *navajuela* (Tagelus dombeii) (Sandoval et al, 2019).

Conservation biologists lamented the deformation of Tubul-Raqui's 'singularity' (Tironi et al, 2021). But Lafkenche people, ancestral Mapuche inhabitants of this territory for more than 12,000 years, mourned something different. For them Tubul-Raqui is a *bude*, an entity that includes what Western knowledge calls a 'wetland' and yet is something utterly different. Different beings, geographies and temporalities constitute a *bude*. From the perspective of the *bude*, what worried Lafkenche communities was not the effect of the tsunami on ecosystems and biodiversity, but the resurgence of *el chocolate* (the chocolate).

I want to think *el chocolate* as a spectre – and as such, about what Lafkenche onto-epistemics has to teach us about ecological reparation and hesitation. As forces that are 'neither present nor absent' (Derrida, 1994: xix), spectres abound in Mapuche worlds (Course, 2011; Bonelli, 2014). Actually, if the spectre is 'an ontological-like category that hovers between being and nonbeing, real and unreal, present and absent' (Klima, 2019: 19), then Mapuche life *is* with the spectral. Social and personal life is in constant interaction with beings that under the form of ghosts, vampires, or monster-like entities inhabit the interstitial space between life and death (González-Galvez, 2016). Their spectrality also entails an uncanny materiality. While

Figure 2.1: Location of Tubul-Raqui

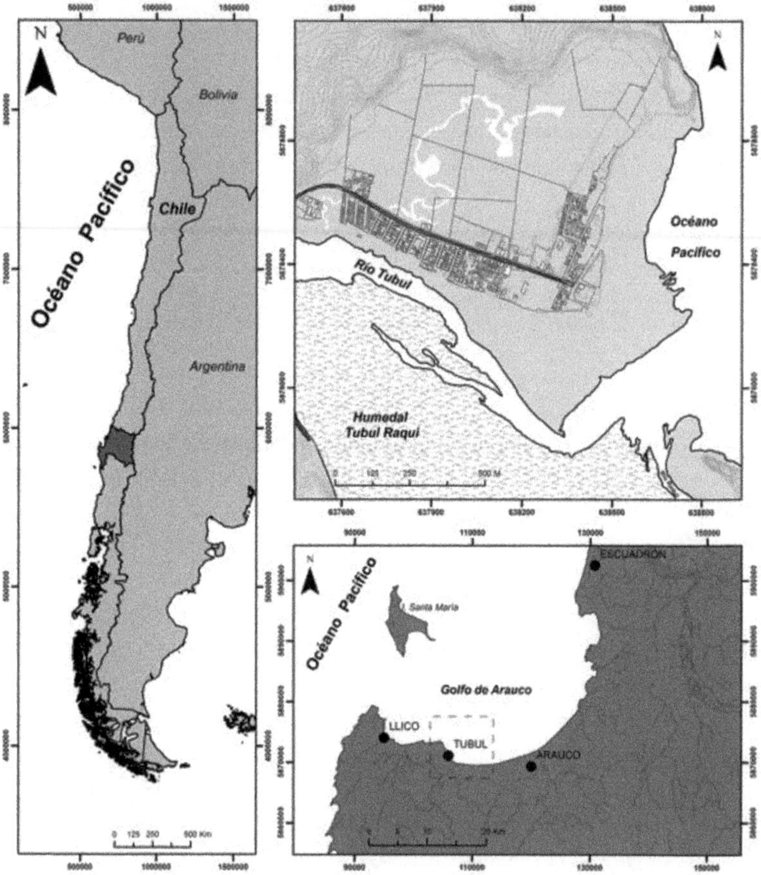

Source: Martínez et al, 2012

real, these eerie beings are perceptible only under certain conditions, erupting into visibility ephemerally and whimsically. A sound connection to the *mapu* (territory) is often a prerequisite for attuning to the spectral and therefore its existence usually defies definitive evidence, especially to *winkas* (Chileans).

El chocolate is a spectre-like being. It is a muddy substance that compounds by the decantation of toxicants produced by the forestry industry in the wetland's bottom-soil. Lafkenche farmers and *pelillo* (*Gracilaria spp.*) recollectors have known about *el chocolate* for decades. It becomes more notorious with winter precipitation and high tides, when messy currents bring it to the surface, and it resurfaced after the tsunami. Accounts of an orange-ish mud entering through the estuary or flaring up from the bottom of the wetland after the earthquake proliferated. Evidence was elusive, however, at least for *winkas*. Biochemical analyses confirmed that levels of iron and manganese increased in Tubul-Raqui after the tsunami, but, the forestry industry claimed, not above concentrations that could pose a risk to human and ecosystem health (Valdovinos et al, 2012). Lafkenche communities had a different diagnosis. Its evasiveness was not a proof of its nonexistence, but actually of its eerie condition. Rather than sanctioning its existence by its undisputable visual presence, Lafkenche people preferred the pragmatic test of its consequences. 'That mud kills the *picoroco* (*Austromegabalanus psittacus*), kills the seaweed that sticks to the boats and it's so strong that it cleans up the boats [*los deja limpios*]', an interlocutor told us as he explained the reality of *el chocolate*.

Accretional and unspectacular damage accumulating slowly and below the threshold of technocratic perceptibility, *el chocolate* was a kind of harm unable to match the dramaturgy of the tsunami – nor the temporality of Western conservation science. For if the wetland was destroyed by the tsunami, the *bude* has been enduring destruction for decades – and destruction not by earthly forces such as the 2010 gigantic oceanic wave, but by extractivism. 'It's the tailings [from Arauco, the forestry company], I have no doubts', the *werken* from the Chilcoco community asserted about *el chocolate*. The substance harming the *bude* had to be therefore evaluated and intervened upon in the perspective of accumulated pollution and the economic, political, and knowledge systems behind its production and maintenance. "The *lafkenmapu* [the sea-land territory] has long been contaminated … they [the forestry company] planted pine trees in agricultural meadows, made ditches and channels seven meters deep and ten meters wide. They even diverted the water to irrigate their pine trees," one collaborator told us while trying to locate the wrong done to the *bude* – including *el chocolate* —— outside the depoliticized eventfulness of the tsunami, unfairly converted into the agent of destruction by White conservationism.

So like a spectre, *el chocolate* brought into the present damage accumulating in long temporalities of harm. *El chocolate* was the ghostly instantiation of the history of dispossession and settler violence over the *bude* and Lafkenche life projects. As a spectre that enunciates in the present histories and wrongs from the past, *el chocolate* prevented from foreclosing destruction in the now. Extractivism, industrialism, settler colonialism, whispered *el chocolate* in Tubul-Raqui. Past actions, as slow, silent, and invisible as they may be to *winkas*, interrupt the present and prefigure the future.

Here, then, ecological reparation has something to learn from the hesitation provoked by *el chocolate* and Lafkenche science at large. Mapuche interrelationality prevents separating the facticity of the actual with the deeds and histories of the ancestors and the land (Quidel, 2016, 2020). As a spectre, *el chocolate* summoned for liabilities that went well beyond the present tense of the tsunami, and was a reminder that 'in order to live, we must learn from the dead' (Klima, 2019: 20). So, if *el chocolate* was an adamant indication that the destruction of the *bude* needed to turn to the trajectory of settler colonialism, it also pointed that from the perspective of Lafkenche struggles against extractivism, ecological reparation is about living with exhaustion and violence made in the past, but from which we have to learn in order to restitute balances and response-abilities for the future. *El chocolate* makes visible the entangled temporality of ecosystemic relations and problematizes the futurity of 'sustainability'. It shows that reparation demands the engagement with a kind of time that, as the Andean chrono-logics theorized by Silva Rivera Cusicanqui, 'moves in cycles and spirals', a non-teleological time in which 'regression or progression, repetition or overcoming of the past are at stake at every juncture' (Rivera Cusicanqui, 2010: 54). In Tubul-Raqui, soils and waters, forests, *ngen* (spirits), and chemicals draw connections and tensions between the deep time of geo-physical formations, the contemporaneity of industrial, colonial, or otherwise anthropogenic harm, and the promise of equilibria-yet-to-come. Temporal braidings that, as manifested by *el chocolate*, force doubts and hesitations.

Thesis #2: Ecological reparation is about (unthinkable) politics

In November 2019, we walked 250 kilometres to heal the Salar de Atacama from extractivism. We were an assorted collective led by doña Sonia, Lickanantay (Atacameño) healer and central figure in the Indigenous defence of the Salar. The walk was a ceremony for reparation (Tironi, 2020). The action, as well, was intended to attract political attention. The last stop of the journey was the delivery of a formal letter to the governor of Antofagasta to demand stopping extractive harm to el Salar. The governor did not receive us. But we managed to arrange a meeting with the Regional Secretary of the Environment.

We met outside the building entrance. The atmosphere was tense. Chile was in the midst of the largest social uprising since the recovery of democracy and Antofagasta had been one of the most intense epicentres. That morning, the smell of tear gas and burned rubble from barricades was still in the air. At the building we were received by representatives of diverse Lickanantay, Aymara, and Quechua collectives from Antofagasta that had heard about the meeting and wanted to support us (Figure 2.2). So instead of the six-people party they were expecting, the secretary and his team were met by 20 people with wiphala flags and a large banner *Pueblos originarios en la lucha* (First nations in the struggle). The secretary's chief of staff swiftly found an apt meeting room and there we were, in this improbable situation: the Regional Secretary of the Environment, in the representation of the Chilean state, meeting with a Lickanantay shaman.

Once in the room and after salutations, doña Sonia explained about the walk and the suffering of the Salar. She explained to the secretary that the

Figure 2.2: Encounter in Antofagasta

Salar was a sentient being and an *abuelo* (grandfather), ancestor and kin. And before handing in the letter, she solemnly explained that while the letter was written by her, it was the Salar speaking; it was the Salar himself who was communicating, through her, about his pain and grief, and about what it needed to heal. This meeting, doña Sonia explained, was summoned by the Salar and she was there to speak on his behalf. She also said that the Salar was not asking for sustainable management or the protection of habitats and ecosystems, not even for the restitution of water rights. The Salar, she explained, wanted his *liberation*: to be freed from all forms of extractivism and industrial intervention. The Salar was not asking for a respectful coexistence nor for any other kind of cohabitation arrangement, but for its complete emancipation and the sovereignty to decide about his own mode of existence.

Then doña Sonia gave the letter to the secretary. He expressed his gratitude, thanked the letter, and told doña Sonia that the Regional Secretary of the Environment was fighting the same fight. He then detailed all the programmes – new conservation areas, forthcoming regulatory amendments, fresh funding for ecological reparation – they were pushing forward.

The secretary did not understand what doña Sonia, el Salar, and the letter were saying. He couldn't. What was uttered in that meeting was inconceivable for him. He was able to engage with doña Sonia only by translating what was being said as a community demand for more environmental protection and participation. For what was being proposed, a saltflat with feelings and voice capacity and the radical ejection of all extractive interventions in the territory, exceeded the limits of discourse and imagination. It was not an issue of good faith – the secretary was genuinely interested in the ecological wellbeing of the salt flat – but about the intractableness of the terms with which the issue was described in a public space. The political proposition presented by doña Sonia entailed a version of the real that he simply could not process, or that he could, but only by locating it in the realm of the aberrant (Povinelli, 2022). It was not intolerance – the framing with which liberalism makes sense of the opposition against alternative forms of critical reasoning – but the sheer impossibility to accept, let alone understand, events and beings that were outside the delimitations of critical reasoning and history.

To borrow from Michel-Rolph Trouillot (2012), the political project proposed to the secretary that morning was *unthinkable* – a proposition 'one cannot conceive within the range of possible alternatives [since it] perverts all answers because it defies the terms under which the questions were phrased' (Trouillot, 2012: 82). And I wonder to what extent ecological reparation in and for Indigenous worlds always invokes a politics of the unthinkable. A politics of the unthinkable because it relates to implausible beings, illogical kin, and abhorrent responsibilities: to a form of publicity and moral obligation based on and pointing to an uncanny alterity. A politics of the unthinkable also because it involves the articulation of a programme around aims that are impossible yet necessary in the face of the challenge at stake. For instance, the ecological reparation of el Salar demands, today, land sovereignty, as doña Sonia told the secretary, despite the fact that territorial autonomy is unthinkable within the contours of Chilean multiculturalism.

And yet this impossibility is the condition for at least imagining a possible impossible. To think ecological reparation as and through a politics of the unthinkable does not mean to remove politics nor to desert from the possibility of politicizing the practice of ecological reparation. Rather the contrary, it means to empower a proper political situation by interrupting the meaning of and discourse about the (in)sensible and the (im)possible (Rancière, 2004). Hence the conundrum of the politics of that which is inconceivable in late-liberalism: to exist in the traffic zone between interrupting politics for its expansion and to account for political beings and projects that by their impossibility will remain unannounced, unnamed, unseen. What we learn from doña Sonia and her unconceivable struggle, if anything, is that ecological reparation as politics of and for a radical otherwise entails an insistence both on that which cannot be accepted by Western onto-epistemologies *and* on ethical goals – land restitution – that, while preposterous, have the power to delineate a thinkable and actionable political agenda for ecological reparation – an intrusion of the impossible within the field of the possible. Perhaps the process of working for those impossible objectives might open fissures in which an ecopolitical possibility can root, and where unconceivable ways of defining what is unconceivable can be cultivated.

Thesis #3: Ecological reparation is the restitution of abundance

The literature on ecological reparation suggests that the success of a repaired ecosystem rests upon four achievements (Clewell and Aronson, 2013): the re-composition of potentially co-adapted species representing all known functional groups; a community structure composed of populations in sufficient quantity, adequately distributed on the site to facilitate the structural development of the biotic community; an abiotic environment with the physical capacity to sustain the biota of the restored ecosystem; and an ecosystem properly integrated into a larger ecological matrix or landscape.

Spending time with Lickanantay communities in the puna makes me wonder if these four conditions are no more and no less – than what emplaced collectives have honoured and strived to maintained for millennia: abundance.

Abundance is an ethical and cosmo-ecological principle in the Andes. The folklorization of Andean shamanism and culture by white systems of knowledge and production, however, has situated abundance in the horizon of expectation, escapism, and desire – a state of fulfilment in a time to come to be achieved through *pagos*. However, in everyday Andean worlds abundance is rather the immanent condition of existence as it is. Ceremonial labour is not mainly intended at petitioning for an otherwise evasive plentifulness, but at honouring and thanking the exuberances with which life is.

Take the long histories of Lickanantay co-habitation with elemental extremes. Take, for example, the Láscar, one of the most active volcanoes in the central arch. The Láscar developed in the western margin of the puna, about 30 km east of el Salar on the basement pre-Cenozoic rocks. Some 60 million years later, the Láscar became a lively node around which human and more-than-human life have flourished. A *maico* or *malku*, a tutelar mountain, the Láscar articulates a territory of ravines, springwaters, plants, and game around which several Lickanantay communities have established their life projects for at least 13,000 years. Archaeologists and palaeontologists recognize the magnitude of this accomplishment (Llagostera, 2004). Situated in the world's driest desert, the climatic conditions have confronted Lickanantay people to the limits of biophysical life. The basin elevates between 2,300 and 4,000 metres above sea level, with the puna occupying the last terrace at 3,400 metres. Dryness, altitude and extreme daily thermal variation pose physiological limitations to the development of vegetation. Geoclimatic events are not uncommon. Altiplano summer rains have washed away villages and cultivation fields for millennia, and the entire basin is surrounded by several volcanoes, forming one of the most complex volcanic mega-calderas in the world. Aucanquilcha, Apacheta-Aguilucho, Putana, Sairecabur, Alitar, Colachi, Acamarachi, Chiliques, Puntas Negras, Miscanti, Miñiques, Caichinque, Pular, Socompa, Llullaillaco, and Escorial are some of the active strato-volcanoes that surround the basin. In the last 120 years, the Láscar alone has had around 30 explosive eruptions.

This geological activity has, for as long as Atacameños occupy the land, frictioned life in the puna. But has also allowed for it. Volcanic powers, both geological and cosmic, have terraformed the Lickanantay territory and defined the possibility of social life in the altiplano. A monstrous force that despite being deemed as hazardous by Western science, enables sociality and thus is in itself an expression of the excessiveness at play: inhuman force that carries death and life, violence and profusion at unison. As a *maico*, the Láscar is the force that enables this complex system of interdependences and exchanges. *Maicos* are ancestors and, also, where ancestors live and express themselves (Grebe and Hidalgo, 1988). The Láscar is kin: a family member, an *abuelo* that *takes care* of the community. Not a spectacular, patriarchal, or patronizing care; not as a deity or temperamental god, but as an immanent presence that *facilitates* life, health and water, particularly water (Grebe and Hidalgo, 1988).

Communities around the Láscar channel water from *aguadas* or springs in Tumbres and Saltar, at the base of the Láscar, to their alfalfa fields and vegetable gardens. The bottom-floor humidity of ravines and gullies that descend from the volcano to the Salar forms *bofedales* or wetlands, floodplains and meadows that cater water for cattle and agriculture. *Cha-cha*, *rica-rica*, *kopa-kopa*, *kume*, *lampalla*, and *chuchekan* are some of the rich variety of plants and bushes in the *puna* that Atacameños have learned to identify and utilize for food, medicine and ceremony (Figure 2.3). Obsidian rocks from the Láscar were utilized to make hunting paraphernalia, while ignimbrite or *liparita*, a pumice-dominated igneous rock formed from the cooling of pyroclastic material ejected from an explosive volcanic eruption, has been utilized as building blocks for the construction of housing, corrals and irrigation canals. The rock eaves left by numerous geological accommodations and climatic exposures since the pre-Cenozoic have offered shelter for hunters and shepherds, while the platforms formed by the Láscar eruptions became the terraces in which Atacameños experimented with agriculture.

It is not volcanic eruptions that disrupt this abundance in the puna, but extractivism. And particularly lithium extraction. Lithium —the 'white gold' of sustainable transition— is industrially unearthed by pumping saline groundwater from beneath the salar. The extraction of dissolved lithium is done by letting lithium brine evaporate in shallow open-air ponds (Bustos-Gallardo et al, 2021). It is estimated that for every kilo of processed lithium,

Figure 2.3: Puna ecologies, Salar de Atacama

two litres of water are utilized, heavily disrupting the fragile hydrogeology of the puna. *Bofedales* and wetlands, the water ecosystems that have sustained for 13,000 years Lickanantay agriculture and cattle, are being severely impacted. Their interruption has effects on the biotic and abiotic communities that share the puna with humans and other beings, including biodiversity losses and the decimation of unique microbial ecologies (Bonelli and Dorador, 2021).

What is to be repaired when lithium extraction breaks down these more-than-human ecologies? Not something but everything, and yet not *everything* but Lickanantay-specific dynamics of reciprocity. Hence the hesitancy. Ecological reparation is threshold-oriented: it aims at restituting the minimal conditions of reproductibility upon which an ecosystem becomes functional. Lickanantay reparation, instead, is oriented towards an economy of excesses in which what has to be repaired is abundance: interventions not aiming at the mechanics of survival but at the reposition of profusion. Profusion of connections, frictions, beings, and forces. Abundance of meaning and time. It is not accumulation. It is not the logic of addition nor the anxiety of securing in the future that which is scarce in the present. Abundance is about situating community and self within the always-expanding copiousness of relations and interdependences (Kimmerer, 2021). Abundance is to account responsibly for the excess by which life is possible: the voluptuosity and radical exuberance of agency, destruction, propagation, change – radiation, metabolism, photosynthesis, energy. Ecosystemic damage by extractivism in the Salar does not call just to the reestablishment of chemical balances, species populations, or hydric flows, but the plethoric production of reciprocities, the indifferent generosity that renders puna life possible. That's the injustice of extractivist harm and the equivocations around ecological reparation in the Salar: extractivism shatters kinship and what has to be restored is redundancy and gratuity.

References

Bonelli, C. (2014) 'What Pehuenche blood does: Hemic feasting, intersubjective participation, and witchcraft in Southern Chile', *HAU: Journal of Ethnographic Theory*, 4(1): 105–27.

Bonelli, C. and Dorador, C (2021) *Endangered Salares: Micro-Disasters in Northern Chile*, Tapuya: Latin American Science, Technology and Society.

Bustos-Gallardo, B., Bridge, G. and Prieto, M. (2021) 'Harvesting lithium: Water, brine and the industrial dynamics of production in the Salar de Atacama', *Geoforum*, 119: 177–89.

Clewell, A. and Aronson, J. (2013) *Ecological Restoration: Principles, Values, and Structure of an Emerging Profession*, New York: Springer.

Course, M. (2011) *Becoming Mapuche: Person and Ritual in Indigenous Chile*, Urban-Champaign: University of Illinois Press.

Derrida, J. (1994) *Specters of Marx: The State of the Debt, the Work of Mourning, and the New International*, London: Routledge.

González-Gálvez, M. (2016) *Los mapuche y sus otros: Persona, alteridad y sociedad en el sur de Chile*, Santiago: Editorial Universitaria.

Grebe, M. and Hidalgo, B. (1988) 'Simbolismo atacameño: un aporte etnológico a la comprensión de significados culturales', *Revista Chilena de Antropología* 7: 75–97.

Kimmerer, R.W. (2021) 'The Serviceberry: An economy of abundance', Emergence Magazine. Available from: https://emergencemagazine.org/story/the-serviceberry/ [Last accessed 28 February 2021].

Klima, A. (2019) Ethnography #9, Durham, NC: Duke University Press.

Llagostera, A. (2004). *Los Antiguos Habitantes del Salar de Atacama, Prehistoria Atacameña*, Santiago: Editorial Pehuén.

Martínez, C., Rojas, O., Aránguiz, R., Belmonte, A., Altamirano, A., and Flores, P. (2012) 'Riesgo de tsunami en caleta Tubul, Región del Biobío: escenarios extremos y transformaciones territoriales posterremoto', *Revista de Geografía Norte Grande* 53: 85–106.

Povinelli, E. (2022) *The Cunning of Recognition: Indigenous Alterities and the Making of Australian Multiculturalism*, Durham, NC: Duke University Press.

Quidel Lincoleo, J. (2016) 'El quiebre ontológico a partir del contacto mapuche hispano', *Chungara* 48(4): 713–19.

Quidel Lincoleo, J. (2020) 'Fey ga akuy ti aht'ü. Entonces el día llegó. Una lectura de la pandemia desde un mapuche rakizuam', Comunidad de Historia Mapuche. Available from: www.comunidadhistoriamapuche.cl/fey-ga-akuy-ti-ahtu-entonces-el-dia-llego-una-lectura-de-la-pandemia-desde-un-mapuche-rakizuam/ [Last accessed 28 August 2021].

Rancière, J. (2004) *Disagreement: Politics and Philosophy*, Minneapolis, MN: University of Minnesota Press.

Rivera Cusicanqui, S. (2010) *Ch'ixinakax utxiwa: una reflexión sobre prácticas y discursos descolonizadores*, Buenos Aires: Tinta Limón.

Sandoval, N., Valdovinos, C., Oyanedel, J.P., and Vásquez, D (2019) 'Impacts of coseismic uplift caused by the 2010 8.8 Mw earthquake on the macrobenthic community of the Tubul-Raqui Saltmarsh (Chile)', *Estuarine, Coastal and Shelf Science* 226(15).

Tironi, M. (2020) '"Rocks have history." Theorizing the contemporary, fieldsights', 22 September 22. Available from: https://culanth.org/fieldsights/rocks-have-history [Last accessed 28 August 2021].

Tironi, M., Vega, D., and Roa Antileo, J. (2021) *Budc Uncommon: Extractivist Endings and the Unthinkable Politics of Conservation in Lafkenche Territory*, Tapuya: Latin American Science, Technology and Society.

Trouillot, M-R. (2012) 'An unthinkable history: The Haitian revolution as a non-event', in A. Sepinwall, A. (ed.) *Haitian History: New Perspectives*, London: Routledge.

Valdovinos, C., Muñoz, M.D., Vásquez, and Olmos, V. (2012) 'Desastres naturales y biodiversidad: el caso del humedal costero Tubul-Raqui', *Sociedad Hoy*, 19: 33–52.

Viveiro de Castro, E. (2004) 'Perspectival anthropology and the method of controlled equivocation', *Tipití: Journal of the Society for the Anthropology of Lowland South America*, 2(1): 3–22.

3

The False Bay Coast of Cape Town: A Critical Zone

Lesley Green and Vanessa Farr

Cape Town, the city that almost ran out of water in 2018, is usually presented in picture-postcard perfection to the global tourist market: Africa's piece of Europe. Like a screenshot from Minecraft, in Google Earth Street View the landscape of the False Bay coast on the periphery of Cape Town is flat, unpopulated and blurred for lack of details that would be available from a ground-level photograph. The southern face of the city's famous UNESCO world heritage landmark, Table Mountain, lies to the north-west in the earlier image. To the right are the hills that ring Table Bay in the Atlantic Ocean. Underfoot is sand that, not too long ago, was under the Indian Ocean, when carbon levels were as high as they are now.

In a city managed for much of the past decade by the centre-right Democratic Alliance, this is a side of the mountain unfamiliar, and off-limits, to the average tourist. This is the side in which the city's many forms of slow violence multiply, not only in the sprawling settlements of the Cape Flats, an impoverished gangland that earns the city its reputation as one of the world's most dangerous, but also in a number of sites of a failing hygiene and sanitation system. This includes a major water treatment works built by the Apartheid state in 1956 in which unlined settlement ponds on sandy soils leach sewage directly into the Cape Flats aquifer. Since the drought of 2018, these are now being lined. In March 2020, when the national government announced a strict policy of home-based quarantine and a ban on travel, the well-heeled tourists who feed the city's public image and coffers disappeared from the northern seaboard. Also banned, in an effort to keep people out of the emergency rooms of hospitals, was alcohol; and so, in the south, mired in daily precarity, rates of street violence plummeted. With the spectacularity of its usual assaults temporarily slowed, the city's attritional but also exponential

Image 3.1

slow violence came sharply into focus. Experienced by 40 per cent of the city's inhabitants before the pandemic, hunger rose even more sharply as millions, crammed cheaply and dangerously into filthy informal shanty settlements and backyard shacks – barely affordable to those who hustle for a daily wage – tried to 'stay home' and 'socially distance'. Hospitals, emptied abruptly of their usual burden of victims of gunshots, domestic violence, traffic accidents and other health impacts of poverty, alcohol and drugs, filled up instead with sufferers of the unknown new pandemic, COVID-19. Within months, scientists would identify the virulent Delta variant of the virus, one unresponsive to the affordable and easily-stored AstraZeneca vaccine on which countries in the Global South were pinning their hopes. Politicians' assumptions of a quick recovery of tourism, the city's economic lifeline, were dashed; but no new economic or social paradigms emerged.

What, now, is the future of this lunar landscape, this online abstraction, flattened out and exiled from ground level view because its violence, hunger and xenophobia make it too dangerous to be traversed by Google Earth Pegman and his Street View Car? How, with the relations of its precarious lives, its soils, air and water agglomerated into such a landscape of unbecoming, does ground level exist in such a periphery? And what might the conditions be that could repair the necropolitical relations that characterize this uber-Anthropocene urban ecology? What kind of reparative ecologies are possible, and if they are to be possible, what rethinkings of ecology are necessary?

Some answers to these questions begin to emerge when we turn attention to the failings built into a contemporary 'social ecological systems' approach in which 'society' (that is, the collective organization of humans) is understood separately, and therefore managed separately, in current urban ecological governance approaches of the city of Cape Town. The failings of the social-ecological systems paradigm are inscribed into this landscape in which dollar values are presumed to connect 'nature' and 'society' that, empirically, have never been separated anyway. Moreover, the flaws of this paradigm are underscored when one recognizes the parallel behind fenced-off nature and apartheid South Africa: a racist society was built on the supposition that 'Black life' constituted 'nature' that had to be governed separately from 'white society'. 'Ethnic Bantustans' were the empirically flawed solution to instantiate in law what did not exist on the ground: a division of nature from society. Thus, apartheid's architects sought to banish Black citizens from 'White' South Africa.

South Africa's environmental policy, like many other countries, perpetuates this same hard divide between 'nature' and 'society': banishing 'nature' (defined as intact biomes) to the other side of a fence outside of 'society'. The net result is that 'society' is governed in terms of lines inscribed in law as private property, and its materialities are governed in terms of rationalities associated with the dollar values (rand values, to use the local currency) defined by 'ecosystem services'. The actual material flows between them remain ungovernable – a situation exemplified in 'the lore of the law of the plastic bag' that is often referred to jokingly by environmental governance scientists in meetings to try to resolve the ongoing problem of watershed governance. The lore goes like this:

> If a plastic bag is in a road, which municipal structure is responsible for picking it up? Answers: If it's on the road, the responsibility is the Transport Department. If it blows into the park: Recreation and Parks. In the ditch on the side of the road, the responsibility for picking it up falls to Stormwater; in the sewer, Sanitation. If it falls into the river, the Environmental Management Department. If the river flows into a nature reserve, the Conservation subdivision; if onto the beach: Oceans and Coasts. If by some chance some distant relative of Franz Kafka picks it up on the beach and puts it in a litter bin, the responsibility falls to Solid Waste Management … unless it blows out of the garbage truck or away from the landfill site. And so on: repeated infinitum, as the bag degrades into microplastics and molecular components.

The Lore of the Law of the Plastic Bag exemplifies the absurdity of governing environments via property law. For the material flows that make up 'nature' know no property boundaries: not in law; not in fences; neither in wind nor rain; not in disciplinary boundaries as they are taught in universities. And yet the assumption that legal boundaries constitute empirically useful boundaries for environmental governance persists unassailably as a founding 'first principle'.

The fenced off 'environment' is therefore not novel, but an expansion of the same logic behind the technologies of containment practiced by the apartheid regime's bantustans.[1] Usefully reconceived as a 'naturestan', conservation areas in this piece of Cape Town are set in concrete and wire fences, backed up by policing (with guns where deemed necessary), and justified by environmental governance policies practiced globally for the past 400 years. In southern Cape Town, however, the nature-society logic of this landscape has permitted the decimation of wildlife and the crushing biodiversity everywhere that is not designated 'nature'. It has also set the wealth–poverty divide at warp speed, aided and abetted by the flawed paradigm of growth economics and its absurd claim that putting dollar values on 'nature' leads to objective, neutral, unbiased decision-making that promotes 'progress'. The resulting ideology of 'ecosystem services' backed by the ideologies of neutral science and neutral governance means that only what can be dollar-valued is worth protecting. So steeped in the illusory efficiency of these ideologies is the city of Cape Town that the 2014 mission statement of its Department of Water and Sanitation states: 'Nature is our silent customer who shall receive an equal share of our services.' Corporate culture – privatization and commodification – here encourages Water and Sanitation officials to imagine themselves as retailers. From this perspective, they are purveyors of what both nature and citizens, framed as 'paying customers', need.

What are we to infer from this capitalist sorcery (Stengers et al, 2011)? The key is to see the sleight of hand for what it is: by designating 'nature' within the framework of 'property', capitalist ontologies of ownership instantiate 'borders and edges' as natural reality – 'nature itself'. Thus captured, earthly spaces are set apart from 'society'; humans are free to live as extraterrestrials, independent of the movement of materials, matter, flows and life. And society's relationship with nature is financialized: in an 'end of pipe solution' via waste flows – sewage and solid waste – that, in Cape Town, land up on this edge of 'society' that happens to be, for the most part, a Black landscape. Thus terraformed, southern Cape Town serves as 'the arse-end' of the rest of the city: the waste of rich and poor alike lands here, in a sacrifice zone of waste and poverty that contains the sacred space of a rapidly vanishing nature (protecting some 15 near-extinct plant species).

On this ground, our exploration of the landscape we both live in finds a more meaningful ecological conversation with critical zone sciences, an approach developed by Earth scientists to understand the generation of conditions that support life at any place on the Earth's surface. We ask whether extending a critical zones approach to human systems can expose the necropolicies of ecosystem services thinking and better guide environmental governance; and whether, by focusing on habitability and wellbeing instantiated by material flows (instead of dollar values), ecological reparation might be possible in this zone of desperation and devastation: exclusions, extinctions and expulsions. In this chapter, therefore, we are writing both descriptively and experimentally to pose this question: how might an approach grounded in material flows direct our attention to remediative strategies in this landscape that exemplifies the Capitalocene?

We begin this work by following the call of anthropologist Eduardo Viveiros de Castro for a new field of 'anthropocenography' – a study of societal geology and its multi-species relations (UC Davis Social Sciences, 2015). To do so, we return to Google Earth: and to the images that, constrained by the risk of on-the-ground violence to Google Earth staff, offer only a god's eye view.

Anthropocenography of a necropolis

Google's archived pixels, in the metadata 177 m above ground, render crammed corrugated iron roofs into a sparkling mosaic, but at 500 m altitude – a shock, as the scale of the shack settlements becomes clear.

In this treeless shanty suburb, street life includes the dark, dank mud that results when sand mixes with effluent, harbouring microbial diseases of all kinds, yet this dystopia has an unimaginable name: Vrygrond, which in English means FreeGround, or Freedom Ground, named thus because it emerged in the immediate aftermath of apartheid in the 1990s. Thirty years later, its fragile shelters adjoin one another, and dozens may take flame in a night from a single fallen candle or illegal electrical connection. The slow

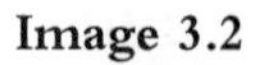
Image 3.2

Image 3.3

Image 3.4

violence of contamination is criss-crossed by daily gang violence: note in the adjoining area of Lavender Hill that park pathways do not testify to soccer games or picnics under trees, but to speed: walk as fast as you can, for conflict between Lavender Hill's gangs yields some of the highest gunshot rates in Cape Town. In Vrygrond, the few taps shared by thousands of people means that frequent handwashing and social distancing is a dream. In Lavender Hill, the struggle for water is a struggle against water management devices that can reduce household water supply to a 6 litre/hour drip-feed if water bills go unpaid.

At 1,000 m up, the extent of Lavender Hill becomes evident. There are no urban farms here. Space for plants, in this, the world's richest floral kingdom known as Cape fynbos, is a luxury. Backyards are places in which to erect more shacks, to earn a rental income or accommodate a family member whose name may be somewhere on the city's social housing list.

To the east, a pocket of the False Bay Nature Reserve, frequently invaded by those desperate for space to build shelter, also comes into view. Removal of 'land invaders' was in 2021 outsourced to a private security company known as the 'Red Ants' – one of the only companies in the country to have been delisted by the security industry for its use of excessive violence. Their 'hardened military vehicles' were branded with their bright logo featuring a red map of Africa; that they were contracted allowed both the city and the province to deny responsibility once the reports were in from Human Rights Commission monitors.

Homelessness is the defining political struggle of our city. In Cape Town, the apartheid state had a 7,000-long waiting list for affordable homes by 1986. The current city has an official waiting list of 17,000 – but in this city, where states of exception have become the rule, a housing activist forum estimates the unofficial housing shortfall to be at least 40,000 homes.

What is not visible, even in the desolation revealed at this elevation, is that many whose desperation has led them to build in this landscape of exclusions, expulsions and extinctions were already climate refugees when they arrived here, displaced by the flood–drought cycles and the wars that arise ever more frequently on an east coast of Africa warmed by rising temperatures in the Indian Ocean. Somalia, Malawi, Zimbabwe, Mozambique; South Africa's own Eastern Cape Province: all are represented.

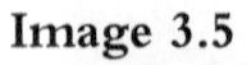

Image 3.5

Our viewpoint soars to 3,000 m; and now, the brutalism of Capitalocene Cape Town is in full view.

On the left, in the west, is Marina da Gama, an upmarket waterfront named for the sixteenth-century Portuguese protocapitalist who first opened Africa's shores to plunder; a high-walled estate trying to protect itself from spillovers of crime and sewage, with electric fences and citizens' environmental activism. One recent credible study of its water quality, undertaken by a freshwater ecologist, cited an *E. coli* count of over 8,000,000 colony-forming units (cfu) per 100 ml. The safe limit for recreation is around 1,000 cfu/100 ml.

In the centre, south of Vrygrond, is Capricorn, the lake that looks like the astrologers' sea goat, surrounded by a feudal-style high-security manufacturing park, branded an 'eco park' and surely earning its tenants green sustainability points towards listing as ESG shares ('environment, social, governance' – that is, those that advance the sustainable development goals). Its shopping centre is anchored by two upmarket food stores, and on the street corner opposite them, migrants sell their wares in the dust. Without commonage for food production, without jobs or documents, and with open lands fenced for either 'nature' or capitalist extraction, this offers yet another instance of the prevailing *conditio inhumana* (Agamben, 1998) of the settler-colonial Capitalocene.

The City's answer to the question of how the poor are to survive once lay in the east of the image, in the patch of ground that looks like a grumpy bureaucrat. This is the Capricorn solid waste dump, growing daily through the mechanized sedimentation of toxic suburban detritus and to which the southern city's waste is trucked. As the City formalizes its recycling policies in recognition of the previously overlooked value that lies in some waste, it is now the 'Green' capitalists, the recyclers – with their privileged access to the City's procurement processes and tenders – that have the first go, after which the poorest may be permitted to pick over the rest, the waste of the waste.[2] A new industry taps off methane from the belly of this beast: another nod to a 'green economy'.

Yet Cape Town's infamous prevailing south-easterly wind swirls the dust of the waste dump into the air of Vrygrond, where respiratory illnesses are legion, and into the surrounding nature reserve, on whose edges we have homes. To the east of the waste dump are the beginnings of the False Bay sewage treatment plant, which, as a favoured destination of birds and birders, forms part of the most industrialized Ramsar site in the world.[3]

★

At 40 km up, another surprise awaits: the Cape Flats is a valley; a channel between mountains and oceans.

And here, at an altitude of 500 km, False Bay is a small alcove tucked in behind the peninsula on which Cape Town lies, on the south-west shores of the southernmost tip of Africa.

Image 3.6

Image 3.7

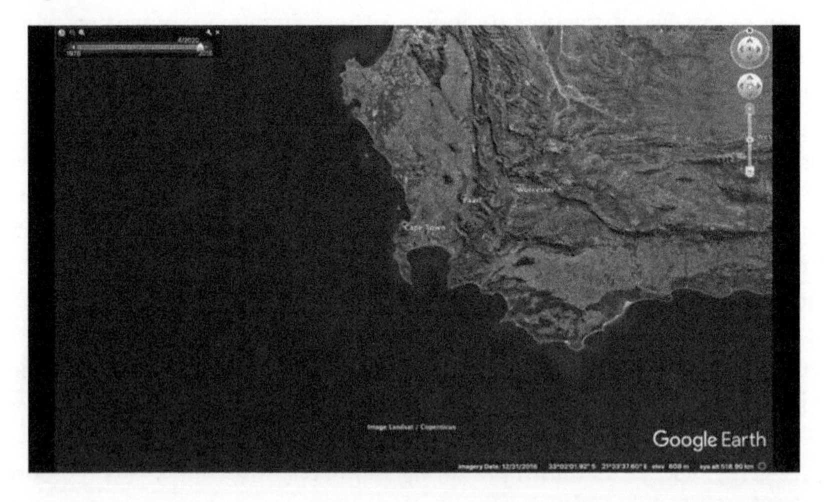

Diving back to a height of only 7 km on a cold winter day, when the icy clouds of the Cape winter reflect in the waters of the Capricorn green business estate, the shallow seasonal pans known as Rondevlei ('Round Lake') and Zeekoevlei ('Hippo Lake') lie just north of the wagon wheel of the sewage plant. Rondevlei is home to a family of six hippos, the only hippos in the greater Cape Town regions, whose fragile lives are stewarded by dedicated nature conservationists in the employ of the City.

At this elevation, the entirety of the Ramsar site is visible. To this protected wetland way-stop for migrating birds, stilts arrive yearly from Siberia, swifts from Europe and southern African birds are abundant: pelicans, flamingos,

Image 3.8

spoonbills, red bishops, bee-eaters, pin-tailed whydahs, fork-tailed drongos, sparrow-tailed hawks, cape barn owls, a pair of fish eagles. The extremely rare Cape clawless otter and the highly endangered Western leopard toad, a colony of butterflies extinct elsewhere and the rare plant on which they breed and feed; water mongoose; cape cobras, puffadders and the *boomslang*, or tree snake; porcupines; the angulate tortoise, genus *Chersina*, endemic to southern Africa. Their lives are made possible by the nourishing, endangered Cape Flats dune *strandveld* (beach scrub), once capable of supporting large herds of grazing animals but now preserved in only four reserves, of which the 290 hectares (720 acres) of mostly permanent wetland of the Rondevlei reserve established in 1952, the small lake to the left of Zeekoevlei, is the oldest.

Ringing the *vleis* and the sewage plant, Vrygrond, Capricorn and the dumpsite is False Bay Nature Reserve, a manifestation of the idea that a paradigm built on financialized ecosystem services will clean up the Anthropocene, the Capitalocene, even when ringed in by its rising tides of waste.

Fed by several small rivers and groundwater that, travelling roughly 100 m per year, takes 30 years to travel through sandstone formations from Table Mountain before surfacing atop a layer of impermeable granite, the *vleis* form the eastern edge of the Cape Flats Aquifer system. The rain that fell on the mountain last week will emerge here in 2051.

At the top right corner is the Philippi Horticultural Area, whose fresh produce – cabbages, carrots, potatoes, spinach, butternut, onions – supplies the whole city, and whose porous sands are on the primary recharge zone of the aquifer system when it rains, polluting it with heavy metals from the waste dump, agricultural toxins and fertilizers, and persistent organic pollutants

Image 3.9

from sewage. As the water cycle changes, the importance of this drought-proof farmland has become increasingly clear to concerned citizens – and for that reason was the focus of a court case brought by activists against the City as it emerged from its infamous drought in late 2019. The court battle, which they won, opposed the planned sell-off of the land to developers who would hard-surface it, risking its salination by seawater, presumably forcing the construction of more expensive infrastructure to pump treated effluent into the aquifer in the future.

Further capitalist sorcery underlies the next two Google Earth images, in which appeared – and disappeared a year later – a desalination plant that failed spectacularly in the drought.

Part of an almost billion-rand 'solution' to the drought's effects, this is one of three desalination plants that failed because the seawater it was meant to process is so polluted by sewage discharges, spills and polluted groundwater runoff that excessive algal blooms overwhelmed the filtration system on which desalination depends. From December to February – the dry, hot summer months in which the city's need for water was at its greatest – the plants could not function. Despite costly maintenance efforts to replace membranes and adjust treatment systems, all three plants closed shop before the end of their two-year contracts. One of them is heading for a court battle over claims that the City failed to disclose the full extent of seawater pollution.

With more than a tenth of the coastline of the country as a whole – 300 kilometres – forming Cape Town's deep bays and long peninsula, the failure of the desalination plants underscores that no 'end of pipe solution' awaits in a technotopian, 'Smart City' future. Moreover, the City's inability to identify a suitable site for desalination plants anywhere along its long seashore tells us that what was presented as a wastewater pipeline solution through which to

Image 3.10

move our unsightly problems to the edge of the periphery has only revealed a new centre. We are not extraterrestrial excretors: our waste stays on this planet. Sailors and scientists long abandoned the idea that the Earth is flat, yet the City's sewage management assumes that sending excrement to the sea makes it magically disappear, no longer affecting humans, animals or the life of the ocean – as if flushed shit falls off the edge of the Earth. Decades later, in the False Bay section of Cape Town designated for urban excreta, that magical thinking continues to fail. Contaminants cannot be corralled by law, policy, property boundaries or neoliberalism's reliance on segregating humans by race and class. Refuse refuses fences: when winds blow and rains fall, heavy metals from batteries and e-wastes and lightbulbs enter the streams. Sewage pollution enters the air when mud dries. Its contents enter the groundwater. Many farmers collect sludge as cheap fertilizer for their fields, even though Cape Town's sewage treatment plants offer no tertiary treatment that would remove chemicals listed by the UN Environmental Programme as 'chemicals of emerging concern'. These will not break down in generational timescales. They will be with us in geological time, perhaps as far forward as the time of the Neanderthals is behind us: 30,000–40,000 years. Out in the open, these contaminants travel with evaporated water high into the atmosphere, to other countries, oceans and continents. They bioaccumulate in fish and birds, in sediments and waters and in humans, no matter where, or with what smart technologies, we live in relation to waste dump and ocean.

*

At an elevation of 7 m, in a February 2000 satellite image, Google Earth showed the waste dump as a sand dune. A mere seven months later, sand

Image 3.11

Image 3.12

miners began to strip the vegetation and haul off loads of sand for building industries, their dune extraction shaping a small lake of groundwater at around 4 m above sea level – an indication of how close to the surface the groundwater lies.

By 2013, 11 years after the sand dunes were trucked away, the waste of the city had been sculpted into terraced slopes of beige and grey.

By 2016, instead of bushes, thousands of dump truck loads polka-dotted the surface, bulldozed into a sediment when they covered an entire contour. Will these be picked over and remarked on by archaeologists, geologists and anthropocenographers of the future or explode as a result of the methane

Image 3.13

Image 3.14

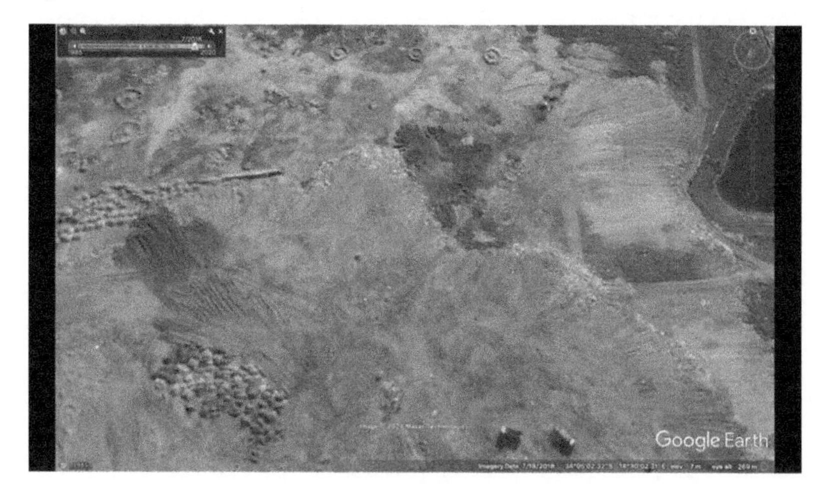

rising from this dump, vented off by engineers brought in to resolve the risk of a deep-belly fire – or will they be washed over the City in storms accompanying the sea level rises predicted when the west Antarctic ice sheet slips into the ocean?

A perfect storm gathers in this wasteland. After the City issued a tender to a private company to recycle waste in 2020, the Vrygrond waste-pickers, evicted from their only source of income, set the dump alight in protest. Did they do so with any knowledge of the multi-decade fires that can ignite in a waste dump, or of the toxins thus released? A few tense days followed their action, marked by thick toxic smoke billowing across the southern shores,

Image 3.15

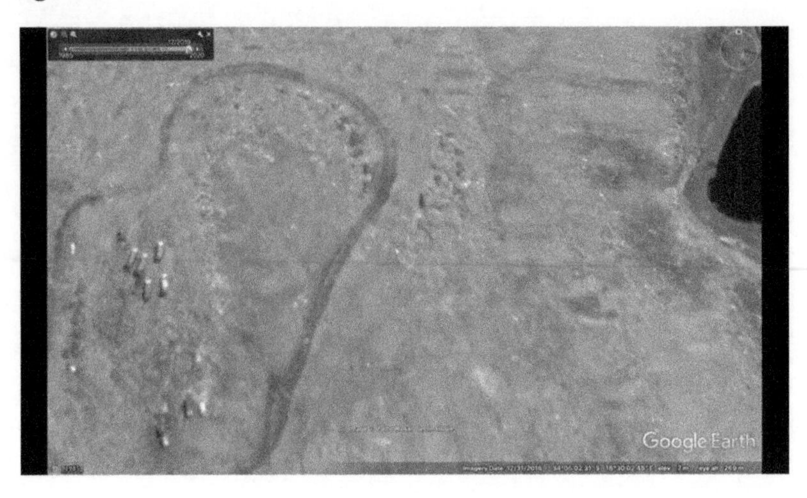

up through the nature reserve. A cape cobra took shelter from the smoke in an underground gutter pipe in Farr's garden in Zeekoevlei. The birds were eerily quiet. Our throats burned as we breathed the chemical compounds of the knowledge economy: did our heads ache from

> cadmium (Cd), chromium (Cr), lead (Pb), mercury (Hg), polybrominated diphenyl ethers (PBDEs), polybrominated and polychlorinated biphenyls (PBBs/PCBs), polychlorinated dibenzo-*p*-dioxins (PCDDs), polychlorinated dibenzofurans (PCDFs), and polycyclic aromatic hydrocarbons (PAHs)? (Dissanayake, 2014)

Which plastics burned our throats and eyes: bisphenol A (BPA) or perfluorooctanoic acid (PFOA) or polychlorinated biphenyls (BCPs) or chlorinated dibenzodioxins (CDDs) or furans or polyvinyl chloride (PVC) or brominated compounds (Young, 2005; Johansson et al, 2016; Verma et al, 2016)? A month before COVID-19 challenged our collective breathing, rare earths displaced from deep underground and persistent organic pollutants sourced from across the planet were settling in our lungs and blood, bones and tissues.

The waste dump, this shrine to the Anthropocene, this 'great derangement' (Ghosh, 2016), testifies to the philosophies of modernity: that fences, concrete and property law can keep solid waste solid, liquid waste liquid and everything in place. Cape Town's policies for water and waste management are designed for solid, liquid and gas, yet the chemical and pharmaceutical industries of our time leach, flow, aggregate, metabolize, sediment and bioaccumulate in bodies and soils, from the dump site and the sewage plant.

Image 3.16

This mingling of states of matter is central to life processes but peripheral to the idea of property boundaries, and so urban design is framed, taught, graduated and rewarded with little thought to flows. Homo sapiens created the knowledge economy and evolved into Homo economicus (Persky, 1995), who, becoming enamoured with the life of pixels – the Earth is known via Google, not soils – has forgotten the root of her humanity: her humus, that is, her soil, whose tending made possible her humanity, and whose abandonment unmakes humanity.

In a great visual joke, hidden in this bird's-eye view of where we live is Anthropos, the 'man of the world' who claims objective control of matter, now revealed as the face of the dump site near False Bay. The image has not been altered. The Zeus of our time, Anthropos shows the face of the great derangement: that of the bureaucratic patriarch who throws not bolts of lightning, but exiles trash from the socius without thought to regeneration, the commons or a circular economy; without thought to the dependency of all life on the re/cycling of matter.

A reparative ecology?

What future is there for the damaged, degraded, necropolitical landscape of the False Bay coast? What reparative ecology is possible?

In the Google Earth image of December 2018, our geological patriarch, Anthropos, faces away from the sewage plant behind him to which every toilet pipe in about a third of the city leads. Exemplifying the mythical 'geology of patriarchy', the ideas behind the ongoing harm to this landscape are: (1) a cosmological belief in the division of nature from society; (2) the belief that that bifurcation necessitates sacrifice zones of pure waste at ends-of-pipes (and ends-of-roads), and sacred zones (of pure nature, also at the ends of roads); (3) that there is no alternative, and that their 'good management' will resolve the nature-society divide; (4) the ontology of private property is naturalized in the belief that fences and legal lines have greater force in nature than material flows; (5) Anthropos believes that ecosystem services (like the nature reserve; the ocean) are a 'business partner' who will clean air, water and soil at minimal cost to the fiscus; (6) as a bonus, this same 'nature reserve' serves as a buffer between rich and poor, White and Black, citizens and undocumented, stateless, surplus people; (7) as a remedy, Anthropos will trumpet the gospel of Homo economicus: the figure of the human economic actor whose actions mediate nature and society; and (8) while denying material flow, Anthropos' public devotion to the dogma of trickle down is absolute (tax havens have not yet been admitted to the canon). The gospel of Homo economicus: the advent of the self-regulating market whose endless growth will spontaneously trickle down, magically releasing the poor from their burdens.

Yet the material trickle down of Anthropos is heavy metals out-of-place; of environmental contaminants that add to the disease burden, the hunger burden and the multiple burdens of slow violence in the City, all of which accumulate in the bodies of the waste-pickers; the undocumented migrants; those forced into apartheid resettlements; those who live (as we do) on the edges of nature reserves, and those who believe that new forms of green feudalism (like the green business park of Capricorn) will save them. Anthropos denies his ecological violence; his schizophrenic ecopathology; he is terrified of the loss of post-colonial communion with the patriarchs of London, Paris and New York. In order to maintain his belief in his mastery of nature and justify slowly violent depositions outside of his patriarchy, Anthropos must permanently create new sacrifice zones where necropolitical relations are subject to constant denial. A toxic politics of toxicity ensues, led by buffoonish figures on the New Right who believe themselves to be masters of land; or urban planning; of justice and of science, and whose account of their political worth rests on the new green accounting that is possible with measures like tapping methane from a rubbish dump (without encouraging recycling), or relying on ecosystem services to do the work of cleaning the commons, or investing in so-called 'green technologies' – like desalination. The world of relations, for such buffoons, is reduced to those evident in the x-y axes of an Excel spread sheet: relations of which they are always the master. Yet the plastic bag blows in the wind.

When the drought crisis hit Cape Town, the emergency desalination plants failed because governance only by financials ignored evidence of the proverbial 'shit-storm' the City had created in its oceans and because no-one, not even the social scientists among whom we count ourselves, had realized the extent to which social order depends on material flows such as water. We shared, at that time, some of the pathological self-delusions of Anthropos. Ecological reparation, we argue, must begin with a paradigm reset in which the delusions of 'social-ecological systems' give way to a science of earthly flows.

Inventorying the material flows in a zone that inscribes both sacred and sacrificial environmental cosmologies makes it possible to understand how and why the environmental governance paradigm of social ecological systems is failing Cape Town and the world in the task of dethroning Anthropos. A transformative environmental governance modality would embrace what anthropologist Eduardo Viveiros de Castro calls an 'anthropocenography' – social sciences as earth sciences and biosciences; not a biogeochemistry but a biogeosocial chemistry. That field does not yet exist. This study seeks to set out a method: an anthropocenography of a landscape of poverty, wealth, sewage, solid waste and conservation, that brings into view the big picture of flows and movements of toxins and nutrients, of solids, liquids and gas, metabolites and soils and society, between life and non-life.

The gaze of Anthropos has for so long told us *There Is No Alternative*. Yet an entire body of work in the environmental humanities has challenged what novelist Amitav Ghosh has termed 'the great derangement'. That approach appears surreal because it challenges the fixity, certainty and solid forms in which the environmental governance science is reduced to a bureaucrat's key performance indicator in an ontology of private property.

The surreal is necessary to bring to the fore that which cannot be seen within perspectives framed by prevalent dogma. Environmental justice work, with its long genesis and international network of environmental justice activists, attends, politically, to the restoration of the commons: not to private property or territory, but, building on biogeosciences, to the flows that make life possible everywhere on Earth. Yet this work alone is unlikely to persuade a city dedicated to proving its international marketability to put knowledge together in a way that solves crises that are core to it, since banishing waste to the periphery is central to the ontology of Rational Man. What is needed in environmental governance is a partnership with earthly flows, landforms and species: a paradigm shift that foregrounds biogeosocial relations and metabolic remediations.

An independent consultant to the City, not wishing to be named, explained that when the Capricorn Waste Dump was first approved, they advised the City to separate e-waste from garden and other waste. The City, citing 'cost', declined to listen – much as it has declined for more than a century to revise its policy of sending sewage to the sea (Overy, 2020). In so doing, the City's government clings to a neoliberal model of expendable-because-cheap nature, cheap work and cheap lives. The situation exemplifies the argument of Patel and Moore (2017) in *A History of the World in Seven Cheap Things* that 'cheapness' has given rise to the ecological and social crises of our time.

Unmaking the Anthropocene requires a fundamental rethinking of knowledge, key to which is to undo the fragmentation and atomization of knowledge to numbers and calculus that has rendered individual scientists, and individual disciplines, mute in the face of the logics of neoliberalism (Mavelli, 2019; Stiegler, 2020). Moreover, some of the City's elected officials responsible for water prior to the election of 2021 claimed to speak in the name of universal science, but sought to silence dissent because post-truth is a normative condition in situations where Anthropos environmental governance has failed but environmental governance cannot admit any alternative. The effect of that silencing, played out via reputational harm, has been to divide the scientific community who could be advising on improving the southern Cape Town 'anthrome' but have not done so, even when the opportunity has arisen.

An example: A recent attempt to synthesize three decades of social-ecological change in False Bay (Pfaff et al, 2019), the bay that adjoins Cape Town south, neither lists chemical contamination nor cites current published research in False Bay by environmental chemist Leslie Petrik, who has consistently contested the City's science and has been subject to ongoing reputational attacks by City officials in response. At no point in the review do the Capricorn Waste Dump and the Cape Flats Sewage Works receive mention, even while the paper's authors note that metal concentrations are most pronounced between Muizenberg and Strand, where these infrastructures are located. Contaminants are summarized in a single sentence: 'Metal concentrations in False Bay are influenced by the meteorology of the area, coastal topography, geomorphology, and hydrodynamics.' Further obfuscation follows: 'These environmental factors also influence the extent of metal contamination caused by anthropogenic activities.' Attesting to the City of Cape Town's effective war on the work of Petrik and other colleagues working on chemicals of emerging concern and persistent organic pollutants in City waste streams, the paper's authors avoid citing any science that discomfits City officialdom. Their prose is obfuscatory. These environmental factors – it's natural, not culpable – also influence the extent – we won't name any governance, management or policy problem – of metal contamination caused by anthropogenic activities – or make any effort to accurately ascribe any blame to the humans who cause the problem. The passive voice, the grammar that displaces responsibility, slips into tautology: Anthropocene contaminants are anthropogenic. This section of the paper concludes: 'recent research has confirmed that concentrations of metals such as cadmium, lead, and manganese in Western Cape marine ecosystems have increased since 1985

Image 3.17

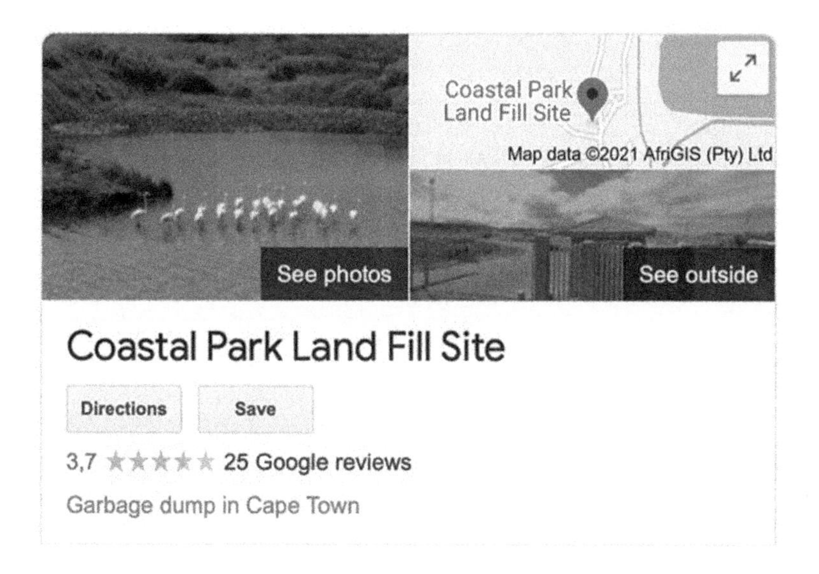

and are influenced by localized sources' and concedes that there is also 'evidence of bioaccumulation of metals such as arsenic, molybdenum, cadmium, copper and zinc in mussels (*M. galloprovincialis*) in False Bay'. 'Further research needs to focus on determining the source of contaminants to False Bay', the writers note, making no mention of the waste dump that sits directly on the aquifer, or of the policies and management practices that encourage their leakage. Invisible in this purportedly social ecological systems analysis of the False Bay biome is any mention of the existence of an 'Anthrome' – the biome of human material exchanges. This 'science' reflects the City's marketing department representation of the site on Google: at Capricorn there is no waste dump – it is a 'coastal park landfill site'! It is not a fetid wetland: it protects flamingos behind a security fence! Meet Brand Cape Town in the age of post-truth, where what is understood to be true, good science is what is marketable (Mavelli, 2019); where the sacrifice zone has become sacred pure nature. One could be forgiven for thinking the authors responsible for that section of the paper (Pfaff et al, 2019) have never visited Vrygrond, Lavender Hill, Philippi or Capricorn.

Unravelling the Capitalocene geology of False Bay requires a line of flight from the confines of disciplinary inquiry established in the age of expansion: the nature-society divide, with its sacred and sacrificial zones. An ecopolitics for living on a damaged planet must begin with the reconfiguration of knowledge through a politics of attention to material flows that make for habitability or inhabitability.

Notes

[1] Bantustans, or 'homelands' were small, unconnected enclaves, first set aside to separate Indigenous African inhabitants from white settlers by the British colonial regime, and then taken up by the National Party administration of South Africa and South West Africa (now Namibia) as part of its policy of apartheid.

[2] Emerging policies around waste management are a new vector for violent clashes between the City and the poor. See Steve Kretzmann (2020) for GroundUp.

[3] Wetland sites designated to be of international importance were first recognized in an intergovernmental environmental treaty, the Convention on Wetlands, or Ramsar Convention, established in 1971 by UNESCO, which came into force in 1975. The False Bay Nature Reserve became South Africa's 22nd Ramsar Site in December 2015, after more than a decade of citizen activism.

References

Agamben, G. (1998) *Homo Sacer: Sovereign Power and Bare Life*, Palo Alto, CA: Stanford University Press.

Dissanayake, V. (2014) 'Electronic Waste', in P. Wexler (ed) *Encyclopedia of Toxicology*, Cambridge, MA: Academic Press.

Ghosh, A. (2016) *The Great Derangement: Climate Change and the Unthinkable*, Chicago, IL: University of Chicago Press.

Johansson, K.O., Dillstrom, T., Monti, M., El Gabaly, F., Campbell, M.F., Schrader, P.E., Popolan-Vaida, D.M. et al (2016) 'Formation and emission of large furans and oxygenated hydrocarbons from flames', *Proceedings of the National Academy of Sciences of the United States of America*, 113(30): 8374–9.

Kretzmann, S. (2020) 'Tragic landfill conflict fouls Cape Town's Air', *Daily Maverick*, 28 February. Available from: www.dailymaverick.co.za/article/2020-02-28-tragic-landfill-conflict-fouls-cape-towns-air/.

Mavelli, L. (2019) 'Neoliberalism as religion: Sacralization of the market and post-truth politics', *International Political Sociology*, 14(1): 57–76.

Overy, N. (2020) '"Pollute the bay and poison the people": A short history of the Green Point Marine sewage outfall, 1882–1992', *South African Journal of Science*, 116(9/10): art. 8173.

Patel, R. and Moore, J.W. (2017) *A History of the World in Seven Cheap Things: A Guide to Capitalism, Nature, and the Future of the Planet*, Berkeley, CA: University of California Press.

Persky, J. (1995). 'Retrospectives: The ethology of homo economicus', *Journal of Economic Perspectives*, 9(2): 221–31.

Pfaff, M.C., Logston, R.C., Raemakers, J.P.N., Hermes, J.C., Blamey, L.K., Cawthra, H.C., Colenbrander, D.R. et al (2019) 'A synthesis of three decades of socio-ecological change in False Bay, South Africa: Setting the scene for multidisciplinary research and management', *Elementa: Science of the Anthropocene*, 7(32).

Stengers, I., Pignarre, P. and Goffey, A. (2011) *Capitalist Sorcery: Breaking the Spell*, London: Palgrave Macmillan.

Stiegler, B. (2020) *The Neganthropocene*, translated and edited by D. Ross, Saint Philip Street Press.

UC Davis Social Sciences (2015) *Anthropocenography: On the Coming Cosmopolitical War*, Video, 18 May. Available from: www.youtube.com/watch?v=ccPbLnTI9B4

Verma, R., Vinoda, K.S., Papireddy M. and Gowda A.N.S. (2016) 'Toxic pollutants from plastic waste – A review', *Procedia Environmental Sciences*, 35: 701–8.

Young, R.A. (2005) 'Dioxins', in P. Wexler (ed) *Encyclopedia of Toxicology*, Cambridge, MA: Academic Press.

PART II

Deskilling<>Experimenting

4

Reflections on a Mending Ecology through Pastures for Life

Claire Waterton

> Above all, repair occupies and constitutes an aftermath, growing at the margins, breakpoints and interstices of complex sociotechnical systems as they creak, flex and bend their way through time. It fills in the moment of hope and fear in which bridges from old worlds to new worlds are built.
>
> Jackson, 2014: 223

I'm a researcher in STS or Science and Technology Studies. And so, I have the privilege of doing what we anachronistically call 'field work' – or, I might say, the chance to encounter people and things in their ordinary settings. I'm interested in witnessing the way in which people are navigating and doing socio-ecological change. In the summer of 2019, I visited 17 different farms, scattered across the UK: from Fife, in Scotland, to Devon in the south-west of England.[1] I had the chance to talk at length to each farmer, or sometimes to a pair of farmers, on each farm, and to take a walk around the farm so that they could show me the work that they were doing. The farms were varied in size and in the diversity of their farm work, but they all had a few things in common. They were all rearing cattle for eventual sale as beef; they were all feeding their animals exclusively on grass pastures; and they were all following, more or less strictly, the guidelines of an organization that supports this practice – the Pasture-Fed Livestock Association (PFLA).

It doesn't seem a very remarkable practice – to feed your cattle on pastures! But in this chapter, I narrate the actions of these farmers in a particular way (and perhaps, problematically, in a way that they may not recognize). I narrate their practices as 'mending ecologies'. Mending ecologies is my preferred term, rather than say repairing ecologies, since mending denotes a mundane, lay activity, not necessarily requiring specialist skills. Mending requires observation (of what has broken down) and a focused amount of care (to mend and re-materialize). I suggest that the PFLA farmers are attempting to mend ecologies as they rear their cattle. As they do so, they contribute to a deep re-materialization of the ecology of the farm, including its socio-technical, non-human and human elements. Mending ecologies are therefore low-key, steady and care-ful actions. But they are also nouns – ecological assemblages that are subtly changing within complex systems. Consisting, therefore, both of human work, and of emerging webs of human/non-human life, mending ecologies contribute in emergent, possibly fragile, human and non-human ways to the ecological reparation that is the focus of this book. In the chapter I aim to show the intricacies of the practice of mending; what it resists and pushes against; how, contrary to common ideas about mending, this practice is innovatory and highly creative; how particular strategies and spaces are required for the work of mending; and how whole ecologies are re-materializing as a result.

Mending stories

The word itself suggests a modest activity. Mending is done. It is not necessarily talked about. It takes place often, without needing a story. But sometimes, a particularly good or bold bit of mending may inspire or even deserve a story. For Steven Jackson (2014) breakdown and repair are not unremarkable and dull aspects of our technoscientific world, but rather 'sites for some of its most interesting and consequential operations' (2014: 227). When things require mending, we can see more – how those things really work, and how they fail to work. Breakdown is seen by Jackson as a fertile 'site' for disclosing the kind of world we have been inhabiting and the kind of values that have been built into that world (2014: 231). And he sees those that fix broken worlds (or the broken ecologies that we will consider in this chapter) as hosts of specific and valuable kinds of knowledge and understanding (Jackson, 2014: 229).

Stories of breakdown and mending can gather (as in this volume) and, when this happens, they can also be stories about change and about scale. This happens in the work I present here. From stories of everyday mending experiments on livestock farms, we will end up thinking about land, soils, forms of life and whole ecosystems. But that is for the end of the chapter. Where does the mending story begin?

Stories, writes John Berger, don't come through searching for them.

> [Stories] come up behind you and tap you on the shoulder. You look round and there's nothing there. Then, another day, they tap again.

> And this time you don't look round, you start remembering. You start listening to something that now won't let you forget it. Something you heard or saw or met which you hardly took any notice of at the time. A story is something that happened (even if only, sometimes, in imagination) and which wants to come back, like a dog, on the jacket of your imagination. (Berger, 2020)[2]

I'm at the farm. We're in a field. But the field is not quite right somehow. It feels odd. Like the oddness you sometimes sense in dreams where things are not quite as some part of your brain expects. Although I have come to see cattle grazing, I can't see any cows at all. Then I see them, but they are very far away. I notice that, rather than being dispersed throughout the field, the cattle are all in one tight group, grazing in a small, fenced off square of grass to the side of the field. The fence holding them there looks flimsy – just one strand of wire held up by some thin posts thrust into the ground. The wire is electrified. The farmer is keen for us to be at this spot at the right time and so we make our way over to this part of the field. "I'm just thinking, I'll show you as they move to the next pasture. What time is it now? They're literally going to go through in the next minute."

Then, CLACK!

"There we go!" The farmer looks happy as a length of electric fence pings free from the corner of the square paddock. The cattle seem to hear the sound and they jostle forward into the next square of pasture. They have become used to the timed release-latch mechanism made by the farmer to let them out of one square paddock into another. The farmer explains to me why what I'm seeing is significant. Rather than allowing the cattle to roam and graze freely across the entire field, his latch mechanism controls their movement, allowing them to graze for specific time periods on confined paddocks of lush, knee-high pasture. This is a low-tech innovation allowing a new kind of grazing. The cattle are enticed forward into fresh, tall pasture four times a day, every day. He delights in watching the cows skip through: "Honestly, I just love it! I set my automatic latches up, but if I'm not busy I'll go down just to watch them go through."[3] The CLACK! reverberates in my memory. It's the soundtrack of a mending ecology. It's a sound emblem, enabling, for this farmer, an altogether new way of farming.

As I travel to different farms in different parts of the country that summer, I see the same thing again and again – the farmer's pleasure in observing the animals as they graze, and (if I am there at the right time) his/her scrutiny as they pass through from one paddock into the next. I notice each farmer gauging, by eye, the height and the quality of the pasture that the cattle are feeding on. The grass and the animals. The same observations each time. How much have they eaten? Are they hungry or full? How much will the next paddock last? When will they next need to be moved on?

But there's another image that is reverberating in my memory. We're on a train. We're travelling south, working together on this research about the PFLA farmers. My colleague is talking about what she imagines other people feel when they look out of the window of the train: "Doesn't this look lovely – lovely green fields!" She imagines their sense of relaxation as the train passes through what seems like miles and mile of lush green vegetation in the north of England. But where her imagined other passengers see nature, she sees miles and miles of pasture dominated by a single species of fast growing, highly nutritious grass.[4] We're speeding across a vast monoculture. She tells me what percentage of the land in England is 'down' to this kind of pasture. She sees a broken ecology.[5] We are travelling across a broken ecology and it needs mending.

A broken ecology

These two recollections, tapping me on the shoulder after a summer of farm visits, provide a kind of framing. We can now see a broken ecology, visible as lush green fields from our passing cars and trains. And we can also hear the CLACK! and the jostle of the cattle as they crowd into a new patch of knee-high pasture. Starting with these two cues, we can begin to imagine them as fragmentary indications of something that is going on, something that is recognizable elsewhere, something that is an '*issue*' (Marres, 2007), at scale, within institutions, within policy, at the level of regions and countries, and even globally. The issue is known, it is in the media, it is being talked about, it is reverberating and it is uncomfortable. As one of the 17 farmers put it:

> 'I grew up in the model of the post-war method of farming which was "Increase production using as much fertiliser, and re-seed, as much as possible. Push, push, push the cows". ... You know, maximise inputs to get the output.' (Farm 1 Interview)

Having grown up with this model, he applied it on his own farm:

> 'The soil just degraded. It was unworkable in summer and too wet to get on in winter. We didn't get the rainfall in summer so the fertilisers you applied didn't do any good. In fact, they did harm. And, looking back, it did real harm to the farm. ... Within six years I realised I was going nowhere, we were starting to lose money by this time. I was exhausted. Er, you know, I just wasn't really coping.' (Farm 1 Interview)

It takes a lot for farmers, producers themselves, to acknowledge that the post-war productivist turn in agriculture came with, and continues to operate with, huge costs.[6] But farmers reflecting on the high-input, high output, chemically intensive, industrialized agriculture that has become a global norm today point not only to the first-hand damage to the farm socio-ecological system, as earlier, but to the deeply capitalistic nature of agriculture and the systemic, financial lock-ins deriving from past and present inequities:

> 'For me it's the realities of economies of scale … so the margins are tight, you know. If you look at Australia now, they're talking about 3000 cows per labour unit/person. And we're going to be in the world market. Yeah, so it's capital intensive. There's no realistic routes into it now, I'd say.' (Farm 4 Interview)

This farmer is referring to the trend, seen in Australia and New Zealand, but also in North and South America, towards massive farm enterprises with high numbers of cattle managed through precision livestock data systems. He's referring to a particular kind of business intensity, and to the corporatization and globalization of farms and markets that create barriers to entry for smaller family farmers. These realities – through their intensity, their risks and their scale – are by their very nature on the verge of breakdown, or are already causing specific kinds of breakdown (fewer routes in for young generations of farmers, for example).

And last, to add to this picture of a broken ecology, or one that is felt by farmers themselves to be at breaking point, farming – especially beef farming – has also had to face the charge that it is actually causing harm to the environmental system upon which it depends. Greenhouse gases from beef production, including pasture-fed systems, hit the UK headlines around 2017, just before the COP23 United Nations Climate Change Conference in Bonn, as journalists picked up on research about the relationship between beef farming and climate (Garnett et al, 2017; Pearce, 2017; *The Guardian*, 2018). Debates have since proliferated, ranging from concerns about animal welfare and the quality of meat to those that are motivated to put an end to the animal agriculture industry altogether (*The Guardian*, 2021). Farmers are thinking through this terrain:

> 'We were always thinking we're doing a job of producing food … we're helping feed the planet in a sustainable way, and then, you're told that actually what you're doing is bad, you're not doing a good thing at all, you're actually part of the problem, all this planet's problems is because of the way that we are farming, and you're going "shit", going "bloody hell, have I been kidding myself all this time? Are we actually doing a negative thing?" And that's made us change things.' (Farm 6 Walk)

Broken world thinking

What we are seeing in these quotations is what Jackson (2014) calls 'broken world thinking':

> an appreciation of the real limits and fragility of the worlds we inhabit – natural, social and technological – and a recognition that many of the stories and orders of modernity (or whatever we choose to call the past two-hundred odd years of Euro-centred human history) is in process of coming apart, perhaps to be replaced by new and better stories, but perhaps not. (Jackson, 2014: 221)

And of course, broken world thinking is the pre-cursor to repair: '*That's made us change things*'. But there are two points of emphasis that will help us understand the work that the farmers are doing to mend ecologies. First, this is not only about fixing, or patching things up. Jackson declares repair – and mending – to be highly innovatory. In studies of mending, he picks out and celebrates the 'innovation at the heart of repair', and the counter-intuitive feeling that menders and repairers are not 'going back'; they do not prefer stasis over change; they are not harking back to some lost or pre-modern harmony; nor are their projects inherently noble or good (Jackson, 2014: 233). Rather, menders are propelled forwards. In an ordinary way, and as many have done before them, they are using what they have to build fallible projects into the future (Jackson, 2014: 237–8).

And second, as Michelle Murphy explores, the menders – those that build new worlds, or new ecologies – are compelled by their circumstances to *experiment* (Murphy, 2006: 176). Once again, they are not looking to go back to a perfect harmony, but to create, out of impossible or damaging body-ecologies, 'concrete and valuable ways of inhabiting the world' (Murphy, 2006: 177–8). Through her study of toxic building syndrome, Murphy shows us what ingredients are needed for mending ecologies to take place. But what is more important for our thinking is that mending demands room for experimentation – specific 'spaces' that help build these new ecologies. Institutional and epistemic spaces to legitimate the work of mending ecologies. Solidarities – increasingly through the spaces of virtual media – to support the angst and determination of going against the grain. And experimental practices for micro-managing the new and emergent ecologies by those who are bringing them into being. The performance of experiments is needed, and these experiments draw on previous cultural narratives and performances. The menders involved are busy, focused, sometimes obsessive. They are re-arranging things, 'drawing on and mutating dominant assemblages, sharing information, and pulling objects out of one arrangement and into another' (Murphy, 2006: 172).

'And that's made us change things'

Each of the farms I visited over the summer of 2019 had a story of a broken ecology, be it a story of 'pushing' production beyond the capacity of the land, a story of degraded ecosystems, a story of human exhaustion, a story of inaccessible economies of scale, or an uncomfortable narrative about contributing to the global climate crisis. Each of these stories, however, propelled new and experimental practices of repair. In this section I look at the innovation inherent in the new mending ecologies taking place on livestock farms, focusing particularly on the experimental practices of the farmers. Again, I want to highlight that these new experimental practices are not striving for a lost harmony; that farmers are aware that their new practices may be unleashing new problems and breakdowns; and that the act of recognizing a broken ecology and 'changing things' runs counter to the still dominant productivist narrative regarding how to raise animals and produce meat (although this is rapidly changing[7]). All of these things point to the fact that mending ecologies, 'changing things', is a risky activity. Farmers are therefore also constructing 'safe spaces' (Murphy, 2006) in which to build better ecologies.

I begin to explore these spaces with the PFLA, founded by livestock farmers. This organization sees itself as a mechanism for bringing together British farmers who want to produce food 'in a more natural way' (PFLA 2016: i). This more natural way, for the PFLA, means feeding beef cattle entirely on grass and forage crops and 'kicking the habit' of fattening animals on grain (PFLA 2016: i). So the PFLA seeks to create community among practitioners and to commit that community to adhere to the one key principle whereby animals eat grass and forage crops their entire life. But if this is seen by some as radical, even 'revolutionary' in the context of conventional livestock farming, the PFLA wears its philosophy lightly. In its various publications it does not go into the pitfalls of the agro-industrial farming system. It does not stick its neck out as a campaigning NGO. Rather it creates a safe space for community, for championing what it defines as 'more natural ways' of rearing ruminant livestock. Rather than portraying itself as a radical counter-cultural farming organization, it seeks to draw on scientific evidence to support the kinds of claims that can safely be made – for example, regarding the nutritional benefits of produce derived from livestock raised solely upon a pasture-based diet.[8]

The PFLA's modest claims, and its single defining standard, whereby animals have only eaten grass and forage crops for their entire life, however, belies much more terrain in terms of sustainable ambition. The PFLA Certification Standards for Ruminant Livestock (2020) 'cover land as well as animal management to ensure that farms with Certified [PFLA] Enterprises deliver environmental goods rather than just avoiding environmental harm'.[9]

The PFLA standards, therefore, pro-actively try to 'change things' in order that 'environmental goods' conventionally externalized by the market (healthy soil, carbon storage, clean water, diversity of species) are brought into the production cycle. While explicitly aimed at ensuring that PFLA Certified animals have only eaten grass and forage crops for their entire life, they implicitly expect: (1) a distinct nutritional quality to its products; (2) environmental benefits; (3) animal welfare benefits; (4) extensive production. In the last of these stipulations, the PFLA make clear that on all farms with PFLA Certified Enterprises the numbers of ruminants in a herd 'reflect traditional mixed farming practice as distinct from specialist intensive production'.[10]

PFLA thus take care in curating a safe space for farmers that want to produce food 'in a more natural way'. They are treading a fine line between the old and the new, convention and innovation, the routine and the novel. This does not mean that they are reverting to an outmoded, a-modern, way of farming, or that they prefer stasis over change. Neither does eschewing specialist intensive production mean that their members are uninterested in making profits from their new mending ecologies. The PFLA website is a public forum through which the organization can showcase farmers' own experiences of experimenting and innovating the way that they rear pasture-fed cattle. Farmer narratives about their own businesses are prominent features and these allow specific aspects of the grass-fed system to come to light – specifically practicality, economic viability and profitability. Further virtual 'safe spaces' are available for members through the PFLA Google Group – an intensely active forum that all 17 of the farmers I visited regularly access. The Google Group was uniformly referred to as the go-to source for reliable, practical, time- and money-saving advice for farmers looking to change their working practices in order to raise their beef on a pasture only diet.

'We had to learn to utilize our grass better'

As a community of practitioners wanting to support each other in the goal to raise, feed and fatten their stock solely on grass and forage crops, one of the most important goals of the PFLA farmers is to 'learn to utilize our grass better' (Farm 6 Interview). But working out how to mend an ecology in order to be able to use pastures 'better' is not as easy as it may sound. We can think back here to the lush green fields that we sped through on the train at the beginning of this chapter. Re-seeding and fertilizing, with the high yielding and nutritious grass *Lolium perenne*, perhaps mixed in with some white clover for nitrogen fixing, has been the standard way of increasing what farmers can gain from a field of pasture since the 1950s. But, as the very first farmer quoted in this chapter realized, and as the PFLA farmers and a growing group of 'regenerative' farmers also increasingly recognize *Lolium* – which is fertilizer-hungry (Williams, 1985), tending to decline and require re-seeding after a few years on poorer soils (Grime et al, 1988) – is no longer the answer.

What are the alternatives? Faced with their green fields, PFLA farmers are undergoing a vast amount of experimentation, trial and error in the way that they go about herding and grazing their animals. As they do so, they are reporting their methods and results, their tricks and tips, their triumphs and their problems, on the PFLA Google Group. In place of 'set stocking' their cattle on a single field for up to a year, they are trying out new methods – variously termed mob grazing, holistic planned grazing, strip grazing, techno-grazing, precision grazing, disruptive and tall-grass grazing. Mob grazing – perhaps the most often used term – has been dubbed 'the new old craze':[11] practitioners are aware that agriculturalists have been experimenting with similar methods since the late eighteenth century![12] Even so, the number of grazing groups, on-farm demonstrations, virtual videos, tweets and discussions on the subject continues to rise exponentially.

Following each other in the PFLA, as wellbeing inspired by some vocal pioneers in the USA, Canada and South Africa, all of the 17 farmers I visited had stopped the practice of set-stocking their cattle. So, rather than leaving their cattle to graze in a fenced off single field for up to a year (or until they have eaten all of the available pasture), the farmers are herding cattle at a rhythm that suits their new working practices – from every few days to as regularly as four times a day – into new, much smaller patches of pasture containing high (often knee- or thigh-high) levels of grass. And rather than allowing the cattle to graze that pasture down to the level of the soil, they allowed them just enough time to eat only the top third of the plants before moving them on to the next paddock. And rather than re-using each paddock of grass for grazing as soon as the grass grew back, they told me that they would let the grass in each small paddock fully recover, 'resting' the trampled grass for between 40 and 100 days before allowing the cattle back onto that patch.

Among the farms that I visited, 'utilizing our grass better' is a key motivation, inspiration and even obsession of the farmers. In contrast to set-stocking, all of these farmers, in subtly different ways are trying out what Tom Chapman, Nuffield Scholar and expert on the subject, calls 'short duration, high density grazing with a longer than usual recovery period'.[13] In a YouTube tutorial Chapman explains that the thinking behind this is 'trying to emulate nature'. To illustrate this he draws on a strangely familiar vision – that of a herd of bison, a compact mass of animals, moving slowly across the prairies in North America. The idea put forward is that the herds move through, eating the nutritious tips of the prairie grasses. They splatter the grass with urine and muck, they trample the grasses down and move onwards. They may not return to that patch of land for many months or even a year. Chapman thus calls on, and entangles, a familiar vision/ideology of wild nature with the new proposed performances of high-density grazing. There is something complex about drawing on this vision, enmeshed as it is in a heady romanticism combined with the memory and remorse of unthinking slaughter (Williams, 1960).[14] He offers a long list of benefits: maximizing the capture of energy from the sun in the pastoral system; improving soil health, improving cattle health, improving mineral, nutrient and carbon recycling; improving the water cycle and improving biodiversity dynamics.

What became clear as I visited the farmers was that a major learning and collaborative process – through reading, watching YouTube videos, visiting each other's farms, hosting and attending farmer-led conferences, gauging advice on PFLA Goggle Group discussions, Facebook and Twitter, and so on – is resulting in a significant shift, not only in the actions of farmers but in their entire vocabulary and way of thinking. As one farmer put it: "I'm a soil farmer and the beef cattle are probably the most important thing on my farm. … In fact, no, I'm not. I'm a sunlight farmer because that's what's feeding everything. You know, without that we're, we're pretty stuffed" (Farm 11 Interview).

The changes on farms are everyday. Decisions that calibrate the health and growth of the cattle against the health and growth of the pasture are being made by each farmer at the micro-scale – how far to move the fence along? What is the best way of sequencing the paddocks? How long should the cattle be in each one? Should I increase or decrease paddock size? What is very apparent is that, as the farmers learn to 'utilize their grass better', the farms look different. The pastures are long – their 'solar panel' enhanced leaf areas give a rough textured look to the fields. Flowering grasses and plants and seed heads change the colour and tone of the fields over the seasons. Once the cattle have passed through a grazing paddock, the stalks of the remaining 60 per cent of the forage are half trampled into the ground, half left standing, splattered with cowpats. To a conventional farming eye this may look 'quite a mess', explains Chapman.[15]

There are uncertainties and discomforts in this learning, then. Farmers talked about getting the hang of managing their cattle and their pastures in this way: "The mob isn't big enough; the rest period isn't long enough; the grazing period isn't short enough. We're actively trying to change that" (Farm 14 Walk).

Farmers are making mistakes and reflecting on them:

> 'You have to experiment, and sometimes you get it wrong, you're just judging how much they can eat. ... I let the other lot in last night into a bit that I thought would last them all day, and I had a look just before

> lunch and, um, they were all pacing around, and making a hell of a mess in the mud, saying they were hungry already.' (Farm 17 Walk)

And farmers are collaborating, not only between themselves, but with their animals, as they refine and fine-tune their paddock boundaries and timings, and as they observe and judge daily the quality of the grazing and the health of the animal. Farmers are 'in partnership with their animals', excited about their new methods, enjoying the new rhythm of the work: 'I get to see the cattle more' (Farm 12 Walk). Some farmers are keeping their mobs large, and multigenerational, so that the animals can express 'their total behaviour' (Farm 5 Walk). Farmers are noticing that watching the cows can be just as useful as using a specific tool to judge when to move on to the next pasture: 'The cows will tell me when they want to go' (Farm 5 Walk). A farmer who had been mob-grazing for five years told me how he could tell that his cows were contented: "Well, they're eating, they're eating, putting on weight. Er, you know, their coats shine. Look at that dark line down their back, you know. They're fat, they're a happy group, their eyes are bright. They're fine" (Farm 3 Walk).

In specific ways these farmers, together with their cattle, and drawing on both old and new performances, are constructing and materializing a new mending ecology. What the YouTube videos, the twitter and Facebook discussions, and the PFLA Google Group have done is support the farmers in specific kinds of experimentation to 'utilize their grass better'. This has enabled them, through setting up new paddock boundaries, changing the rhythm and the pace of grazing, even sometimes changing their animal breeds, to materialize a new ecology. This is a slow process – a mending ecology – involving a revised assemblage of technologies including solar powered electric fences, especially made reels for rolling out fences, automatic latches, specific kinds of mobile or static water trough, and so on. Some parts of the old assemblage gradually become less useful – overwintering barns, storage for hay and silage, tractors and loaders, feed areas and feeding systems – since farmers are keeping their cattle outdoors for longer periods over winter. The changes ripple through all parts of the socio-agro-ecology that is the farm.

For most of the 17 farmers I visited, the materialization of this new mending ecology was something relatively new. Their excitement at the learning, the constant gauging of grass growth against the needs of the soil and the health of the animals was palpable. Their day-to-day work had changed. Their identities as farmers have changed: "We're hoping for maximum production as well as improving the quality of the land at the same time" (Farm 17 Walk). Farmers were observing vital changes on their land: in the course of preparing this chapter I have come across 22 references to different bird species that farmers have observed on their farms, 23 references to plant species and 13 mentions of different invertebrates observed, including insects such as moths and butterflies. Farmers are meeting for coffee to discuss dung beetles! The farmers' profits were also reported to be healthy: "If we can obviously just do it off grass we're more in control of the costs" (Farm 7 Interview); "We're just seeing if we can strip out the inputs you know" (Farm 6 Interview). Of mob grazing: "so not only was it much cheaper but they [the animals] did better" (Farm 17 Interview).

Learning to mend their farm ecologies in the ways I have just described thus seems to be an immersive, collaborative and even transformative process. The farmers I visited seem to be shifting the way they work: rather than working on the farm, they seem to be working with, and becoming part of, the emerging ecology of the farm. As they do this, their senses are enlivened; their observations become acute; their enthusiasm and willingness to tinker and experiment is heightened and shared. But they are also aware that this is not just about them: they are learning, in a relational and interdependent way, to attend to a multi-species ecology that is on-the-mend, a mending ecology that is vulnerable but that does have the potential to flourish.

Reflections on a mending ecology

This mending ecology thus contributes, through specific shifts in human labour and complex multi-species interactions, to the ecological reparation that is the theme of this volume. The mending ecology charted here is characterized by what our editors call the unfolding, relational and interconnecting movements of ecological reparation, whereby human and non-human others patch up and reclaim the socio-environmental fabric.

Following Jackson (2014) and Murphy (2006) I want to say that this mending involves a gradual rematerialization of the whole body of the farm in a way that counters the mode of production of intensive industrialized livestock agriculture. As in Murphy's example, this has been carried out in specific spaces and with specific supportive and collaborative mechanisms in place. It is clear that this involves a re-making of the old and the new in experimental and possibly fallible ways. It is risky, but it is also hopeful and exciting for those involved. Much of the new system is built from rejection of the 'push, push, push' of intensive agriculture, yet many of the PFLA farmers are still keen to optimize production, to grow more grass, increase their mob sizes and to see how much their pasture can produce while simultaneously improving the ecological health of their land. At the same time, the debates around livestock and their relationship to environmental harm are loud and contested. The stakes are high – for livestock farmers, for diets, for the land and for our atmosphere. There may be mistakes. As Jackson (2014) suggests, the dynamics of breakdown and repair do not guarantee noble outcomes.

But what is clear is that the farmers I visited in the summer of 2019 are energetically engaged in a process of self-directed and farm-centred learning to try to mend an entire socio-agro-ecology that has come under strain and that, in turn, is putting huge strains upon the dynamics of our planet. This at least gives us the possibility of looking with more discriminating and fresh eyes as we pass through lush green fields on a southbound train. Are mending ecologies at work on this land? Might we pass through a patchwork of pastures indicating a new patterning of grazing going on? Might we catch the CLACK of a mechanism designed to move the cattle on, keep the pastures long, harvest the sunlight, grow the roots, and turn what was a monocrop into a lively, life-giving place? This mending is non-innocent. It is fallible. But, like all processes of reparation, it is hopeful.

Notes

1 I would like to thank Dr Mark Toogood who accompanied me on some of these farm visits and took part in the interviewing.

2 John Berger's note on stories and storytelling were incorporated into a guide for teachers of drama assembled by the Theatre Company Complicité. www.complicite.org/media/1439372000Complicite_Teachers_pack.pdf

3 Farm Walk 1: 19.

4 We were seeing the dominance of the grass *Lolium perenne* in British pastures.

5 As an ecologist, she sees what Aldo Leopold (1972: 165), the early twentieth-century ecologist, calls 'damage inflicted on the land'.

6 There is a vast literature on this. A recent example arguing that farming needs to change radically – meaning literally 'down to its roots' is Tudge (2021).

7 A double spread on the benefits of 'mob grazing' in *Farmers Weekly* in 2019 spurred on debates on livestock farming methods among British farmers: 'How beef herd is delivering £350/cow net profit with deferred, mob grazing'. *Farmers Weekly* 22 February 2019. Available from www.fwi.co.uk/livestock/grassland-management/how-beef-herd-is-delivering-350-cow-net-profit-with-deferred-mob-grazing

A more recent cluster of articles on regenerative farming added to these debates, 'Livestock: Soil Special' in Farmers Weekly *Farmers Weekly* 23 October 2020, pp 30–41. Available from www.fwi.co.uk/livestock/grassland-management/how-dairy-farmer-gets-14t-ha-dm-of-grass-with-minimum-inputs

A report by the Food, Farming and Countryside Commission (2021) 'Farming for change: Mapping a route to 2030', arguing that livestock have a key role in sustainable farming will no doubt add to the debate. Available from https://ffcc.co.uk/assets/downloads/FFCC_Farming-for-Change_January21-FINAL.pdf. It was also reported in *Farmers Weekly*: www.fwi.co.uk/livestock/beef/orfc-2021-livestock-have-key-role-in-sustainable-farming

There are counter-arguments in the current agricultural press that are cautious about the excitement around regenerative agriculture. New Zealand-based soil scientist Doug Edmeades, for example, expresses caution: "I'm concerned that the term regenerative agriculture is undermining the confidence of many farmers who may now be questioning if their practices are bringing about environmental harm." At the WA Livestock Research Council (WALRC) and Australian Association of Animal Science conference 'The Great Livestock Industry Day Out (GLIDO)', 1 February 2021 he promised to 'have a good hard look at that.' www.farmweekly.com.au/story/7067968/big-topics-on-livestock-industry-agenda/

8 www.pastureforlife.org/why-pasture/

9 PFLA Certification Standards for Ruminant Livestock, Version 4.0, May 2020: p 8. www.pastureforlife.org/media/2020/08/PfL-Standards-Update-Version-4.0-FINAL-v2.pdf

10 PFLA Certification Standards for Ruminant Livestock, Version 4.0, May 2020: p 8. www.pastureforlife.org/media/2020/08/PfL-Standards-Update-Version-4.0-FINAL-v2.pdf

11 www.farms.com/news/mob-grazing-the-new-old-craze-90285.aspx

12 John Anderson, a Scots Enlightenment thinker and farmer stated in 1777, for example: 'To obtain this constant supply of fresh grass, let us suppose that a farmer who has any extent of pasture ground, should have it divided into 15 or 20 divisions, nearly all of equal value: and that, instead of allowing his beasts to roam indiscriminately through the whole at once, he collects the whole number of beasts that he intends to feed into one flock, and turns them all at once into these divisions; which, being quite fresh, and of sufficient length of bite, would please their pallet so much as to induce them to eat of it greedily, and fill their bellies before they thought of roaming about, and thus destroying it with their feet.

And if the number of beasts were so great as to consume the best part of the grass of one of these enclosures in one day, they might be allowed to remain there no longer; giving them a fresh park every morning, so as that same delicious repast might be again repeated.

And, if there were just so many parks as there required days to make the grass of these fields advance to a proper length after being eat bare down, the first field would be ready to receive them by the time they had gone over all the others; so that they might be thus carried round in a constant rotation.'

13 www.soilassociation.org/our-work-in-scotland/scotland-farming-programmes/mob-grazing/what-is-mob-grazing/

14 John Williams' novel drives home the cultural complexities of wild buffalo slaughter in the late nineteenth-century West of the United States.

15 www.soilassociation.org/our-work-in-scotland/scotland-farming-programmes/mob-grazing/what-is-mob-grazing/

References

Garnett, T., Godde, C., Muller, A., Röös, E., Smith, P., De Boer, I.J.M., zu Ermgassen, E., Herrero, M., Van Middelaar, C.E., Schader, C. and Van Zanten, H.H.E. (2017) 'Grazed and confused? Ruminating on cattle, grazing systems, methane, nitrous oxide, the soil carbon sequestration question-and what it all means for greenhouse gas emissions', *Food Climate Research Network*, 708.

Grime J.P., Hodgson J.G. and Hunt R. (1988) *Comparative Plant Ecology*, London: Unwin Hyman.

The Guardian (2018) 'Rising global meat consumption "will devastate environment"', The Guardian, 19 July. Available from: www.theguardian.com/environment/2018/jul/19/rising-global-meat-consumption-will-devastate-environment

The Guardian (2021) '"Let's get rid of friggin' cows" says creator of plant-based "bleeding burger": Impossible Foods working on milk and fish substitutes as Patrick Brown pledges to put an end to animal agriculture industry', *The Guardian*, 8 January. Available from: www.theguardian.com/environment/2021/jan/08/lets-get-rid-of-friggin-cows-why-one-food-ceo-says-its-game-over-for-meat-aoe?CMP=Share_iOSApp_Other#img-1

Jackson, S.J. (2014) 'Rethinking repair', in T. Gillespie, P. Boczkowski, and K. Foot (eds) *Media Technologies: Essays on Communication, Materiality, and Society*, Cambridge, MA: MIT Press.

Leopold, A. (1972) *Round River*, Oxford: Oxford University Press.

Marres, N. (2007) 'The issues deserve more credit: Pragmatist contributions to the study of public involvement in controversy', *Social Studies of Science*, 37(5): 759–80.

Murphy, M. (2006) *Sick Building Syndrome and the Problem of Uncertainty: Environmental Politics, Technoscience, and Women Workers*, Durham, NC: Duke University Press.

Pearce, F. (2017) 'Grass fed beef is bad for the planet and causes climate change', *New Scientist*, 3 October. Available from: www.newscientist.com/article/2149220-grass-fed-beef-is-bad-for-the-planet-and-causes-climate-change/

PFLA (2016) *Pasture for Life: It Can Be Done, PFLA*. Available from: www.pastureforlife.org/news/pasture-for-life-it-can-be-done/

Tudge, C. (2021) *The Great Re-think: A Twenty-First Century Renaissance*, Pari: Pari Publishing.

Williams, E.D. (1985) 'Long-term effects of fertilizer on the botanical composition and soil seed population of a permanent grass sward', *Grass and Forage Science*, 40: 479–83.

Williams, J. (1960) *Butchers Crossing*, New York: Macmillan.

5

Fab Cities as Infrastructures for Ecological Reparation: Maker Activism, Vernacular Skills and Prototypes for Self-Grounding Collective Life

Atsuro Morita and Kazutoshi Tsuda

Introduction

On a chilly morning in November, one of the authors (Morita) was listening to a roundtable talk at an international conference on sustainability transition and economic localization held at a university conference hall on the outskirts of Yokohama, Japan. While the participants discussed various issues in their effort to localize production and consumption and restore the environment, one of the participants, a renowned Japanese permaculture designer, directed the conversation to a rather unexpected topic: the flush toilet and sewer.

Referring to his early career experience in soil analysis, he noted the serious degradation of Japanese soils. He pointed out toilets and sewers as the major causes of this degradation, because they are literally flushing human and other organic waste, which, before the introduction of these infrastructures, circulated thorough the local environment, and out to sea. As a potential solution to this problem, he then suggested a return to the compost toilet that would turn human waste into manure.

The permaculture designer's talk seemed to have successfully impressed on the audience a rather unexpected relationship between aspects of everyday life and environmental impoverishment through the workings of mundane infrastructures. If such uneventful infrastructures can be so detrimental, the relationship between our current civilization and environment seems to be fundamentally broken. He continued speaking on his take on circulation at another session Morita attended in that same afternoon. In this session he started his talk with a simple and impressive statement: 'microbes make soil just by the way they live'. He continued:

> 'But we are now focusing so much on making money and staying away from "day-to-day living" (*kurashi*). If we *lived properly* (*kichin to kurasu*), this would create cycles (*junkan*). You can nourish nature just by the fact that you live there. … [In order to live sustainably], you must actively create a day-to-day life (*kurashi wo tsukuranaito ikenai*). … You cannot change everything at once. You need to make soil in order to make food. In order to change, you have to make a detour. So, you have to learn how to make soil, how to use tools. … We have to make a new social system in which our day-to-day life can regenerate our environment.'

This chapter takes up this statement as a starting point to explore one of the key concerns in ecological reparation, that is, the transformation of material flows from linear systems to the circular. Circularity is an essential characteristic of ecosystems and the biosphere at large, where materials and waste generated by organisms are recycled by other organisms. It is precisely this recycling capacity that is lacking in contemporary industrial systems. Earth system scientists note that the proposed geological epoch of the Anthropocene is characterized by a new geological stratum of the technosphere – the vast collection of artifacts, infrastructures and waste that has accumulated on the Earth's surface. They warn that this ever-expanding technosphere that consists of non-biodegradable and often toxic artificial matter is now threatening the biosphere upon which the species survival of humans depends (Zalasiewicz et al, 2016). Such scientific findings seem to confirm this view of the environment as fundamentally broken alluded to in the permaculture designer's talk mentioned earlier.

Responding to this crisis, circularity has become a common theme in various transition imaginaries, from the EU's circular economy plan and circular business models to techno-activism for sustainability (Diez, 2018). However, permaculture sees circularity not only as a functional condition of sustainable systems but also as an ethical principle central to people's relationships with each other and with other species (Puig de la Bellacasa, 2017). In the aforementioned talk the permaculture designer also followed this tradition when he drew a close link between the ethics of everyday life (*living properly*), circular material flow, and multispecies flourishment. For him, vernacular skills for making and repairing are a key strategy in the shift towards such an ethical life: 'you can nurture nature just by the fact you live there', just like microbes do.

This view of vernacular skills is not limited to permaculture, as these skills feature in many circular imaginations, particularly in grassroots proposals for sustainability transitions. The Fab City Global Initiative is one of them. The Fab City was launched in 2014 by the international community of Fab Labs in order to remake cities that are at least 50 per cent self-sufficient by the year 2054. The thrust of this initiative is turning global linear material flows toward the circular by encouraging people's participation in the manufacturing process by recycling the waste that cities generate (Diez 2018). Here the Fab City shares permaculture's ambition in socio-technical transformation through nurturing vernacular skills.

The Fab City began within the kind of grassroots maker activism based in a particular kind of maker space called a 'Fab Lab'. As a cornerstone of the maker movement, maker spaces are equipped with digital fabrication tools that allow and facilitate people's engagement with making, tinkering and repairing. Inspiring what is now called DIY activism (Shantz, 2013; Ratto and Boler, 2014), the maker movement advocates for technological empowerment of non-experts and community-building by increasing participation in the process of making and repairing (Davies, 2017; Gauntlet, 2018). Its key claim is for technological autonomy, where ordinary people could gain sufficient skills and knowledge to take care of the various technical affairs in their daily life without depending on experts (Gershenfeld, 2007).

We explore how this vision of autonomy configures the way circularity is imagined as a means of ecological reparation. To elucidate this, we focus on one important difference from permaculture. The circular vision of permaculture is grounded on the circularity of the biosphere. Here, the ultimate goal of ecological reparation is harmonizing the patterns of human livelihood and productive activities with ecological cycles (Lockyer and Veteto, 2015). On the other hand, following the maker movement's insistence on technological autonomy, the Fab City highlights a certain kind of autonomy made possible by collectively building communal infrastructures using DIY. Achieving increased circularity is a major condition for and result of this autonomy.

The anthropologist Christopher Kelty points out a similar socio-technical vision of the free software movement with the term 'recursive public' (Kelty, 2008). Kelty notes that the free software movement concerns the development and maintenance of the technical infrastructures of the internet that, in turn, enables the activists to communicate with each other which then forms that public such communication is based on. Here, the internet is the ground for the formation of the public while at the same time the very product of the collective practices of this public. Thus, the public and its infrastructure recursively constitute each other. Similarly, the Fab City movement aims to build DIY urban infrastructures that make possible a new form of collective life, which in turn the movement itself prefigures. In this vision circularity not only represents circular material flow but also the recursive constitution of DIY infrastructures and the autonomous collective life, in which vernacular skills are essential.

We elucidate the way these two forms of circularity relate with each other by drawing on an anthropological take on prototypes. Empirically speaking, prototypes of circular infrastructures are so far the major products of the Fab City initiative. Even more, the idea of the Fab City itself can be seen as a prototype of a future urban life. On the other hand, the prototype as an anthropological concept helps to understand the way the formation of a recursive public and achievement of circular material flow come together in a certain sort of knowledge making practice.

Alberto Corsin-Jimenez (2014) characterizes the prototype as suspended 'object-knowledge' (object-oriented knowledge or a knowledge embodying object) that prefigures the future while at the same time stays open to further modification and modulation. As prototypes, provisional designs for testing functionalities and scalability, most of the Fab City innovations still stay partial, rather than being full-scale solutions to urban linear material flow. However, it is precisely this characteristic as a design for testing that allows activists to further explore into urban infrastructures and material flows to gain fresh views of the way they mediate everyday life and environmental degradation. Such explorations in turn allude to possible forms of collective life where vernacular skills and circularity serve as conditions for the reparation of the damaged environment.

Fab Labs and the political vision of self-manufacturing

In the early 2000s, hackers, educators, scholars and journalists discovered that the increasing availability of digital machine tools such as 3D printers and laser cutters would allow for a new form of citizen participation in manufacturing. This gave rise to a new techno-social movement – the maker movement, which utilizes digital machine tools for hobbies, repairing, innovation and community-building.

Digital machine tools such as 3D printers and laser cutters can manufacture a wide range of objects from digital data that is generated by software on a personal computer. According to the maker activists, these digital fabrication tools would allow individual makers to develop and manufacture a variety of hardware just like the personal computer decades ago enabled individuals to develop their own software. By making analogies with the rise of software start-ups and new forms of sociality based on the internet, maker activists have claimed the coming-of-age of a new industrial revolution, where production would shift from large factories to community-based small enterprises, or even to individuals (Anderson, 2012). Thus, the key concern of the maker movement is introducing data-centred digital manufacturing modelled after the work organization of the free software movement, which claims individual and grassroots freedom to innovate new technologies.

Aside from this ambition, what is conventionally called 'the maker movement' is politically heterogenous, culturally diverse, and organizationally plural. It consists of a politically radical tradition in Europe and a less political one in the US (Davies, 2017), as well as many variations in Asia, Africa and Latin America. Alternatively called hacker spaces, maker spaces or Fab Labs, shared spaces equipped with shared tools are essential for maker activists to facilitate a communal exchange of ideas and skills (Papadopoulos, 2018). For the Fab Lab movement that we take up in this chapter, the format of this shared workspace is particularly important.

In the early 2000s, a few years before the rise of the maker movement, the original Fab Lab was founded by the computer scientist Neil Gershenfeld as an outreach activity of MIT's Center for Bits and Atoms. The idea of the public workspace originated from the course called 'How to Make (Almost) Anything' that Gershenfeld taught at MIT. The course instructed students how to use digital fabrication technologies for their own projects (Gershenfeld, 2007). Stimulated by students' creativity in making and designing new artifacts, Gershenfeld came up with the idea of building a citizen workshop equipped with digital fabrication tools, and named it 'Fab Lab (Fabrication Laboratory)'.

The Fab Lab is a sort of maker space, where people can use shared tools and machines to pursue their own projects. The Fab City Foundation that coordinates the Fab Labs across the world requires a Fab Lab to be open to the public at least one day per week, and to be equipped with standard tools such as a 3D printer, laser cutter, milling machine and various hand tools.[1] In addition, the Fab Labs are required to conform to the Fab Charter, a brief document that defines what the Fab Lab is, and to join the international network.

The Fab Labs can be considered as what Kelty (2008) calls a 'modulation' of the free software movement, where activists seek to apply the principles of open-source software to other fields. The fact that the internet has come to serve as a common infrastructure for many different activities, from distant education to hardware innovation, plays a central role in this transfer. A well-known Japanese Fab Lab activist told one of us that "the Fab Lab is a new infrastructure for making that is built upon the infrastructure of the internet" (Gershenfeld, 2007; Anderson, 2012).

However, the maker movement imagines this future industry as not just a mere digitization of the material world. Rather, it entails a particular political vision. When he started the course 'How to Make (Almost) Anything', Gershenfeld observed that students' creative use of digital tools indicated a future of manufacturing: 'Such a future really represents a return to our industrial roots, before art was separated from artisans, when production was done for individuals rather than the masses. Life without the infrastructure we take for granted today required invention as a matter of survival rather than as a specialized profession' (2007: 8).

As he links the future of industry with its roots in non-alienated labour of artisanal production, Gershenfeld's vision is futuristic and nostalgic at the same time. Gershenfeld claims that digital fabrication technology would revitalize the artisanal mode of production that was once defeated by the rise of mass production and consumption. In this sense, the Fab Lab movement shares a criticism of mass production with permaculture. However, the Fab Lab's vision that relates digital technology with earlier forms of production is hardly new. Indeed, the maker movement's vision for a new industrial revolution (Anderson, 2012) repurposes the economic theory about the earlier digital restructuring of manufacturing systems in the 1980s developed by Michael Piore and Charles Sable (1984). Their book *The Second Industrial Divide* takes on a sort of historical fiction when the authors discuss the historical contingency that brought about the victory of the mass production system:

> Under somewhat different historical conditions, firms using a combination of craft skills and flexible equipment might have played a central role in modern economic life – instead of giving way … to corporations based on mass production. Had this line of mechanized craft production prevailed, we might today think of manufacturing firms linked to particular communities rather than as an independent organizations. (Piore and Sable, 1984, cited in Anderson, 2013: 69)

Rather than Gershenfeld's view on artisanal production that highlights the ingenuity of ordinary people, Piore and Sable focus more on the communal nature of skills and technical knowledge transmitted through communities. The vision of the new industrial revolution of the maker movement is an explicit revival of this vision, as another advocate of the maker movement, Chris Anderson, noted that "digital desktop fabrication has indeed introduced a sort of 'mechanized craft production' that Piore and Sable could only dream of" (2013: 69).

The roots of this technological imaginary might stretch back even earlier. Langdon Winner (1980) indicates that Lewis Mumford saw the history of technology as a constant oscillation between two poles of possibilities: the authoritarian and centralized system, and the democratic and distributed one. Winner also notes that this vision even "mirrors concerns voiced earlier in the works of Peter Kropotkin, William Morris, and other nineteenth century critics of industrialism" (Winner 1980: 121). Similar to these visions, the Fab Lab community envisions a future of industry where technical knowledge is horizontally shared by autonomous makers that forms, together with digital fabrication machines, a productive collective rooted in a particular locality. In the Fab City initiative, this vision became tied with a particular form of techno-social imagination of ecological reparation by autonomous cities.

The Fab City: the vision of material and data circulation

In 2011, the Fab City Global Initiative was launched by the Institute of Advanced Architecture of Catalonia, MIT's Center for Bits and Atoms, the Fab Foundation and the Barcelona City Council. In the inauguration ceremony, the mayor of Barcelona pushed the button to start the countdown to 2054, when the city will have achieved at least 50 per cent self-sufficiency in terms of energy, food and products (Diez, 2018). The Fab City has quickly grown into a network of cities. The Fab City global network now has 28 member cities across the globe. Kamakura, Japan's former capital from the twelfth to fourteenth century, and located about 50 km away from Tokyo, joined the network in 2018.

The Fab City Whitepaper describes the initiative as a response to the increasing pace of global urbanization and environmental crises (Diez, 2018). With increasing numbers of people moving to cities – the UN predicts that 75 per cent of the world population will be urbanites by 2050 – cities continue to consume an overwhelming amount of resources and emit large amounts of CO_2. The *Whitepaper* argues that to tackle this sustainability challenge requires a fundamental rethinking of the ways cities operate.

As a 'modulation' (Kelty, 2008) of open-source activism, the Fab City aspires to tackle this challenge by repurposing digital technologies. In this vision, the key to circular manufacturing rests in the shift from what they call the 'Products In, Trash Out' (PITO) system to the 'Data In, Data Out' (DIDO) system. The PITO system represents the linear material flow of our current industrial systems where cities import most of products they consume from other countries. The supply chains of these products stretch out globally, often originating in the extraction of resources in the Global South, manufacturing and assembling these products in China and Southeast Asia, and then transporting via container ships and trucks over a large distance to their destinations, to then eventually be consumed in cities in the Global North. In those cities, most of the products imported enjoy a relatively short lifespan due to rapid product cycles facilitated by recurrent obsolescence. Finally, they move out from the city as waste to their final resting place in a landfill (Diez, 2018).

The Fab City tries to change this linear flow by developing what they call the DIDO system, which is essentially a network of locally organized circular production systems connected by the internet. On the one hand, the DIDO system is a distributed circular manufacturing system that consists of Fab

Labs, small and medium enterprises (SMEs) and public organizations. In this system, 'more production occurs within the city itself, as does recycling and urban mining of materials to be reinserted in supply chains' (Diez, 2018: 82). However, while production would be localized by a massive recycling of the waste the city generates, the DIDO system is also a globally networked production system built on the internet.

The DIDO model rests on the current mode of making practised in the Fab Labs. As we have seen, Fab activists see Fab Labs as a new infrastructure of manufacturing built upon the infrastructure of the internet on which information freely circulates. In this vision, digital fabrication technologies are seen as devices to translate digital design information to material forms (Gershenfeld, 2007). The circulation of design data draws on the open-source principle, and following this, the Fab Labs share design data on the internet as part of the digital commons. Individual users can download and modify these data in order to make things, taking into account local conditions. They can then upload modified data again for further modification and use elsewhere. The Fab City aims to scale this practice up to achieve urban self-sufficiency by linking Fab Labs, SMEs and other relevant actors. In other words, in the DIDO system, production would be based on globally shared data, while manufacturing happens with local and recycled materials, and thus becomes part of circular urban material flows that would significantly reduce cities' ecological footprint. One interesting aspect of the DIDO system is that it serves as a way to achieve localization of production and circular resource use, a common concern shared with permaculture, while maintaining the global network of data circulation, which is essential for the digital activism born on the internet.

Prototypes: being part of the yet-to-come

In this globally networked activism, individual initiatives are often seen as prototypes, which opens up new possibilities. The architect of the Fab City, Thomas Diez (2017), explicitly notes in the project's proposal to the European Commission's call for Circular Economy Large Scale Demonstrators that the Fab City consists of prototypes for a large-scale shift toward circular economies. In this proposal, Diez delineates a Fab City as a network of Fab Labs, neighbourhoods, artists, SMEs and so on that develops prototypes for circular economies and shares these prototypes with the member cities of the Fab City initiative.

One of those prototypes is evident in De Ceuvel,[2] a circular office park founded by Metabolic,[3] a circular economy and urban metabolism consultant company, located in a waterfront area near the bay of Ij in Amsterdam. Restored from a polluted shipyard, De Ceuvel 'hosts a sustainable restaurant and 16 upcycled houseboats which have been placed on land to serve as offices for some 30 companies' (Monaghan and Black, 2018: 164). De Ceuvel's buildings are made of upcycled materials and fitted with more than 150 solar panels. Waste water from the café's kitchen and offices are filtered through a grey water treatment system 'made of sand, gravel, shells and plants', and each building constructed from the houseboat is equipped with a compost toilet that allows them to recover nutrients from human waste (Monaghan and Black, 2018: 167). In addition, the Metabolic Lab in the park houses an aquaponic system where fish excrement is recycled as fertilizer for growing vegetables. This restoration of the polluted land as well as the construction of the facilities are completed in such a DIY manner by the makers and activists who gathered at Metabolic.

De Ceuvel is perhaps one of the largest Fab City prototypes, as it hosts a number of companies, and has successfully installed an almost full set of circular infrastructures. However, it also shares characteristics with other smaller-scale prototypes such as plastic recycling and composting toilet prototypes developed by Dynamic Lab, a Fab Lab located in Daizaka, a small depopulated village in a mountainous area in the southern part of the island of Kyushu in Japan.[4] Run by the environmental activist known by his nickname 'Tender', Dynamic Lab is hosted in a building repurposed from a closed elementary school. Being sympathetic to the Fab City's vision of circularity, Tender has invented a number of new processes and equipment designed to recycle waste – from ocean plastic to human waste – by using recycled materials from other waste and scraps. For example, their latest invention includes a new open-source design that enables composting toilets to decompose human waste[5] at an accelerated rate.

These prototypes prefigure the future of an envisioned circular city through a metonymic relationship, in which a prototype points to a larger whole of the future circular city of which it is intended to be a part. Such partiality is visible for the compost toilet, a small-scale innovation that is to work as a part of a larger infrastructural configuration. Similarly, the circular office park De Ceuvel, which might look like a self-sustaining circular community, is still surrounded by existing linear urban infrastructures of Amsterdam, and its maintenance and expansion still draw on them.

This dual ontology of these circular prototypes produces a particular effect. Alberto Corsin Jimenez notes that one of the salient effects of the prototype in contemporary techno-culture is to 'keep sociality in suspension', both in its material form and through the social relationship it engenders (Corsin Jimenez, 2014: 385). In the case of the Fab City initiative, such suspension is produced by the duality of prototypes that belongs both to the future circular and the current linear infrastructural configurations.

For example, the Precious Plastic project,[6] a globally coordinated effort to develop alternative plastic recycling processes, which is practiced in Dynamic Lab and other Fab Labs across the world, is concerned with developing technological processes and devices in order to recycle discarded plastic without using as much energy. However, testing such prototype technologies entails an inquiry into existing practices related to discarded plastic. For example, one Japanese Fab Lab activist told us that developing a prototype system of plastic recycling involves not only developing different technology to process these plastics, but also changing the way people discard plastics so that the Fab Lab activists do not have to spend the time meticulously sorting out bulk mixtures of plastic with widely different quality. The current dominant practice of discarding plastic does not take into account quality, cleanliness and the colour of the plastic, all of which significantly come to matter when one tries to produce a product out of recycled plastic. By their very function of recycling discarded plastics, the precious plastic prototypes revealed these entanglements of plastic packages and products with the daily routines in contemporary consumerist society, and thus the project's limited success as a circular system.

To put another way, as prototypes, existing circular Fab City artifacts and processes lack the social and technical infrastructures that would sustain its envisioned smooth functioning. In this particular case of plastic recycling, the invented process does not have a functioning input that could provide appropriately sorted-out plastic waste. Thus, installing Fab City prototypes inevitably reveals gaps between not-yet-realized circular infrastructural configurations and already-existing linear ones. These gaps consist of technical problems such as the lack of proper input, but also social ones, such as the dominant practices of discarding that hinder recycling. As Corsin-Jimenez notes, these prototypes suspend existing infrastructural formations both in terms of their materiality and sociality. What is more, these gaps invite activists to further explore the way current infrastructures organize material flow. These prototypes lead activists not only to question the way people dispose of plastic products and packages, but also to further explore a larger infrastructural configuration that deals with discarded plastics and other waste. Such an exploration often results in other prototypes, such as a new digital infrastructure to match the timing of discarding and recycling.

Importantly, the issues that circular prototypes further invite activists to explore range across diverse domains. For example, in a workshop about circular innovation one of the authors attended, a representative of a waste management company collaborating with Fab Lab activists noted that increasing reuse and recycling in business-to-business transactions requires a temporal coordination of disposal and purchase, unless the cost of transportation and storage becomes prohibitive. This requires the close coordination of activities between companies that dispose waste and its potential buyers, who often have no previous relationship in the current system. The company's new project entails developing a platform to match up these two parties in a timely manner. This case suggests that the changes required for circular material flow not only concern discarding practices, but also the rhythm of organizational routines and relationships between organizations that constrain and constitute the discarding practices.

One of the salient effects of circular prototypes of Fab Labs is such a gradual revelation of entanglements of infrastructures and everyday routines. These entanglements often cut across domains and organizations and thus tend to remain invisible. Importantly, the realization of such complex entanglements often results in highlighting the role of technical skills. In this regard, the circular vision of the Fab City resembles that of permaculture. The Fab Lab activists we spoke to often mentioned the nurturing of the average person's technical knowledge about the material properties of discarded objects and recycling technologies as a key issue. For example, joining a local form of the Precious Plastic project and having hands-on experience in dealing with discarded plastic would greatly improve the way people sort different kinds of plastic when discarding. Such skills are important not only as a capacity to operate recycling processes, but also as an embodied knowledge that help one navigate complex entanglement between infrastructures, materials, organizations and everyday routines. In this sense these skills are indeed vernacular, meaning rooted in and idiosyncratic to particular situations.

Nurturing the production skills and knowledge of lay people is also an essential part of the envisioned sociality of the Fab City. The Fab Lab movement emphasizes the technological autonomy of people gained by participation in manufacturing as key to the transformation of society – from the one characterized by uneven distribution of technical knowledge between experts and lay people to a more equal one. Thus, in the Fab City vision, people equipped with technical skills and knowledge are an essential part of future circular infrastructures. Here, makers, those who can take care of some part of technical infrastructure on which they themselves depend, become prototypes for the circular city, prefiguring a new infrastructural configuration in which people's technological literacy serves as an essential part.

Conclusion: self-grounding public and infrastructures for ecological reparation

The Fab City shares with other initiatives for sustainability transition the view of circularity as a major means for repairing the damaged environment. With permaculture, it also highlights the significance of a new form of sociality in which vernacular skills of making and repairing play a central role. For the Fab City, building skills and new technologies such as digital fabrication become a cornerstone of the networked distributed production system as well as the future form of autonomy in urban life.

Unlike permaculture that grounds itself on the circularity of ecological processes, for the Fab City the figure of the circular denotes not only circularity of material flows but also the recursive relationship between the public and infrastructure. Permaculture aspires to ground its 'regenerative social system' upon natural cycles of ecosystems. Here what is highlighted is harmonization of human livelihood and productive activities with the cycles of multispecies ecologies. However, the Fab City defines and models itself after the recursive public of internet activism, and thus focuses on a recursive relationship of urban collective life, and the effort to remake urban infrastructures, the very condition of this collective life, circular. However, there is one important difference.

On the one hand, the public of the free software movement largely draws on the internet, the infrastructure of their own making, and thus their view of self-grounding rests on the techno-activists' capacity to design and program it. On the other hand, for the Fab City activists, they must also take into account and draw on existing infrastructures such as waterworks, electricity grids and supply chains in order to build their circular prototypes. In other words, they inevitably face the fundamental entanglement between the Fab City public and the linear infrastructures that they aspire to replace in the first place.

This difference stems from the scope of the publics in terms of what Kelty calls the *depth* of self-grounding. He notes that each public considers in its own way the depth of infrastructural layers on which it grounds itself. The free software movement considers its technicality of infrastructures as located down to the level of software and hardware specifications for digital information exchange, but not to the level of the national grid and oil infrastructures that enable the construction and maintenance of these machines in the first place. In this sense, the Fab City further establishes the depth of its own self-grounding by looking into the very basic strata of infrastructural layers such as logistics, energy and waste treatment infrastructures. The more they explore the infrastructural layers to realize the circular city, the more it becomes clear and visible that they draw on existing linear infrastructures. Here the Fab City reveals itself as a prototypical recursive public, an only partially realized recursive public exploring the depth and width of urban infrastructures in the hope of realizing circular material flow.

In this sense the Fab City's primal focus is not so much on re-embedding human activities in ecological cycles as bringing about a new relationship between urban collective life and its infrastructures. It aspires to render urban collective life a practice of creating its own ground, autonomously, without depending on the state or external material flows. Regeneration of the damaged environment is a corollary of this recursive endeavour by curtailing the environmental impact of cities. In one sense this results in the separation of the city from outside ecological processes, something that resembles the modern separation of nature and culture (Latour, 2017). However, this vision of self-grounding does not rely upon the modern view of nature as the ultimate ground for humanity. Rather, the Fab City's recursive effort for self-grounding continuously explores the complex entanglement of industrial infrastructures, linear material flows and the everyday. The Fab City activists see them not only as problems waiting for a solution, but also as resources potentially repurposed and recycled for building circular infrastructures, just like Dynamic Labs endeavours to recycle scraps for building prototypes. In other words, the figure of the circular here substitutes the public's effort to create its own infrastructure for nature as a resource and the ultimate ground. Here, the Fab City's recursive endeavour materializes as responsible acts of repairing and redressing our broken relationship with multispecies ecologies. In this vision of the circular city, the Fab City as an infrastructure for autonomous urban life appears as an infrastructure for ecological reparation.

Notes

1. See more details at https://fabfoundation.org
2. Details of De Ceuvel can be found at https://deceuvel.nl/en/about/general-information/
3. See more details at www.metabolic.nl
4. Details of Dynamic Lab can be found at htpps://sonohen.life
5. Details of this project can be found at https://sonohen.life/koyassy-highend-model/?fbclid=IwAR0DgpmhDEetCBfE4DFqqSPkeZ0231_yzQ0cT_orfTrkqVvLpymOIMmT1R4
6. See more details at http://preciousplastic.com

References

Anderson, C. (2012) *Makers: The New Industrial Revolution*, London: Random House Business.

Corsín Jiménez, A. (2014) 'Introduction', *Journal of Cultural Economy*, 7(4): 381–98.

Davies, S.R. (2017) *Hackerspaces: Making the Maker Movement*, Cambridge: Polity.

Diez, T. (2017) *Fab City Prototypes* [online]. Available from: https://blog.fab.city/fab-city-prototypes-designing-and-making-for-the-real-world-e97e9b04857 [Last accessed 22 September 2021].

Diez, T. (2018) *Fab City: Mass Distribution of (Almost) Everything*, Barcelona: IaaC.

Gauntlett, D. (2018) *Making Is Connecting: The Social Power of Creativity, from Craft and Knitting to Digital Everything*, 2nd edn, Cambridge: Polity Press.

Gershenfeld, N.A. (2007) *Fab: The Coming Revolution on Your Desktop--from Personal Computers to Personal Fabrication*, New York: Basic Books.

Kelty, C.M. (2008) *Two Bits: The Cultural Significance of Free Software*, Durham, NC: Duke University Press.

Latour, B. (2017) *Down to Earth: Politics in the New Climate Regime*, Cambridge: Polity.

Lockyer, J. and Veteto, J.R. (eds) (2015) *Environmental Anthropology Engaging Ecotopia: Bioregionalism, Permaculture, and Ecovillages*, New York: Berghahn Books.

Monaghan C. and Black, K. (2018) 'The circular urban metabolism' in T. Diez (ed) *Fab City: Mass Distribution of (Almost) Everything*, Barcelona: IaaC, pp 162–9.

Papadopoulos, D. (2018) *Experimental Practice*, Durham, NC: Duke University Press.

Piore, M.J. and Sabel, C.F. (1984) *The Second Industrial Divide: Possibilities for Prosperity*, New York: Basic Books.

Puig de la Bellacasa, M. (2017) *Matters of Care: Speculative Ethics in More Than Human Worlds*, Minneapolis, MN: University of Minnesota Press.

Ratto, M. and Boler, M. (2014) *DIY Citizenship: Critical Making and Social Media*, Cambridge, MA: MIT Press.

Shantz, J. (2013) *Commonist Tendencies: Mutual Aid Beyond Communism*, Brooklyn: Punctum Books.

Winner, L. (1980) 'Do artifacts have politics?' *Daedalus*, 109(1): 121–36.

Zalasiewicz J., Williams, M., Waters, C.N., Barnosky, A.D., Palmesino, J., Rönnskog, A.-S. et al (2016) 'Scale and diversity of the physical technosphere: A geological perspective', *Anthropocene Review*, 4(1): 9–22.

6

The Cosmoecological Workshop: Or, How to Philosophize with a Hammer

Martin Savransky

Introduction: a metaphysical catastrophe

And suddenly the world began to quake. The ground trembled. Soils liquified under the smooth covers of asphalt. Modern infrastructures were bent, dented, upended. Buildings moved to the rhythm of their crumbling dance. Those inside ran out, when they could. Others succumbed to the collapse. The rumbling noise filled the air. Amidst the turmoil, hundreds of metres of coastline fell off a cosmic cliff, dropping vertically by over half a metre. The Earth itself trembled, shaken, literally knocked off its axis. Days have been shorter ever since. The Pacific Ocean shrivelled a little. And Japan, whose northeastern region of Tohoku was closest to the epicentre, was moved 13 feet closer to North America. With a magnitude of 9,1 M_w, it was the biggest earthquake to have struck the archipelago, and only the fourth most powerful in the history of seismology. And yet it only took six minutes. Six. The planetary blink of an eye, making it present that most things happen in the break, through the cracks, *with the tides of time*, through the resonance of events: 'point of view on a point of view, displacements of perspective, differentiation of difference' (Deleuze, 2004: 200). For indeed, the earthquake that struck northeastern Japan on 11 March 2011 turned out to be but a prelude, a foreshock of its own, a call to another kind of intensity whose response washed it all away. This was what seismologists call an underwater megathrust earthquake – a name that could belong just as well to geology as to poetry – and the 39-metre tsunami that it summoned flooded the entire area, ravaging it all in a 200-square mile range. The dark wave truly devoured everything: almost

20,000 people died, countless other critters saw their lives brought to a sudden end, and over 45,000 buildings were destroyed. Among those who survived, 4.4 million households were left without electricity, over 340,000 people were displaced, and suffered from food, water, shelter, medicine, and fuel shortages for a long period of time afterwards, even when considerable material efforts were deployed to restore infrastructures in the wake.

The tsunami precipitated ripples of its own. In the first instance, of course, the devastation of energy infrastructures brought about by the tsunami triggered a total blackout at the Fukushima Daiichi Nuclear Power Plant, disabling the reactors' cooling systems, thereby resulting in the meltdown of their nuclear cores which in turn led to hydrogen explosions and the attendant radioactive plume that, together, constituted the seminal events of the nuclear disaster that still persist up to this day. The collapse of infrastructures – and of political imaginations – in turn strained the state's disaster response, desperately seeking to reassert the most tenuous semblance of modern sovereignty as it scrambled to repair or confine the broken reactors under pressure from shareholders of the energy sector itself, who insisted on the imperative of saving the reactors as assets for future use. Unable to mend what had been broken beyond repair, the Japanese government cordoned off a 12-mile radius around the plant as an evacuation zone which led to the displacement of 154,000 residents. And yet, the catastrophe of radioactive contamination knows no sovereign borders, whether spatial or temporal. It only took about ten days for radioactive substances to become detectable in Tokyo's tap water, thereby implicating incalculable numbers of people, animals, plants and manifold ecological relations in different manners and varying intensities.

Indeed, we must tragically go further. For if the ripples of this ongoing and unfinished breakdown upended human and other-than-human modes of living in and around the archipelago in such a way that, to quote writer and activist Sabu Kohso (2020: 19), 'living is becoming equal to struggling against such hazards and the regimes that impose them', it is equally the case that they affected manifold *modes of dying* as well. Six months after the tsunami, the region of Tohoku experienced a veritable profusion of ghost stories, sightings, experiences, possessions and hauntings in and around the devastated area. 'A young man' – reporter Richard Lloyd Parry (2017: 99) writes in his wonderful *Ghosts of the Tsunami*, which chronicles some of the personal, social and political effects of the tsunami – 'complained of pressure on his chest at night, as if some creature was straddling him as he slept. A teenage girl spoke of a fearful figure who squatted his house. A middle-aged man hated to go out in the rain, because of the eyes of the dead, which stared out at him from puddles.' Alas, these were not isolated events:

> A civil servant in Soma visited a devastated stretch of coast and saw a solitary woman in a scarlet dress far from the nearest road or house, with no means of transport in sight. When he looked for her again, she had disappeared.
>
> A fire station in Tagajo received calls to places where all the houses had been destroyed by the tsunami. The crews went out to the ruins anyway, prayed for the spirits of those who had died – and the ghostly calls ceased.
>
> A taxi in the city of Sendai picked up a sad-faced man who asked to be taken to an address that no long[er] existed. Halfway through the journey, the driver looked into his mirror to see that the rear seat was empty. He drove on anyway, stopped in front of the levelled foundations of a destroyed house and politely opened the door to allow the invisible passenger out at his former home.
>
> At a refugee community in Onagawa, an old neighbour would appear in the living rooms of the temporary houses and sit down for a cup of tea with their startled occupants. No one had the heart to tell her that she was dead; the cushion on which she had sat was wet with water. (2017: 99–100)

Indeed, this wasn't simply a 'natural' calamity, the total breakdown of infrastructures, or the inadvertent destruction of an environment. If it constituted a veritable *ecological* disaster it is not least because, as Félix Guattari (2001) reminded us, environmental ravages can never be dissociated from the destruction of social ecologies through which modes of living and forms of dying are woven together, just as the devastation of such social ecologies cannot be separated from the domain of subjectivity, the personal and collective experiences of grief and grievance, of loss and desolation that the wave left in its wake.

And in the case of the Great East Japan Earthquake and Tsunami, through the ripples of the Fukushima Nuclear Disaster, it was the world itself that quaked: upending ecological as much as cosmological arrangements, it tangled humans, animals, plants and soils, but also of the spirits of the living and the dead, in a metamorphic process that left no one and no thing unscathed (Savransky, 2021a). Indeed, it is precisely in this sense that a group of activists writing in the aftermath of the various metamorphic processes set in motion by this ongoing and heterogeneous catastrophe characterized the disaster as nothing short of a 'metaphysical incident', one whose consequences have transformed 'the very fabric of our reality' and upended not only myriad habitats but their very modes of habitation, the composition of the world and their ways of living and dying. 'How', they powerfully asked, 'to compose a collective after such an event? How is the relationship with the cosmos affected? How to go on living after such a catastrophe?' (Les éditions des mondes à faire editorial collective, 2018: 19). Thinking in the wake of this disaster, in the presence of these and other resonant questions, and alongside some of the practices and experiments that were born in response, in this chapter I hope to experiment with some of the lessons they together might be capable of yielding when it comes to thinking about questions of 'ecological reparation'. How to think the ecological work of reparation when it is the very *arts of living*, the collective modes of valuation by which lives become worth living and deaths become worth living for, that are in question? How is a cosmos recomposed?

I wager that exploring these questions itself demands a transformation of our philosophical habits, perhaps rediscovering something of what Nietzsche (1997) called 'philosophising with a hammer': not in order to prolong the critical gesture of smashing our idols but so as to craft thought immanently, with a tuning fork, experimenting with stories and practices whose resonances and reverberations make other ways of living and other modes of earthly habitation perceptible. Which is to say that, if as Nietzsche would argue, we always have the thoughts, beliefs and values that we deserve given our way of being and our modes of living, perhaps the task in the wake of breakdown is that of consenting to the genuinely pragmatic challenge, not just of making other modes of living thinkable, but of experimenting immanently with living one's way into other modes of thinking (Savransky, 2019, 2021b). In order to elaborate on this speculative proposition in the wake of the 3.11 disaster and in spite of the slow violence of Japan's own process of 'modernisation' – which saw both the psychologization of ghosts and the centralization of care practices as part of the Meiji government's project incorporation of Western structures and institutions of statecraft (Ivy, 1995; Figal, 1999) – I relay the story of a number of practices developed by collective of priests of various denominational faiths who gave themselves over to an experimentation with forms of 'spiritual care work' in an effort to tend as much to the survivors of the tsunami as to the ghosts of the dead. Their efforts, which revolved around the creation of a mobile café that would offer various forms of spiritual care to those affected by the wave, involved what I would characterize as generative, partial, improvisational forms of cosmoecological reparation: not the attempt to restore the modern terms of order, to render life resilient to disaster, but that of carefully reweaving modes of living and of dying otherwise from the interstices of the catastrophe, of experimentally recomposing a cosmos on a ravaged earth.[1]

'Consolation for the Spirits': repair and the aporetics of reparation

What perhaps makes efforts to attend to the subtle acts of care that constitute repair especially generative today is precisely their commitment to think from the perspective of a 'broken world', affirming the ways in which breakdown 'disturbs and sets in motion worlds of possibility that disappear under the stable or accomplished form of the artifact' (Jackson, 2014: 230). In their oft-cited, neo-Heideggerian approach to the question of repair, Stephen Graham and Nigel Thrift (2007) note that Heidegger's philosophy presupposes a world given ready-to-hand that implicates human beings in tasks and processes around things, such that it is the thing itself that 'discloses a world', that shapes the manner in which its own material, those that are concerned with and implicated in its use, and the environment in which these various forces are emplaced, become assembled together. But of course, just as Heidegger himself did, Graham and Thrift are quick to note that things aren't always so 'ready'. In fact, it is precisely when that readiness isn't given, when the worlds they disclose break down, that the thing itself reveals its powers and as such becomes the object of attention, that it demands to be thought. Breakdown, on this account, is the event of a figure/ground reversal where the previously smooth and unproblematic background of a universe of things becomes disturbed and is thereby foregrounded, simultaneously calling on the work of repair while opening up, in the course of the operations that are assembled in response, a new distribution of affordances and alternative modes of reassembling the world. 'It is in this space between breakdown and restoration of the practical equilibrium', they argue, 'that repair and maintenance, makes its bid for significance' (Graham and Thrift, 2007: 3).

Yet everything changes when breakdown implicates not only the world-disclosing thing (a technological device, a body, a form of infrastructure, or even an environment), but the very fabric of reality as such – that which a multiplicity of beings collectively and differentially compose, in which they divergently partake whether they like it or not. For here no reversal is possible: both figure and ground become implicated in a radical metamorphosis that leaves no being, no relationship and no established set of values untouched. It involves collective modes of living and dying in active processes of decomposition no amount of *Dasein* can transcend, in losses no amount of mourning can atone for, in a runaway transformation no practical search for equilibrium can hope to restore. What might 'repair' mean then, when the world itself falls apart, when it is no longer a question of readying it back to hand, when the very collective modes of valuation and earthly habitation become both instrument and object of an irreversible collapse? Perhaps it is then, here, that the notion of 'repair' reencounters the aporetics of 'reparation', that of the radical incommensurability between destruction and redemption, between loss and compensation. 'How does one compensate for centuries of violence', ask Stephen Best and Saidiya Hartman (2005: 2), 'that have as their consequence the impossibility of restoring a prior existence, of giving back what was taken, of repairing what was broken?' Best and Hartmann ask this question in relation to and in the interstices of the ruin and afterlife of slavery, but something of this question – which makes the crucible of a life lived in loss perceptible – resonates with those questions raised by the living and dead in the wake of the waves of sorrow that flooded the region of Tohoku in the months following the devastation.

For indeed, to say that most things happen in the break, through the resonances of events, is also to say that the earthquake and tsunami were their own aftershocks to other foreshocks, that the devastation was not so much originated as compounded by them. If the profusion of ghosts in the wake of the tsunami was an unsettling event, this was not least because, despite ghosts enjoying a long cultural and symbolic history in the archipelago, the Japan in which ghosts made themselves present in such powerful ways was one which had already undergone its own processes of social, cultural and ecological attrition and slow violence (Nixon, 2011). Indeed, since the Meiji period, Japan underwent a process of 'modernisation' that involved not only an ambitious financial and infrastructural development programme, but also the invention of new knowledges and other belligerent operations that turned ghosts and their stories into signs of superstition and backwardness, into obstacles that the modern state sought to eliminate through a mix of legislation, education and intervention (Ivy, 1995). As the responsibility and possibility of care was wrested from the practices of local practitioners and became centralized, administered by governmental authorities guaranteeing, through modern knowledges, the promise of individual wellbeing and the maintenance of the health of the population, ghosts were refashioned, de-realized, turned into elements of a nervous symptomatology 'that could be explained and thus controlled with a newly coined language that represented a newly constructed knowledge' (Figal, 1999: 29).

As a result, many of those afflicted by loss, grief and the emergence of ghosts in the wake went on to consult doctors, psychologists and social workers in search for support, to repair what had been broken, to restore their selves to a prior existence, to bring themselves back to health. To no avail: modern health care practices had nothing on the ghosts whose presence became patently felt in the wake. This in turn led people to try and quell unhappy spirits by seeking help from priests and healers of various denominational faiths. And while their responses were marginally better, many of them did not know how to care for ghosts or for those who contended with them, and some did not even care (Takahashi, 2016). Instead, their concerns revolved mostly around theoretical and theological questions associated with what in scholarly circles became known as the 'ghost problem': questions as to whether these ghosts really existed; what correspondence, if any, they bore with the Buddhist realm of the *hungry ghost*; or whether these stories could simply be the product of superstition or delusion, a substantiation of the Devil or apparitions of the Holy Spirit (Lloyd Parry, 2017). Confronted with the impossibility of effective forms of care, their concerns turned instead to the attempt to articulate forms of theoretical repair: seeking explanatory frameworks that could restore the conceptual brokenness out of which ghosts' unsettling presence emerged, to a prior state of good theological sense.

Yet it is also out of this broken cosmos, in the incommensurable break between the irredeemable event of grief and the imperative impossibility of reparation, that a singular collective of priests from Buddhist, Shinto and Protestant faiths, led by Reverend Taio Kaneta, risked a form of experimentation with stories and practices that sought to affirm other ways of living and other modes of earthly habitation in the wake of the devastation.[2] This wasn't easy. Above all, it required not explaining the metamorphic breakdown away but instead daring to take the catastrophe seriously, with all its grief and sorrow, as the very broken world they now inhabit, and the very means by which to revaluate living and dying on these islands. The catastrophe had changed everything. As such, theoretical questions about the nature of ghosts were not the right questions to pose, Kaneta said, for

> what matters is that people are seeing them, and in these circumstances, after this disaster, it is perfectly natural. So many died, and all at once. At home, at work, at school – the wave came in and they were gone. The dead had no time to prepare themselves. The people left behind had no time to say goodbye. … The dead are attached to the living, and those who have lost them are attached to the dead. It's inevitable that there are ghosts. (Lloyd Parry, 2017: 100)

The collective of priests came to this realization only in the wake of their own direct confrontation with the aporetics of reparation: by slowly learning to affirm the incommensurability of the immensity of loss and the imperative impossibility of careful consolation. After being consulted by several people with cases of ghost possession, they decided to perform a ritual march to the coast, through the devastated town of Shizugawa. During their march, 'the landscape through which they walked was broken, and corrupt with decay. Bulldozers had cleared ways through the rubble, and piled it into looming mounds of concrete, metal, wood and tile.' But the heaps had not been completely searched, and the area was suffused with the smell "of dead bodies, and of mud. There was so much rubble," Kaneta recalled, "and mementoes of people's lives still lying around on the ground. We had to take care where we stepped to avoid trampling on photographs" (2017: 224). The procession of priests roamed through the devastation, carrying a placard bearing the characters meaning 'Consolation for the Spirits'. As they passed through workers and machines trying to sort out the debris, they began to worry that, rather than helping in the efforts of restoration, they had become 'an unwelcome obstruction to the clean-up operation' (2017: 224). Yet there were ordinary people there as well, looking for the bodies of their loved ones. And when "they saw us marching past," Kaneta told Lloyd Parry (2017: 224), "they turned and bowed their heads. They were praying desperately to find their loved ones. Our hearts were so full when that happened. I have rarely been more conscious of suffering." It had been the aim of these priests to sing sutras and hymns as they marched, but amidst the rubble, their voices failed them – none of the hymns seemed right, the sutra came out in screams and shouts:

> 'And when we got to the sea,' said Kaneta, 'when we saw the sea – we couldn't face it. It was as if we couldn't interpret what we were seeing.' … We realised that, for all that we had learned about religious ritual and language, none of it was effective in facing what we saw all around us. This destruction that we were living inside – it couldn't be framed by the principles and theories of religion. Even as priests, we were close to the fear that people express when they say, 'We see no God, we see no Buddha here.' I realised then that religious language was an armour which we wore to protect ourselves, and that the only way forward was to take it off. (2017: 225)

The cosmoecological workshop: spiritual care out of a broken world

Taking the armour off was not a matter of abandoning faith, however. It was not a question of consenting to the catastrophe of a secular reckoning, or of resigning themselves to the sheer impossibility of crafting a response. On the contrary, taking the armour off precisely meant giving to this catastrophic breakdown the power to affect them, to transform their thoughts, values and practices, to let themselves become affected in such a way that might render them capable of activating a regenerative response from the very incommensurability of the break that was now their present. Indeed, if I here make a case for the need and possibility of learning to 'philosophise with a hammer', of not seeking theoretical answers that might tell us how to live but experimenting with revaluating our values by living our way into other modes of thinking, it is because this is precisely what these priests found themselves engendering in their unflinching efforts to recompose modes of living and modes of dying worthy of the broken situation to which they sought to respond. Faced with a devastation that was as much ecological as it was cosmological, what the collective of priests improvisationally composed was nothing short of a *cosmoecological workshop*: experimenting pragmatically, across the worlds of the living and the dead, with the possibility of nurturing collective ways of living and dying otherwise from the interstices of the catastrophe, of seeking to recompose a cosmos on a ravaged earth.[3]

Indeed, what they set in motion was a mobile café which, in a triple play with words, they decided to call *Café de Monku*: *Monku* being the Japanese word for 'complaint'; *Monk*, the English term for priest; and Thelonious Monk, the name of the jazz pianist and composer whose improvisational music would accompany their practices. 'The dissonant and loose tempo of Monk's music', Kaneta (2015) suggested, 'accurately expresses the disaster victim's hearts.' The initial goal was nothing more but also nothing less than to make 'a space where people can relax in the midst of the rubble!' And so would the *Café de Monku* travel around the devastated area, nurturing a milieu in which those who had survived the wave could gather together around tea, coffee and cakes, share their stories and sorrows, raise their 'complaints', and collectively remember the dead. In addition to tea and cakes, the priests would offer exercises, humour and massages; music would be played, flowers would be laid; sutras and hymns were offered to those who benefited from them; and they would infuse the air with the scent of incense. The survivors, many of whom were at that stage living in temporary shelters, were invited to sit and string Buddhist rosaries as the priests inscribed and blessed memorial tables with the names of the dead. And of course, they would occasionally perform rituals and techniques devoted to caring as much for those who had been possessed and as for the ghosts who made themselves felt through them (Yōzō, 2015).

It was in the indeterminate course of fabricating this cosmoecological workshop that the collective of priests came to learn how to tend to the living and the dead, to compose modes of living and dying otherwise in the wake. Indeed, they learned that, though death awaits us all, it matters what modes of dying precipitate one's end, such that, when 'people die violently or prematurely, in anger or anguish, they are at risk of becoming *gaki*: hungry ghosts, who wander between worlds, propagating curses and mischief' (Lloyd Parry, 2017: 103). And thus, they also learned to discern the presence of ghosts in the suffering of the living, which in turn enabled them to develop practices thanks to which they could hold a veritable conversation with the dead. Strikingly, they discovered that not all ghosts that made themselves present in the wake of the tsunami were of those the wave had washed away. For some, the problem was that the tsunami had taken *their living* away, leaving no one in the world to look after them in their death. And not all of them knew they were dead. Among those that Kaneta sought to help was Rumiko Takahashi, a young woman that had been repeatedly possessed by ghosts. Indeed, in his spiritual care work with Rumiko, Kaneta consoled over 25 ghosts that spoke and made themselves present through her. Among the first was the ghost of a middle-aged man who, through Rumiko, 'despairingly called the name of his daughter. "Kaori!" said the voice. "Kaori! I have to get to Kaori. Where are you, Kaori? I have to get to the school, there's a tsunami coming"' (2017: 247). The man was on his way to pick Kaori up from school when the – very late – tsunami warning was triggered, and the wave caught him driving along the coast, carrying him and much else to the bottom of the sea. As Lloyd Parry (2017: 248) relays their exchange:

> The voice asked, 'Am I alive or not?'
> 'No,' said Kaneta. 'You are dead.'
> 'And how many people died?' asked the voice.
> 'Twenty thousand people died.'
> 'Twenty thousand? So many?'
> Later, Kaneta asked him where he was.
> 'I'm at the bottom of the sea. It is very cold.'
> Kaneta said, 'Come up from there to the world of the light.'
> 'But the light is so small,' the man replied. 'There are bodies all around me, and I can't reach it. And who are you anyway? Who are you to lead me to the work of the light?'

The conversation, Lloyd Parry tells us, 'went round and round for two hours' until at some point Kaneta asked him to consider the consequences of his own mournful wandering on the suffering of his host:

> 'You are a father. You understand the anxieties of a parent. Consider this girl whose body you have used. She has a father and a mother who are worried about her. Have you thought of that?' There was a long pause, until finally the man said, 'You're right' and moaned deeply. Kaneta chanted the sutra. He paused from time to time when the voice uttered choked sounds, but they faded to mumbles and finally the man was gone.

It is this improvised configuration of doings and modes of care that the collective of priests developed in response to the imperative impossibility of reparation in the wake of cosmoecological devastation. Practices, as they called them, of 'listening to the heart': diligently and attentively enabling stories of suffering to be told, even when they profoundly challenge the thoughts, beliefs and values that one holds dear, while carefully reweaving those tales and addressing them to the future. 'Don't think,' says Kaneta (2015), in an emphatic call to arms that dramatizes the gesture at the heart of their practices. 'Feel!! Creation and Act!!' Indeed, it was only by feeling their way into another mode of thinking, by attending to stories of suffering as much from a personal as from a cosmic perspective, that they were able to reweave and repurpose a Buddhist concept, *jita funi* – which translates as 'self and other: undivided' – to reweave relationships among the living and the dead. And they did so by making it perceptible that this 'universe wraps everything up inside it, in the end. Life, death, grief, anger, sorrow, joy. There was no boundary then, between the living and the dead' (Lloyd Parry, 2017: 222).

Café de Monku proved highly successful, making the spiritual care work an essential emergency response to the invisible devastation of a life lived in loss in the wake of ecological disaster. But neither practices of listening to the heart nor the reactivation of this Buddhist concept were capable of repair if by that we understand a piecing back together, a restitution of what broke down, a return to a world that had collapsed.[4] The dead did not come back to life, and neither did the living go back to theirs. The aporetics of reparation pressed on in the incommensurable break between grief and restoration. 'This', Kaneta (in Lloyd Parry 2017: 242) said, 'is consolation. This is understanding. We don't work simply by saying to people, "Accept." There's no point lecturing them about dogma. We stay with them, and walk with them until they find the answer on their own. We try to thaw the frozen future.' Out of the brokenness of their world, the priests' cosmoecological workshop began to repair something else. Philosophizing with a hammer amidst the resonance of events, they and those that joined them at the *Café de Monku* began to learn how to tell their stories otherwise, to regenerate their collective capacities to think, feel and imagine, to reweave modes of living and dying in the wake of the catastrophe, to pragmatically recompose a cosmos on a ravaged earth.

Notes

1 For a resonant and very generative take on the politics of repair and recuperation in the context of post-crash politics in Portugal, see the thoughtful reflection by Sánchez Criado (2020).

2 They, of course, were not alone in this. For other experiments in the wake of the Fukushima Nuclear disaster see the accounts by Sabu Kohso (2020) and Mari Matsumoto and Sabu Kohso (2017).

3 For an account of the radical pragmatism dramatized by this collective of priests, see Savransky (2021a).

4 There are in this sense possibly generative resonances to be explored, between the practices nourished by *Café de Monku* and a traditional craft technique for broken ceramics, known in Japan as *Kintsugi* and dating from the late sixteenth century, which is characterized by rendering both the object's history of brokenness and the transformative operations of repair visible and sensible rather than pursuing restoration through erasure. While a full discussion of *Kintsugi* and any possible connections to *Café de Monku* exceeds the scope of this chapter, see the article by Keulemans (2016) for a thoughtful conceptualization of the craft.

References

Best, S. and Hartman, S. (2005) 'Fugitive justice', *Representations*, 92: 1–15.
Deleuze, G. (2004) *The Logic of Sense*, London: Continuum.
Figal, G. (1999) *Civilization and Monsters: Spirits of Modernity in Meiji Japan*, Durham, NC: Duke University Press.

Graham, S. and Thrift, N. (2007) 'Out of order: Understanding repair and maintenance', *Theory, Culture & Society*, 24(3): 1–25.

Guattari, F. (2001) *The Three Ecologies*, London: Continuum.

Ivy, M. (1995) *Discourses of the Vanishing: Modernity, Phantasm, Japan*, Chicago: University of Chicago Press.

Jackson, S. (2014) 'Rethinking repair', in T. Gillespie, P. Boczkowski and K. A. Foot (eds) *Media Technologies: Essays on Communication, Materiality, and Society*, Cambridge, MA: MIT Press, pp 221–40.

Kaneta, T. (2015) 'Listening to the heart: Among the people in the Great East Japan Earthquake Disaster Area.' Paper presented at the third *United Nations World Conference on Disaster Risk Reduction*. Sendai, Japan, March 14–15, 2015. Available from: http://drr.tohoku.ac.jp/system/wp- content/uploads/2015/01/aaf2cd443b060aaaf3090227d870528a.pdf [Last accessed 4 April 2019].

Keulemans, G. (2016) 'The geo-cultural conditions of *Kintsugi*', *The Journal of Modern Craft*, 9(1): 15–34.

Kohso, S. (2020), *Radiation and Revolution*, Durham, NC: Duke University Press.

Les éditions des mondes à faire editorial collective (2018) *Fukushima & Ses Invisibles*, Vaulx-en-Velin: Les Éditions Des Mondes A Faire.

Lloyd Parry, R. (2017) *Ghosts of the Tsunami*, London: Vintage.

Matsumoto, M. and Khoso, S. (2017) 'Rages of Fukushima and grief in a no-future present,' in C. Milstein (ed) *Rebellious Morning: The Collective Work of Grief*, Chico, CA: AK Press.

Nietzsche, F. (1997) *Twilight of the Idols: Or, How to Philosophise with a Hammer*, London: Hackett.

Nixon, R. (2011) *Slow Violence and the Environmentalism of the Poor*, Cambridge, MA: Harvard University Press.

Sánchez Criado, T. (2020) 'Conclusion: Repair as repopulating the devastated desert of our political and social imaginations', in F. Martínez (ed) *Politics of Recuperation*, London: Bloomsbury.

Savransky, M. (2019) 'The bat revolt in values: A parable for living in academic ruins', *Social Text*, 37(2): 135–46.

Savransky, M. (2021a) *Around the Day in Eighty Worlds: Politics of the Pluriverse*, Durham, NC: Duke University Press.

Savransky, M. (2021b) 'After progress: Notes for an ecology of perhaps', *Ephemera: Theory & Politics in Organization*, 21(1): 267–81.

Takahashi, H. (2016) 'The ghosts of the tsunami dead and *Kokoro no kea* in Japan's religious landscape', *Journal of Religion in Japan* (2016): 176–98.

Yōzō, T. (2015) 'Chapliancy work in disaster areas', in C. Harding, I. Fumiaki, and Y. Shin'ichi (eds) *Religion and Psychotherapy in Japan*, London: Routledge.

PART III

Contaminating<>Cohabiting

7

Multispecies Mending from Micro to Macro: Biome Restoration, Carbon Recycling and Ecologies of Participation

Eleanor Hadley Kershaw

Amid the manifold social and ecological crises of the Anthropocene,[1] diverse understandings of the entanglement of human life with more-than-human species and entities have emerged across and between disciplines, alongside attempts to intervene in these relations. Fields such as Earth system science and sustainability science have envisioned, charted and emphasized the interrelatedness of (socio)ecological processes and phenomena across spatial scales and geopolitical borders, while science and technology studies has explored how these global environmental knowledges are co-produced with global political orders (Miller and Edwards, 2001; Jasanoff, 2004). Posthuman and new materialist theory, often indebted to Indigenous ontologies and epistemologies (Sundberg, 2014; Giraud, 2019), has challenged Anglo-European anthropocentrism and dualisms (of nature/culture, mind/body, social/material), by considering the human to be always, relationally and mutually constituted with often agential or lively nonhumans, whether animals, technologies, energies, landscapes or elements (Haraway, 2007; Bennett, 2010; Povinelli, 2016).

In environmental, life and health sciences, the symbiotic roles of bacteria and other microbes in constituting animal lives and creating or maintaining individual, ecological and planetary health has become a key area of investigation (McFall-Ngai et al, 2013), shaping popular imaginaries of ecologies within and beyond human bodies (Turney, 2015; Yong, 2016), and producing new forms of microbiopolitics (Paxson, 2008). By the early twenty-first century, dominant scientific and popular representations of microbial life had 'shifted from an idiom of peril to one of promise' (Paxson and Helmreich, 2014: 165), in which microbes have come to represent both a 'not yet tamed or fully known nature' and model ecosystems that might prescribe the ways in which relations between organisms and humans '*could*, *should* or *might* be' (p 168). Meanwhile, in feminist science studies and anthropology, the resurgence of life forms such as fungi in the 'capitalist ruins' (Tsing, 2015), and the multispecies, material relationalities in practices such as permaculture (Puig de la Bellacasa, 2017) have been taken as much-needed symbols of hope and examples of the possibility of care and renewal.

Microbial capacities for ecological and material transformation are increasingly mobilized within what geographer Jamie Lorimer (2019, 2020) refers to as 'the probiotic turn', a shift occurring across a range of science and policy domains in parts of the WEIRD (Western, Educated, Industrialised, Rich, Democratic) world. Probiotic approaches 'use life to manage life', deliberately changing the composition of biophysical systems to control, restore and enhance ecological dynamics in order to address or cure pathologies resulting from earlier 'antibiotic' efforts to 'eradicate, control, rationalize, and simplify life' (2020: 2). Lorimer's examples span a wide range of organisms and environments, not only microbes (to be discussed later), but also the reintroduction of other keystone species (for example large herbivores or beavers) in rewilding initiatives to address the detrimental effects of modern natural resource management and intensive farming, and the use of ocean seeding or afforestation for geoengineering, among others. While these can be read as attempts towards ecological reparation that practice interconnectedness and afford more reciprocally caring relations between all beings (Puig de la Bellacasa, 2017), they can also (and simultaneously) be seen as Anthropocenic, ecomodernist, productivist, reductionist or biocapitalist fantasies of 'a technofix planet' where human control is reinscribed over unruly ecologies (Hird and Yusoff, 2018: 268).

Notwithstanding the shifts in human-microbial relations that have occurred during the COVID-19 pandemic (McLeod et al, 2020), microbes play – and will likely continue to play – a key role in narratives and practices of ecological repair, promise and hope in a range of contexts, from intimate bodily processes (McLeod et al, 2019) to global systems and planetary geological formation (Hird and Yusoff, 2018). This chapter explores two approaches to mending damaged ecologies with microbes. First, therapeutic efforts to 'rewild' human microbiomes through the restoration of mutualistic microbes (hookworms and bacteria) to treat diseases of microbial absence and imbalance. Second, synthetic biology endeavours to engineer microbes to assimilate greenhouse gases and produce sustainable chemicals in carbon circular economies. Drawing on scientific, social scientific and popular literatures on helminth and faecal microbiota transplant therapies (biome restoration), and ongoing ethnographic engagement with synthetic biology for gas fermentation (carbon recycling), the chapter explores what might (and might not) be afforded by ecological, multi-scalar thinking, fixing and caring in these contexts. In concluding, the chapter turns briefly to the analytical lens of 'ecologies of participation' (Chilvers et al, 2018) to consider its value in accounting – and perhaps also making space – for diverse modes of engagement with ecological mending. This concept, while coined to consider the systemic and co-productive nature of varied forms of public engagement with sociotechnological change such as energy transitions (Chilvers et al, 2018) or algorithms in public services (Pallett et al, 2021), can usefully be extended beyond the human to bring into focus the political, constitutive aspects of efforts towards ecological reparation – such as biome restoration or carbon recycling – that harness more-than-human capacities for biophysical, material and relational transformation to mend damaged (or damaging) corporeal, industrial and planetary ecologies. This perspective allows us to shift beyond reification – or, equally, obfuscation – of 'the microbe' to consider reparation as a multiplicity of contingent multispecies processes (entailing diverse entities and forces) co produced with and co-producing political imaginaries, cultures and constitutions.

Mending microbiomes

Ecological understandings of the human microbiome – the communities of organisms[2] living on and in the human body (and their collective genome) – emerging in the life sciences describe deep co-evolutionary, symbiotic and constitutional relationships between microbiota and their humans (or their non-human animals) (Blaser, 2014; McFall-Ngai et al, 2013). This work suggests that microbiota shape a range of vital bodily processes and functions, including immunity, metabolism and cognition, through 'intimate biochemical chatter' (Sachs, 2008, cited in Lorimer, 2016: 67; Blaser, 2014; Wammes et al, 2014). In particular, as explored by Lorimer (2016, 2017a, 2019, 2020) as a key exemplar of the probiotic turn, hookworms, among other helminths (parasitic worms), were once a common part of the human microbiome across many more geographical regions than today.[3] Ecoimmunologists propose that, through co-evolutionary relations spanning millennia, hookworms learned to disguise their presence and create hospitable conditions in the human gut by training or modulating the immune system in communication with commensal bacteria, suppressing the host's normal immune response (Wammes et al, 2014; Zaiss et al, 2015; Lorimer, 2020). While a heavy infection can result in hookworm disease (with symptoms such as abdominal pain, diarrhoea, weight loss and anaemia), a light infection can be tolerated and may even be beneficial (Wammes et al, 2014).

The 'hygiene hypothesis' (Strachan, 1989) and 'biome depletion paradigm' (Parker, 2014) posit that antibiotic trends in healthcare and sanitation (not only antimicrobial drugs but also hygiene infrastructures and practices such as flush toilets and antiseptic cleaning regimes) have distanced people in industrialized regions from diverse microbial exposure and created absences of 'old friends' in the microbiome (Rook and Brunet, 2005). These benign commensal bacteria and other organisms such as hookworms are no longer present to 'educate' or calibrate immune systems to tolerate their presence, and, without this regulation, immune responses run amok, resulting in the increasing prevalence in affluent countries of allergies and autoimmune diseases such as asthma, inflammatory bowel disease and multiple sclerosis (Wammes et al, 2014). In these 'epidemic[s] of absence' (Velasquez-Manoff, 2012), the microbiome is seen as a damaged ecology, set off-balance by technologies (for example anthelmintic drugs) and lifestyles that separated humans from the microbial life needed for commensal and mutualistic relations to keep bodily systems in check. This calls into question antibiotic ways of (controlling and eradicating) life, including the ethics of ongoing deworming programmes in areas of the world where hookworms are still endemic (Wammes et al, 2014).

These developments in microbiology, parasitology and ecoimmunology have resulted in new approaches to treating dysbiosis (imbalances in microbial ecologies leading to pathological situations); not only the absence of mutualistic helminths, but also, for example, the over-abundance of pathogenic, opportunistic and antimicrobial-resistant bacteria such as *Clostridium difficile*, a sometimes lethal lower gastrointestinal infection often acquired during or after antibiotic treatment. 'Biome restoration' has emerged from these ecological ways of understanding health, taking the form of proactive management or 'rewilding' of microbiomes through deliberate helminth infection or the replacement of gut microbiota through faecal microbial transplant (FMT) (Bono-Lunn et al, 2016; Lorimer, 2020). In both cases, this has occurred not only through formal clinical trials but also through informal networks of expert patients sharing and circulating biological matter (hookworms or faeces) and knowledge about their therapeutic administration and potential. Lorimer (2020) characterizes the emergence of this 'hookworm underground' (Velasquez-Manoff, 2012) – operating (sometimes illegally) outside of regulatory structures and counter to capitalist regimes of resource management – as a type of multispecies commoning (Lorimer, 2020): helminth providers 'grow and distribute worms for free or at cost … [and go] to some lengths to make sure the probiotic value of their worms stays in the public domain' (Lorimer, 2020: 195–6). The hookworm underground could be seen to espouse a more-than-social movement of ecological reparation (Papadopoulos, 2018), attempting to maintain some kind of (for some humans[4]) open access to multispecies cooperation for microbiome repair.

However, the political imaginaries and alter-ontologies of DIY biome restoration are challenged by the emergence of biotechnological attempts to privatize the probiotic value of helminths and faeces, turning from 'the ecological work done by bugs as drugs' to the synthesis of 'drugs from bugs', and from biobanks of healthy, donated faeces to the synthesis of the active ingredients therein to produce microbiome therapeutics for profit (Lorimer, 2020: 199–200). Here, researchers attempt to decouple the beneficial ecosystemic effects of hookworms and microbiota from their ecological embodiment and embeddedness, making them more amenable to existing routes to biocapitalization, by, respectively, identifying molecules secreted by worms in the hopes of synthesizing them as pills, and delivering microbial consortia in capsule form rather than in faeces (Ditgen et al, 2014; Seres Therapeutics, nd; Lorimer, 2020).

Across these multiple modes of microbiome repair, a range of microbial roles emerge. The hookworm or microbe as an 'ecosystem engineer' or immune system manager, working to build, transform and preserve its habitat and manage bodily balance (Jones et al, 1996; Laforest-Lapointe and Arrieta, 2018). Hookworms (or even bacteria) as 'gut buddies' and 'colon comrades'[5] that enter into reciprocal relations of care with their human hosts who do what they can to provide ideal conditions for microbial flourishing. And hookworms or microbes as sources of biomimetic inspiration or 'more-than-human intellect' ripe for capitalist enclosure (Johnson and Goldstein, 2015, cited in Lorimer, 2020: 192). While each draw on ecological ontologies to understand and enact relations between human and microbiome, these roles differ in the hope they can offer for moving beyond productivist stories, economies of service, or 'entrenched relations to natural worlds as "resources"' (Puig de la Bellacasa, 2017: 221). Ecological reparation in the sense of repairing microbiomes entails varied visions of ideal human-microbe relations and thus varied possibilities for more-than-human agency and reciprocal care.

Mending carbon cycles

Developments in gas fermentation are a contrasting instance of microbial mending, where ecological thinking informs sociotechnical imaginaries, but does not always shape human conceptions of microbes. In the context of a carbon cycle deeply affected by land use change, industrial scale mining and burning of fossil resources, emissions from the production of materials such as cement and steel, and the mass-manufacture of persistent carbon compounds (for example plastics), industrial biotechnologists and researchers in the field of synthetic biology endeavour to engineer microbes to (more efficiently) assimilate greenhouse gases and produce chemicals, fuels, materials, feed and food. This work is oriented in relation to addressing climate change through the dual benefits of capturing waste gases (for example carbon dioxide, carbon monoxide, methane) that otherwise would be released into the atmosphere,[6] and using (exploiting or valorizing) these gases as a carbon source to make chemicals, fuels and materials that otherwise would be produced using fossil resources, thus reducing 'societal reliance on petrochemicals-based feedstocks' (SBRC-Nottingham, nd). In practice, this could mean situating gas fermentation facilities (comprising bioreactors and other processing technologies) in high-emitting industrial contexts such as steel mills, cement works, power generation or waste treatment plants, or potentially operating at a smaller or more distributed scale (to be discussed later).

While much research is still in development at the laboratory scale, US-based LanzaTech was the first company to operate a proprietary microbial gas fermentation process converting steelworks off-gases to ethanol at commercial scale, with their first commercial plant launched in Shougang, China in 2018, and further facilities established, in construction or planned in Belgium, India, South Africa, the UK and the US, all aiming to 'use nature to heal nature', whether through optimized processes that enhance the productivity of non-engineered bacteria or by also 'optimizing' the microbes themselves.[7] In November 2020, the Shougang carbon recycling facility reached 150 days of full capacity continuous operation, and since its start up has produced more than 20 million gallons of ethanol, 'the equivalent of keeping 100,000 tons of CO_2 from the atmosphere' (Burton, 2020). Some of this ethanol was used in late 2018 to part-fuel a transatlantic flight (The Guardian, 2018). Other companies and research projects operating at a range of 'Technology Readiness Levels' and scales (from lab to commercial) are exploring the potential of gas fermentation for a range of other contexts, feedstocks and products (Humphreys and Minton, 2018; Carus et al, 2019), using gases such as carbon dioxide from industrial waste streams or municipal waste treatment plants, syngas (from a range of sources including gasified solid waste) or methane for products such as animal feed,[8] bioplastics[9] and (via CO_2 to acetate/acetic acid) omega-3 fatty acids (Sherrard, 2020).

These forms of biological 'carbon recycling' (alongside chemical CO_2 conversion processes) are gathering pace and garnering attention though coordination efforts such as the Renewable Carbon Initiative. This international alliance of companies, led by German private research institute the nova-Institute,[10] aims to promote a transition from fossil carbon to 'renewable carbon' in the chemical and materials industries:

> Renewable Carbon entails all carbon sources that avoid or substitute the use of any additional fossil carbon from the geosphere. Renewable carbon can come from the biosphere, atmosphere or technosphere – but not from the geosphere. Renewable carbon circulates between biosphere, atmosphere or technosphere, creating a carbon circular economy. (Renewable Carbon Initiative, nd)

Renewable carbon (AKA carbon recycling or carbon (capture and) utilization (CCU)), is positioned as the equivalent of decarbonization of the energy industry:

> Organic chemistry cannot be decarbonised because it is defined by the use of carbon. The same applies to the plastics industry, without whose versatile polymers the modern world would be inconceivable – or only with a considerable renunciation and higher greenhouse gas emissions. (Carus et al, 2019: 10)

This work is frequently framed in terms of circular economy or circular bioeconomy (Venkata Mohan et al, 2016; Hadley Kershaw et al, 2021); interpretively flexible, contested concepts that have foundations in 'ecological' ways of conceiving economic and material resource flows, arising from ecological economics and industrial ecology (Inigo and Blok, 2019):

> Industrial ecology (IE) is an emerging framework for environmental management, seeking transformation of the industrial system in order to match its inputs and outputs to planetary and local carrying capacity. A central IE goal is to move from a linear to a closed-loop system in all realms of human production and consumption. In this and other ways the industrial world can move closer to an ecological model in its dynamics. Industrial ecosystems embody a concrete strategy for developing closed-loop systems locally, in industrial parks or regions. (Lowe and Evans, 1995: 47)

While perhaps not embodying the full interrelatedness of industrial ecosystems (and cognate ideas such as industrial symbiosis and industrial metabolism) imagined in some sectors, many of the developments in carbon recycling rely on co-location with and valorizing waste streams of existing industry, as in the case of LanzaTech's facilities. Organizational and spatial imaginaries such as bioclusters and public-private partnerships have become key to bioeconomy implementation generally (Philp and Winickoff, 2017), and smaller-scale carbon recycling projects have ambitions for distributed biorefinery options, for example enabling farms or consortia of farms to valorize waste gases from existing or future anaerobic digestion and biogas facilities; alternative political and sociotechnical imaginaries of the uses and scales of these technologies beyond centralized industrial parks towards smaller-scale, rural and agricultural community use.

Carbon recycling could represent the becoming ecologically obliged (Papadopoulos, 2022) of chemical, material and waste management industries, in the sense of acknowledging and attempting to rectify extractivist, productivist, and polluting approaches to feedstock sourcing, chemical production or waste processing by aiming to move towards industrial systems that are 'restorative and regenerative by intention and design' (Ellen MacArthur Foundation, 2013: 7), working across species to achieve carbon fixation and salutary (bio)physical transformation. Smaller-scale imaginaries also open up the potential for wider, more distributed or decentralized access to these technologies across a broader range of sectors and actors. However, despite the framing discourses linking to ecological (or ecologically informed) ontologies (such as industrial symbiosis), and the material constitution of sociotechnical infrastructures and transformational substance flows that are in some ways shaped by these ideas, carbon recycling is at risk of techno-optimist, ecomodernist and reductionist limitations in its approaches to mending damaged carbon cycles, cleaning up industrial systems, and managing (or 'optimizing') the microbial life that is deployed to help.

Climate change and fossil resource dependency are highly complex, wicked problems (Hulme, 2009), not easily amenable to technofix solutions (Levidow and Raman, 2020). There is a danger that pursuing carbon recycling technologies without attending to their sociopolitical dimensions and wider contexts diverts attention or urgency from the crucial project of rethinking, restructuring or repairing the institutions, relations and consumption patterns that produced current pathologies, potentially even further locking in carbon-intensive industries and practices through rebound effects (whereby greater efficiency or valorization of waste leads to greater demand for resources and products) or other unintended outcomes (Inigo and Blok, 2019; Genovese and Pansera, 2021). Developments in carbon recycling must be accompanied by ongoing investigation and contextualization of the problem(s) they intend to address, the pathways they are taking and the worlds they are making. Circular economy strategies in general, and carbon recycling in particular, frequently echo the ecomodernist ambition to reconcile the resolution of environmental problems with economic growth, based on renewed belief in the possibility of human 'mastery' and control (Hajer, 1997: 34; Genovese and Pansera, 2021). This challenges the possibility of ecological reparation through carbon recycling, not only due to the destructive legacies and limitations of capitalist market economies (Benessia and Funtowicz, 2015; Genovese and Pansera, 2021), but also through its anthropocentrism and treatment of nonhuman entities as service providers or resources. The synthetic biology that microbial CO_2 conversion routes often require has been critiqued for its modernist and reductionist attempts to simplify and decontextualize living organisms by isolating and building with 'parts' (Calvert, 2010; de Lorenzo, 2011).

In these contexts, bacteria that are able to grow on gases such as carbon monoxide and carbon dioxide are engineered using synthetic biology techniques, either to produce higher yields of chemicals, fuels and materials than they would produce 'naturally' at lower yields, or to produce different – or a wider range of – products (Humphreys and Minton, 2018). The microorganisms involved include acetogens (microbes that generate acetate), chemoautotrophs such as *Cupriavidus necator*, which accumulates polyesters within its cell as energy storage, and photoautotrophs that draw on light as an energy source. The microbiologists, modellers, chemists and engineers working with these organisms often refer to them in terms of 'chassis' (as in a car frame or other mechanical structure to house, or on which to build, a range of metabolic pathways) and 'factories' (for example Humphreys and Minton, 2018; see also McLeod and Nerlich, 2017). The factory metaphor is employed at multiple scales, figuring both individual bacterium and whole bioreactors (including their microbial communities) as production facilities.

The implications of factory and machine metaphors have been explored in social, philosophical and linguistic studies of synthetic biology (Boudry and Pigliucci, 2013; Boldt, 2016) (and more widely in relation to plants or animals (Harrison, 2013; Berry, 2018)). Boldt (2018) argues that expressions taken from car manufacturing (such as 'chassis') detract attention from the ecological aspects of these living beings, such as evolutionary change and interdependencies in ecosystems. In the context of gas fermentation, the notion of bacterium as a chassis or mini-factory and bioreactors as microbial factories simultaneously figures and obscures microbes as resource-makers and labourers, recategorizing them from lifeforms to disembodied, mechanistic technologies or anonymous workers whose function is to transform waste to worth or incalcitrant non-life (inorganic carbon in the form of CO_2) to something more lively (organic carbon in the form of chemicals). Perhaps the exploited carbon[11] dreams of returning to its long, deep rest sequestered in soil or sediment.

Recent work in social studies of synthetic biology has explored shifts in some parts of the field towards the recentring of the agential organism as an important figure, where it had been obscured previously through machine metaphors and a focus on genetic pathways, manipulatable parts, and commercially viable products. Szymanski and Calvert (2018) explore the potential of directed evolution approaches (and metaphors that 'activate' the organism (Szymanski, 2018)) as a means of co-design between microbes and synthetic biologists, in which some control on the part of scientists is relinquished and microorganisms may play a more active role – a shift from rational design towards more reciprocal (or even open-ended) multispecies collaboration, where unexpected sources of value might be allowed to evolve or emerge (Boldt, 2016). Joining other accounts of multispecies participation (for example Waterton and Tsouvalis, 2015; Bastian et al, 2016; van Dooren et al, 2016), this strikes a hopeful note for moving past the reductionism and instrumentalism of a rational design ethos in synthetic biology for carbon recycling and its key metaphors of the chassis and the factory to consider microbes – and perhaps other aspects of the more-than-human world – beyond their 'functions' as resource makers or service providers. However, it also raises questions about the conditions that might enable participation and activation in multispecies collaborations, and the broader political economies and ecologies that they shape and are shaped by. For example, open-endedness may not always be a sustainable or justifiable quality for collaborations in the context of output-focused research and innovation systems and capitalist market economies.

Ecologies of participation

The cases of biome restoration and carbon recycling present various visions of – and challenges to – ecological reparation. Helminth therapy and faecal microbiota transplantation can be seen as probiotic approaches to repairing microbiomes ravaged by antibiotic regimes, restoring human health by reintroducing microbial diversity and re-establishing mutualistic, companionable human-microbe relations. This has been enacted through formal clinical routes and informal patient networks, though its potential for (re)instating reciprocal ecological relations might be challenged by biocapitalist attempts to synthesize the active components of hookworms and bacteria to achieve beneficial effects (for some humans) independent of embodied microorganisms. Carbon recycling technologies can be understood as efforts to repair – or replace – disrupted and damaged carbon cycles, depleted fossil resources, and to redress deleterious carbon-intensive industries by preventing (or delaying) some carbon emissions and substituting them for petrochemical feedstocks to produce chemicals, plastics, fuels and other products. This might be achieved through centralized industrial ecosystems or more distributed, localized arrangements accessible to a wider range of sectors and actors. However, these technologies could be at risk of neglecting the multidimensional character of the problems they seek to address – or even serve to justify or maintain ongoing overconsumption – without careful interrogation and reformation of the ecomodernist, techno-optimist, reductionist or capitalist narratives and systems which they might otherwise perpetuate. The synthetic biology integral to some carbon recycling technologies might further reinforce entrenched exploitative relations between humans and 'natural' resources or entities by treating microbes as machines, factories or mechanisms to be manipulated and improved, and thus simplifying or even negating their ecological dimensions and, arguably, their agency. Means to address this might include recentring or reactivating microbes in reciprocal multispecies collaboration.

These cases reveal multiple ways in which more and less ecological understandings of the world (the microbiome, the carbon cycle, industrial systems, organisms) are shaping attempts to mend damaged ecologies across different contexts and scales (from intimate human to expansive planetary health). Scale is produced in these stories in a range of ways: as size and visibility (microscopic to macroscopic); geographical space (local to global); and 'Technology Readiness Level' – interpreted as bioreactor volume and context of operation (from smaller volumes in the lab to large volumes in commercial settings). The cases show multiple modes of participation and collaboration in mending, and different ways of valuing the significance of more-than-human activities and contributions. Political imaginaries varied from reproducing, extending and maintaining existing biocapitalist systems and technofixes to enabling decentred, distributed or open access forms of bioproduction and probiotic health management. Kinds of relations between humans (scientists, industry leaders, expert patients, etc.) and more than humans (microbes, carbon, chemicals, sociotechnical infrastructures, etc.) included industrial exploitation, labour and service delivery; companionable comradeship; and multispecies co-design. Key questions arose around who has access to and can participate in these sociotechnical nature-cultural ways of repairing, being and living, and what the terms of engagement might be.

These questions are perhaps usefully considered through the lens of 'ecologies of participation' (Chilvers et al, 2018), an analytical approach to public participation in sociotechnical change[12] that aims to think beyond discrete instances of engagement to consider participation practices (and subjects and objects) as multiple, relational and systemic, co-constitutive of each other and of broader political cultures and systems. While not explicitly posthuman, this constructivist relational framework draws on the language of new materialism (Bennett, 2010) to see change as the 'outcome of multiple swarming vitalities' (ecologies of interconnected sociomaterial collectives and practices of participation) rather than as a result of extending, replicating or 'scaling up' pre-defined, individual and bounded engagement exercises or processes (Chilvers et al, 2018: 201). Incorporating more-than-human publics in this framework may enable further consideration of diverse practices, relationalities and multiplicities of efforts towards ecological reparation (such as biome restoration and carbon recycling), bringing into focus their political, constitutive dimensions, exploring their interconnections and, importantly, the conditions that give rise to and co-produce generative, caring and hopeful multispecies collaborations and transformations.

Mapping diverse forms of microbial mending through this lens foregrounds urgent questions of who (is able to) participate(s), in what ways, around which issues (to what end), as well as how these emergent subjects, modes and objects of participation are co-produced with each other and with(in) broader systems and contexts. It draws attention to the worlds, actors and practices (re)made through these activities. This contextualization and focus on emergence might also facilitate a shift beyond reification or obfuscation of 'the microbe' – a human category mediated by human understanding, whether protagonist of our salvation or invisible labourer – to consider reparation as a multiplicity of contingent multispecies processes and achievements, entailing diverse entities and forces, not only a range of microbes and humans, but also chemicals, materials, sociotechnical infrastructures, energies, and more. This approach offers hope when considered in conjunction with Papadopoulos' notion of ecologically obliged chemical practice as reparation (2022), where the global, all-encompassing, intractable scale of Anthropocenic challenges (such as pervasive chemical toxicity, or fossil-dependence of late capitalist societies) might be addressed through multiple, situated, distributed and diverse acts of ontological, sociomaterial healing. Diverse participation – and understanding how myriad experiments of multispecies mending might relate to, depend on and shape each other and their contexts, identifying how to repair without repeating historical and contemporary patterns of destruction – is essential to ecological reparation that begins to redress unevenly distributed benefits and harms of human ways of life and fosters reciprocally caring relations between all beings.

Notes

1. The term 'Anthropocene' was first used in Earth system science to denote an epoch in which humans have taken a central role in shaping geologies and ecologies (Crutzen and Stoermer, 2000). It has generated wide debate – and critique – within and across disciplines (Lorimer, 2017b), not least for its anthropocentrism and universalization of the 'human' (Lövbrand et al, 2015). Here it is employed as shorthand for a diverse range of environmental, ecological and social change, including species loss, climate change, ocean acidification, proliferation of plastic and chemical waste, air pollution, land use change, displacement and degradation of communities, and more.
2. Comprising bacteria, archaea, viruses, fungi and other uni- and multicellular eukaryotes (for example protists and helminths) (Laforest-Lapointe and Arrieta, 2018, NIH Human Microbiome Project, nd).
3. Although hookworms are now largely eradicated in the Global North, they remain endemic in some rural areas of the Global South (for a relational geography of hookworms and human health see Lorimer, 2017a, 2020; for a consideration of the implications of these differential geographies for understandings of and responses to abundance in the Anthropocene, see Giraud et al, 2019).
4. Just as the current spread of pathological hookworm infection and hookworm absence is geographically and socioeconomically differentiated, so is the distribution of antimicrobial and biome restoration therapies, with the latter predominantly developed in North America, Europe and Australia (Lorimer, 2017a) – and available primarily to those able to enrol in clinical trials or access and navigate informal expert patient-led networks.
5. Lorimer (2016, 2017a, 2020) notes that these were the titles of two blogs run by early helminth users. See http://web.archive.org/web/20220420082245/https://coloncomrades.wordpress.com/ and www.facebook.com/people/Gut-Buddies/100063971292667/
6. Though they may still be released into the atmosphere at a later stage.
7. https://lanzatech.com
8. For example, see https://deepbranch.com/

[9] For example, see https://engicoin.eu/

[10] https://renewable-carbon-initiative.com/. See also the CO_2 Value Europe association (www.co2value.eu/), the Global CO_2 Initiative (www.globalco2initiative.org/), the Circular Carbon Network (https://circularcarbon.org/), and UK Biotechnology and Biological Sciences Research Council (BBSRC) funded Network in Industrial Biotechnology and Bioenergy (NIBB), The Carbon Recycling Network (https://carb onrecycling.net/).

[11] https://engicoin.eu

[12] Such as low-carbon energy transitions.

Acknowledgements

Many thanks are due to colleagues in the Nottingham BBSRC/EPSRC Synthetic Biology Research Centre and the ENGICOIN (Engineered Microbial Factories for CO_2 Exploitation in an Integrated Waste Treatment Platform) project. This work was supported by the Biotechnology and Biological Sciences Research Council and the Engineering and Physical Sciences Research Council [grant number BB/L013940/1], as well as the European Union's Horizon 2020 Research and Innovation Programme [Grant Agreement No. 760994].

References

Bastian, M., Jones, O., Moore, N. and Roe, E. (eds) (2016) *Participatory Research in More-than-Human Worlds*, Abingdon: Taylor & Francis.

Benessia, A. and Funtowicz, S. (2015) 'Sustainability and techno-science: What do we want to sustain and for whom?', *International Journal of Sustainable Development*, 18(4): 329–48.

Bennett, J. (2010) *Vibrant Matter: A Political Ecology of Things*, Durham, NC: Duke University Press.

Berry, D.J. (2018) 'Plants are technologies', in J. Agar and J. Ward (eds) *Histories of Technology, the Environment and Modern Britain*, London: UCL Press, pp 161–85.

Blaser, M.J. (2014) *Missing Microbes: How Killing Bacteria Creates Modern Plagues*, London: Oneworld.

Boldt, J. (ed) (2016) *Synthetic Biology: Metaphors, Worldviews, Ethics, and Law*, Wiesbaden: Springer.

Boldt, J. (2018) 'Machine metaphors and ethics in synthetic biology', *Life Sciences, Society and Policy*, 14(1): article 12.

Bono-Lunn, D., Villeneuve, C., Abdulhay, N.J., Harker, M. and Parker, W. (2016) 'Policy and regulations in light of the human body as a "superorganism" containing multiple, intertwined symbiotic relationships', *Clinical Research and Regulatory Affairs*, 33(2–4): 39–48.

Boudry, M. and Pigliucci, M. (2013) 'The mismeasure of machine: Synthetic biology and the trouble with engineering metaphors', *Studies in History and Philosophy of Science Part C: Studies in History and Philosophy of Biological and Biomedical Sciences*, 44(4, Part B): 660–8.

Burton, F. (2020) 'CarbonSmart™ milestones for LanzaTech'. Available from: www.lanzatech.com/2020/12/16/carbonsmart-milestones-for-lanzatech/ [Last accessed 9 January 2022].

Calvert, J. (2010) 'Synthetic biology: Constructing nature?', *The Sociological Review*, 58(1_suppl): 95–112.

Carus, M., Skoczinski, P., Dammer, L., vom Berg, C., Raschka, A. and Breitmayer, E. (2019) *Hitchhiker's Guide to Carbon Capture and Utilisation*, Hürth: nova-Institute. Available from: www.bio-based.eu/nova-papers [Last accessed 9 January 2022].

Chilvers, J., Pallett, H. and Hargreaves, T. (2018) 'Ecologies of participation in socio-technical change: The case of energy system transitions', *Energy Research & Social Science*, 42: 199–210.

Crutzen, P.J. and Stoermer, E.F. (2000) 'The "Anthropocene"', *Global Change Newsletter. The International Geosphere-Biosphere Programme: A Study of Global Change of the International Council for Science*, 41: 17–18.

de Lorenzo, V. (2011) 'Beware of metaphors: Chasses and orthogonality in synthetic biology', *Bioengineered Bugs*, 2(1): 3–7.

Ditgen, D., Anandarajah, E.M., Meissner, K.A., Brattig, N., Wrenger, C. and Liebau, E. (2014) 'Harnessing the helminth secretome for therapeutic immunomodulators', *BioMed Research International*, vol. 2014, article 964350.

Ellen MacArthur Foundation (2013) *Towards the Circular Economy, Vol. 1: Economic and business rationale for an accelerated transition*, Cowes: Ellen MacArthur Foundation. Available from: https://emf.thirdlight.com/link/x8ay372a3r11-k6775n/@/preview/1?o [Last accessed 9 January 2022].

Genovese, A. and Pansera, M. (2021) 'The circular economy at a crossroads: Technocratic eco-modernism or convivial technology for social revolution?', *Capitalism Nature Socialism*, 32(2): 95–113.

Giraud, E.H. (2019) *What Comes after Entanglement? Activism, Anthropocentrism, and an Ethics of Exclusion*, Durham, NC: Duke University Press.

Giraud, E.H., Hadley Kershaw, E., Helliwell, R. and Hollin, G. (2019) 'Abundance in the Anthropocene', *The Sociological Review*, 67(2): 357–73.

The Guardian (2018) 'First commercial flight partly fuelled by recycled waste lands in UK'. Available from: www.theguardian.com/business/2018/oct/03/first-commercial-flight-partly-fuelled-by-recycled-waste-lands-in-uk [Last accessed 9 January 2022].

Hadley Kershaw, E., Hartley, S., McLeod, C. and Polson, P. (2021) 'The sustainable path to a circular bioeconomy', *Trends in Biotechnology*, 39(6): 542–5.

Hajer, M.A. (1997) *The Politics of Environmental Discourse*, Oxford: Oxford University Press.

Haraway, D.J. (2007) *When Species Meet*, Minneapolis, MN: University of Minnesota Press.

Harrison, R. (2013) *Animal Machines*, Wallingford: CABI.

Hird, M.J. and Yusoff, K. (2018) 'Lines of shite: Microbial-mineral chatter in the Anthropocene', in R. Braidotti and S. Bignall (eds) *Posthuman Ecologies: Complexity and Process after Deleuze*, London: Rowman & Littlefield, pp 265–82.

Hulme, M. (2009) *Why We Disagree about Climate Change: Understanding Controversy, Inaction and Opportunity*, Cambridge: Cambridge University Press.

Humphreys, C.M. and Minton, N.P. (2018) 'Advances in metabolic engineering in the microbial production of fuels and chemicals from C1 gas', *Current Opinion in Biotechnology*, 50: 174–81.

Inigo, E.A. and Blok, V. (2019) 'Strengthening the socio-ethical foundations of the circular economy: Lessons from responsible research and innovation', *Journal of Cleaner Production*, 233: 280–91.

Jasanoff, S. (ed) (2004) *States of Knowledge: The Co-Production of Science and Social Order*, London: Routledge.

Johnson, E.R. and Goldstein, J. (2015) 'Biomimetic futures: Life, death, and the enclosure of a more-than-human intellect', *Annals of the Association of American Geographers*, 105(2): 387–96.

Jones, C.G., Lawton, J.H. and Shachak, M. (1996) 'Organisms as ecosystem engineers', in F.B. Samson and F.L. Knopf (eds) *Ecosystem Management: Selected Readings*, New York: Springer, pp 130–47.

Laforest-Lapointe, I. and Arrieta, M.-C. (2018) 'Microbial eukaryotes: A missing link in gut microbiome studies', *mSystems*, 3(2).

Levidow, L. and Raman, S. (2020) 'Sociotechnical imaginaries of low-carbon waste-energy futures: UK techno-market fixes displacing public accountability', *Social Studies of Science*, 50(4): 609–41.

Lorimer, J. (2016) 'Gut buddies: Multispecies studies and the microbiome', *Environmental Humanities*, 8(1): 57–76.

Lorimer, J. (2017a) 'Parasites, ghosts and mutualists: A relational geography of microbes for global health', *Transactions of the Institute of British Geographers*, 42(4): 544–58.

Lorimer, J. (2017b) 'The Anthropo-scene: A guide for the perplexed', *Social Studies of Science*, 47(1): 117–42.

Lorimer, J. (2019) 'Hookworms make us human: The microbiome, eco-immunology, and a probiotic turn in Western health care', *Medical Anthropology Quarterly*, 33(1): 60–79.

Lorimer, J. (2020) *The Probiotic Planet: Using Life to Manage Life*, Minneapolis, MN: University of Minnesota Press.

Lövbrand, E., Beck, S., Chilvers, J., Forsyth, T., Hedréna, J., Hulme, M., Lidskog, R. and Vasileiadoug, E. (2015) 'Who speaks for the future of Earth? How critical social science can extend the conversation on the Anthropocene', *Global Environmental Change*, 32: 211–18.

Lowe, E.A. and Evans, L.K. (1995) 'Industrial ecology and industrial ecosystems', *Journal of Cleaner Production*, 3(1): 47–53.

McFall-Ngai, M., Hadfield, M.G., Bosch, T.C.G., et al (2013) 'Animals in a bacterial world, a new imperative for the life sciences', *Proceedings of the National Academy of Sciences*, 110(9): 3229–36.

McLeod, C. and Nerlich, B. (2017) 'Synthetic biology, metaphors and responsibility', *Life Sciences, Society and Policy*, 13(1): article 13.

McLeod, C., Nerlich, B. and Jaspal, R. (2019) 'Fecal microbiota transplants: emerging social representations in the English-language print media', *New Genetics and Society*, 38(3): 331–51.

McLeod, C., Hadley Kershaw, E. and Nerlich, B. (2020) 'Fearful intimacies: COVID-19 and the reshaping of human–microbial relations', *Anthropology in Action*, 27(2): 33–9.

Miller, C.A. and Edwards, P.N. (eds) (2001) *Changing the Atmosphere: Expert Knowledge and Environmental Governance*, Cambridge, MA: MIT Press.

NIH (National Institutes of Health) Human Microbiome Project (nd) 'About the human microbiome'. Available from: www.hmpdacc.org/overview/ [Last accessed 9 January 2022].

Pallett, H., Burall, S., Chilvers, J. and Price, C. (2021) *Public engagement with algorithms in public services. 3S Research Group Briefing Note: 8*, Norwich: University of East Anglia. Available from: https://uea3s.files.wordpress.com/2019/10/public-engagement-with-algorithms-in-public-services.pdf [Last accessed 9 January 2022].

Papadopoulos, D. (2018) *Experimental Practice: Technoscience, Alterontologies, and More-than-Social Movements*, Durham, NC: Duke University Press.

Papadopoulos, D. (2022) 'Chemicals, ecology, and reparative justice', in D. Papadopoulos, M. Puig de la Bellacasa and N. Myers (eds) *Reactivating Elements: Chemistry, Ecology, Practice*, Durham, NC: Duke University Press, pp 34–69.

Parker, W. (2014) 'The "hygiene hypothesis" for allergic disease is a misnomer', *British Medical Journal*, 349: g5267.

Paxson, H. (2008) 'Post-Pasteurian cultures: The microbiopolitics of raw-milk cheese in the United States', *Cultural Anthropology*, 23(1): 15–47.

Paxson, H. and Helmreich, S. (2014) 'The perils and promises of microbial abundance: Novel natures and model ecosystems, from artisanal cheese to alien seas', *Social Studies of Science*, 44(2): 165–93.

Philp, J. and Winickoff, D.E. (2017) 'Clusters in industrial biotechnology and bioeconomy: The roles of the public sector', *Trends in Biotechnology*, 35(8): 682–6.

Povinelli, E.A. (2016) *Geontologies: A Requiem to Late Liberalism*, Durham, NC: Duke University Press.

Puig de la Bellacasa, M. (2017) *Matters of Care: Speculative Ethics in More than Human Worlds*, Minneapolis, MN: University of Minnesota Press.

Renewable Carbon Initiative (nd) Renewable Carbon Initiative (RCI). Available from: https://renewable-carbon-initiative.com/ [Last accessed 9 January 2022].

Rook, G.A.W. and Brunet, L.R. (2005) 'Old friends for breakfast', *Clinical & Experimental Allergy*, 35(7): 841–2.

Sachs, J.S. (2008) *Good Germs, Bad Germs: Health and Survival in a Bacterial World*, New York: Farrar, Straus and Giroux.

SBRC-Nottingham (nd) About Synthetic Biology Research Centre. Available from: www.sbrc-nottingham.ac.uk/about/about.aspx [Last accessed 9 January 2022].

Seres Therapeutics (nd) Our platform, approach, and strategy. Available from: www.serestherapeutics.com/our-platform/ [Last accessed 9 January 2022].

Sherrard, A. (2020) 'IndianOil and LanzaTech poised to scale-up CO_2 to food, feed and fuel', Bioenergy International. Available from: https://bioenergyinternational.com/biofuels-oils/indianoil-and-lanzatech-poised-to-scale-up-co2-to-food-feed-and-fuel [Last accessed 9 January 2022].

Strachan, D.P. (1989) 'Hay fever, hygiene, and household size', *British Medical Journal*, 299(6710): 1259–60.

Sundberg, J. (2014) 'Decolonizing posthumanist geographies', *Cultural Geographies*, 21(1): 33–47.

Szymanski, E.A. (2018) 'Who are the users of synthetic DNA? Using metaphors to activate microorganisms at the center of synthetic biology', *Life Sciences, Society and Policy*, 14: article 15.

Szymanski, E.A. and Calvert, J. (2018) 'Designing with living systems in the synthetic yeast project', *Nature Communications*, 9(1): 2950.

Tsing, A.L. (2015) *The Mushroom at the End of the World: On the Possibility of Life in Capitalist Ruins*, Princeton, NJ: Princeton University Press.

Turney, J. (2015) *I, Superorganism: Learning to Love Your Inner Ecosystem*, London: Icon Books.

van Dooren, T., Kirksey, E. and Münster, U. (2016) 'Multispecies studies: Cultivating arts of attentiveness', *Environmental Humanities*, 8(1): 1–23.

Velasquez-Manoff, M. (2012) *An Epidemic of Absence: A New Way of Understanding Allergies and Autoimmune Diseases*, New York: Scribner.

Venkata Mohan, S., Modestra, J.A., Amulya, K., Butti, S.K. and Velvizhi, G. (2016) 'A circular bioeconomy with biobased products from CO_2 sequestration', *Trends in Biotechnology*, 34(6): 506–19.

Wammes, L.J., Mpairwe, H., Elliott, A.M. and Yazdanbakhsh, M. (2014) 'Helminth therapy or elimination: Epidemiological, immunological, and clinical considerations', *The Lancet Infectious Diseases* 14(11): 1150–62.

Waterton, C. and Tsouvalis, J. (2015) 'An "experiment with intensities": Village hall reconfigurings of the world within a new participatory collective', in J. Chilvers and M. Kearnes (eds) *Remaking Participation: Science, Environment and Emergent Publics*, Abingdon: Routledge, pp 201–17.

Yong, E. (2016) *I Contain Multitudes: The Microbes within Us and a Grander View of Life*, London: The Bodley Head.

Zaiss, M.M., Rapin, A., Lebon, L., et al (2015) 'The intestinal microbiota contributes to the ability of helminths to modulate allergic inflammation', *Immunity*, 43(5): 998–1010.

8

Involvement as an Ethics for More than Human Interdependencies

Nerea Calvillo

Restoring 200 sq km of kelp forest off the Sussex coast, creating new habitat for heat-sensitive butterflies and connecting fractured wetlands for the reintroduction of beavers – these are some of the 12 UK Wildlife Trusts schemes funded last month to recreate wetlands, restore peatlands and reintroduce beavers, all of which aim to help the UK tackle climate change.[1] They are just some of the thousands of projects that are being deployed throughout the world to address the devastating effects of climate change, extinction, biodiversity reduction, systemic pollution, sea level rise and global warming. These interrelated processes are rapidly changing the physical and affective conditions of the planet, increasing inequality, mass migration and resource colonialism, among other social crises. Therefore, social and ecological relations require radical improvements. The protection, conservation and remediation of existing ecologies is at the forefront of the strategies to compensate for, delay or adapt to these challenges.

Projects like the ones funded by the UK Wildlife Trust are relevant contributions, and the diversity of approaches reflect the complexity of doing so. However, terms like restoration and repair are becoming exchangeable and often unquestionable buzzwords that give little information about the aims of the intervention or what exactly will be done. In addition, at least from my experience as an architecture studio design professor, concepts like repair and remediation seem to justify any intervention because they are 'solutions' to 'global challenges'. They are approached as seamless and easy processes: 'we put this here, change this there, these others will transform each other' and 'voilà'. Repaired. Ecologies are repaired as if they were buildings, with the same techno-scientific mindset: trees substitute bricks and animals substitute mortar. But are concepts like reparation and restoration best suited to address the complex ecological reality we are living in?

Reparation (and repair, remediation, restoration ...)

In common language, words like reparation, remediation and restoration have different meanings. Reparation speaks about making an entity function again, or stopping it being broken. Remediation comes from remedy, to cure a body (human or not). Meanwhile, restoration implies a transformation to achieve a similarity with the past: to make something look or function as it did before. The type of intervention also varies depending on the object of reparation, whether it is violence against Indigenous peoples, a pond, a house façade or a 13th-century painting.

When thinking about repairing ecologies, projects that aim to repair, remediate or restore ecologies vary in their scale, timeframe, technologies and relations with the past, to the point that often they cannot be distinguished from each other. Another reason for these overlaps might be the different uses and connotations that these terms have in different disciplinary contexts. In anthropology, social justice and international relations, reparation and restoration tend to focus on the violence and injury inflicted on discriminated individuals or collectives and stopping or reversing the damage. Along these lines, Patel and Moore (2017) have suggested the notion of 'reparation ecology' as a form of repair that rectifies the social past as a fundamental step to move into the future. In engineering, science or architecture, reparation implies to fix or repair an instrument or a structure – sometimes an ecosystem (Jackson, 2014; Ureta, 2014; Callén Moreu and López Gómez, 2019). The fixing mechanism tends to be technological, and as a consequence, it is criticized by scholars and activists for being too naïve, reductive and technocentric (Levidow and Raman, 2020). In any case, repair presupposes that the object's function or how it is aimed to work in the future is clear. But what is the function of an ecosystem, or how can it be decided when it works properly? In addition, conservation and preservation projects have shown how benefiting one species can have disastrous consequences for others (Büscher and Fletcher, 2020). So, what is going to be eliminated/eradicated/expelled/let die? This then begs an ethical question: what is worth repairing and at the expense of what else? And what exactly is being repaired? For instance, climate goals might be achieved at the expense of livelihoods and biodiversity. Buck argues that "It's the relationships that must be repaired and restored, not just the climate" (2019: 244).

As an alternative, bio-remediation projects mobilize ecologies as the agents – and not only the objects – of reparation. However, in a highly connected world where matter circulates constantly, where does an ecology end and how can it be controlled? Bioremediation projects like reforestation and rewilding, regardless of the scale, presuppose that what both was there and the incoming bio are good. But what counts as a forest? What exactly in the forest is going to be repaired/planted/introduced? There are also assumptions from repair metaphors that do not apply well when the object of reparation is an ecology, because repair presupposes that the solution or remedy is known, presenting some sort of reassurance about a positive result. This lack of uncertainty is strengthened by the reference to a – better – past, as there is a known baseline to go to. This dismisses the fact that any intervention in an ecology creates a new one.

There is another problem with metaphors related to repair. As Moore and Moore argue (2013: 1),

> The metaphors that are currently used in reference to ecological restoration are commonly the languages of healing and repairing (Keulartz 2007). … In each of these cases, the lens shapes our understanding of restoration by placing it in the context of familiar activities that are generally skillful, successful, small-scale, benevolent, and beneficent, carrying positive (sometimes positively cozy) associations with the kindly family doctor and the garage mechanic.

Repair feels like a hard but safe space, which is congratulatory because of its intentions, but which says little of what will be done, how and more importantly, which power relations will be rearranged.

Many community approaches to bioremediation and repair already have a more complex and socio-ecological conception, mobilizing a conceptualization of repair aligned with one of the editors of this volume: reparation 'as the unfolding of relations, movements and interconnections among human and non-human others, in the patching up and reclaiming of the socio-environmental fabric' (see Introduction to this volume). These projects are obviously relevant and important contributions. And yet, what it takes to repair and its consequences are hardly discussed.

Air reparation

These questions become even more complex when thinking about the ecology of the air. A multitude of projects have been designed and deployed to repair or remediate the different forms of pollution and toxicity that take place at multiple scales, locations and altitudes of the atmosphere. The displacement of factories, fossil-fuelled traffic reduction, reforestation through carbon markets and the protection of green areas have managed to locally reduce the concentration of some components, like coal or nitrogen dioxide, while other forms of air pollution like CO_2 are on the rise.

In addition, if we speak about bio-remediation, what counts as the 'bio' and how technological the intervention is becomes blurry. For instance, oxygen is used in regenerative thermal oxidizers to remediate hazardous air pollutants and volatile organic compounds (Manufacturer, nd). CO_2 is dissolved in water to be injected into basalt rock for carbon capture and sequestration – a new method developed in Iceland (Nunez, 2016). Titanium dioxide paint is being promoted in construction to break down NO_2 (Palmer, 2011). Water cannons, also called anti-smog guns, are used to remediate particulate matter, replicating the traditional watering of the streets used for centuries in many parts of the world (Suri, 2017).

These repair and remediation projects work at different scales, from a grand scale or industrial remediation scale to the street or garden. And yet, one of the challenges is that the element, compound or ecology that is used to repair – the 'bio' in bioremediation – is considered a resource. For instance, when watering the streets with water canyons, trucks or even domestic hoses, water is used as a resource to remediate the air. This has two implications. When considering something a resource, it is assumed that it is at the service of humans. It becomes an abstract entity, dislocated in space and time from the ecosystem it is part of. Therefore, it can be extracted, depleted or hurt. So, what is extracted, polluted or exhausted to repair an element or an ecology? The second implication is that resources are usually addressed independently of each other – for example, water, air, a forest or a peatland. Then, when trying to repair one, its interconnections with other ecologies are not taken into consideration, or some other ecologies are put to work for them, damaging others. The remediation/reparation of air as it is currently done may damage other ecological relations, maybe elsewhere.

What if the bio considered the object of reparation – the 'bio' in bioremediation – is considered an ecology instead of a resource? In other words, what would ecological repair look like if the process of ecological repair were conceived from an ecological lens?

Ecological (interdependency and involvement)

One of the main contributions of ecological thinking has been to highlight the interdependency between entities. To reflect the interdependencies between plants and air, philosopher Emanuelle Coccia suggests a metaphysics of mixture (Coccia, 2018), where plants and air cannot be conceptualized on their own because they belong to the same mixture of gas exchanges and co-configure each other. Coccia illustrates his argument through Joseph Priestley, the experimenter who discovered the breathing of plants. Priestly wrote in 1772: 'I have accidentally hit upon a method of restoring air which has been injured by the burning, and that I have discovered at least one of the restoratives witch nature employs for this purpose. It is vegetation' (Priestly, cited in Coccia, 2018: 44). As a consequence, Coccia argues,

> the air we breathe is not a purely geological or mineral reality … but rather the breath of other living beings. It is a byproduct of 'the lives of others'. ... Breath is already a first form of cannibalism: every day we feed off the gaseous excretions of plants. We could not live but off the life of others. (2018: 47)

This interdependency implies that any intervention in one entity affects the others to different degrees. Therefore, ecological interventions have multiple effects beyond their aim, which tend to be neglected or untold. They are technically known as overflows or adverse habitat modifications. This begs the questions: who benefits, and who might be left to die as a result of a reparation?

To detect and unpack how these interdependencies work in practice, I suggest the notion of *involvement*[2] as an analytical tool to think about human-non-human entanglements that may exceed or unsettle conventional notions of environmental agency, responsibility and ethics. Involvement, as a polysemic word, describes an umbrella of relations: from active to passive, between humans and more than humans. For example, one might say, 'I'd like to be involved in your discussion group', or 'I was involved in an accident'. Involvement also allows for describing an emotional or physical association between entities – human and otherwise – as well as the unwanted or unaware connections among them. Etymologically, involvement also refers to a mode of taking part in something, to envelope or surround, as well as a connection, interest, intimacy and attachment.[3] So involvement is a form of engagement or participation in environmental issues, active but not always chosen. It also speaks for the enthusiasm that one feels when caring deeply about something –although that enthusiasm might not always be positive. In sum, there are many ways of being involved, with different humans and more-than-humans, ranging from affective, to political, social and spatial.

I use involvement to unpack how elements, compounds and ecologies are intertwined. To be more specific: to unpack *the modes* in which they are intertwined in particular settings. Because what an intervention achieves – or not – can only be known through these details. To be clear, the objective is not to deconstruct or critique relevant attempts at repairing or remediating the air by reducing pollutants' concentrations. I wish to focus on the forms of relation between elements to suggest more just – and hopefully more effective – forms of intervention, by acknowledging the interdependencies and vulnerabilities that are at stake. So, *what* exactly and *how* it is done is what matters when considering what it takes, and to consider and evaluate its consequences. In sum, to think through involvement to make every intervention an ethical – as well as technical – problem, from its conception to its consequences.

Thinking through an experimental air pollution remediation project

I draw on Yellow Dust to explore what and how an experimental air pollution remediation project was done, and to unpack the specific modes of involvement that took place. Yellow Dust is an installation that I designed and built with collaborators for the Seoul Biennale of Architecture and Urbanism 2017 (Figure 8.1).[4] The theme of the Biennale was 'urban cosmologies – the four ecology commons of air, water, fire and earth and six technological commons'. To my surprise, the curator's proposal to engage in a posthuman urbanism was to 'solve the problem' through technoscientific and geoengineering projects: 'cleaning air, moving polluted air away from city streets … is central to preserving the rights of urban citizens to common resources' (Zaera-Polo and Pai, 2019: 48). But who and what cleans the air? Where is the pollution displaced to and, as a consequence, who is now breathing the lethal air?

Against the proposed technofixes, Yellow Dust aimed to explore other forms of engagement with air pollution beyond cleaning and risk

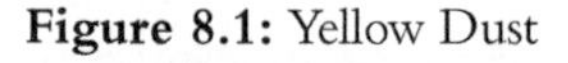
Figure 8.1: Yellow Dust

management. Considering that air pollution levels are consulted by Seoul citizens through apps, Yellow Dust aimed to move away from displaying numbers towards making visible the social and political implications of particulate matter in Seoul, to test the impact of overlapped – and even excessive – sensorial experiences of data. We configured a misty canopy that informed visitors about the concentrations of particulate matter in the air through the density changes in a water vapour cloud. The cloud was tainted in yellow in reference to Hwangsa (yellow dust in Korean), the sandstorms that take place every year in the spring, which the government and the media blame for South Korea's particulate matter pollution. To avoid being only a narrative device, we took the opportunity to explore the remediation capacity of water vapour, as water deposits suspended particles.

Having analysed the initial human involvement with the installation after the opening (Calvillo and Garnett, 2019), here I focus on the process of construction and installation of the piece – often disregarded as a site of inquiry. In particular, I pay attention to the unexpected states, scales and locations in which air and water became materially intertwined with each other, as well as with other materials, spaces, technologies and humans (including myself). While the project's space of intervention was the air, the intense and multiple roles that water acquired made tangible the relevance and power of the interdependencies between elements, and the consequences of taking the bio-remediator – water – as 'just' a resource. Thinking about the construction process of Yellow Dust through involvement permits me to unpack the web of interrelations, and some of the consequences of turning water into a resource instead of an ecology.

Building Yellow Dust, or how water took over

In early August 2017, when we arrived in Seoul three weeks before the opening to finish the production and set up of the installation, the monsoon had not ended – as predicted and as usual. For more than a week, the rain and humidity involved us, not metaphorically, but in the most literal sense. Air and water were the same thing. The thick drops pouring from the sky seemed to play an impossible trick: falling down with roaring strength, while remaining suspended in the air, increasing the humidity to levels of 90 per cent (Figure 8.2). This water-air kept my body soaked from my hair to my toes, despite all of my attempts to prevent it, to the point where I could not distinguish between the water droplets in the air and my own sweat. It was not possible to think about the environment as something 'out there'. The atmospheric resource we needed to build the air installation with – water – greeted us with strength. But not only that. The unfinished floor of the Biennale turned into a sea of mud, and mosquitoes bit every inch

Figure 8.2: Monsoon rain in Seoul

of my skin. It was difficult to breathe, let alone do physical work. It was an enactment of Coccia's metaphysic of mixture.

★

Almost one year earlier, in the first proposal for Yellow Dust sent to the curators and the Biennale production team, we specified water, electricity and Wi-Fi as the project's technical requirements. As common supplies available in large cities' institutional buildings, the request was easily approved. When we arrived in Seoul, our exhibition space did not yet have water or an electricity supply. The refurbishment works of the museum had been delayed, so we occupied our time finding the last materials needed for the construction of the installation across the city – an adventure considering that we did not speak Korean and did not share any common languages with most of the shop salespeople.

After 10 days of chasing the builders – and four days before the opening –- we finally got a water supply. It did not come from our exhibition building, but through a very long hose connected to a water source somewhere far away. Relieved by the fact that water was coming out of the tap, we started testing. But the following morning, when we arrived early to speed up the set-up process and make up for the time lost, we opened the tap and nothing came out of the water vapour system. Following the hose, I discovered that the other end was inside another pavilion. I found the tap from which the hose had been disconnected. At the time I thought that it must have been the water pressure, or that someone else had needed to connect another hose for something else. So, I reconnected it and we continued adjusting the water densities. We were stressed by the proximity of the opening day, but grateful that we finally had all of the supplies we needed.

The following day, to our despair, the water was not running. I went to the pavilion and found the hose, which looked like it may have been cut. We could not reconnect it so on the morning of the opening we were still without a water supply.

★

Participation in the Biennale carried a responsibility: to have the project up and running for the VIP tour at 4 pm on 30 August, the day of the opening. The contractors started searching for alternatives. They connected the hose to a bathroom far away and one storey below the square, so – obviously – the pressure was not enough and it started leaking. Builders were called one after the other to solve it (Figure 8.3). They discussed it and tried different connections, patching it up with any material or object at hand. It felt like the Marx Brothers' cabinet.

In the meantime, some pavilions were already enjoying their opening drinks. The builders tried until we were all soaked again – including the toilet – and the chief manager decided to try a different location. The curator kept sending me phone messages for an update. Two minutes before the VIP cohort showed up at our site, the chief manager connected the hose to a secure water access point on another pavilion's rooftop. To our indescribable relief, the water vapour started covering the skin of the visitors with the data of the particles that they were breathing.

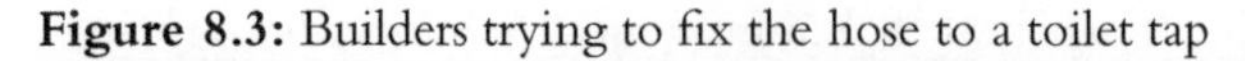

Figure 8.3: Builders trying to fix the hose to a toilet tap

As I was working towards remediating the air, I expected to have problems with air-related issues like the air pollution sensors, for instance. But, why was I having to deal with, fight for and spend all of my time on water? It could be argued that all of these involvements were a random coincidence, and it would have been different had we arrived when the monsoon was over and the village had been finished. But that is not my point. What is useful to me is that paying attention to these involvements permits me to see the interdependency between elements, even far away, and who and what is put to work. In the design phase we had barely considered water. We hadn't properly asked: What is needed to repair? What is depleted, and polluted? Where? For whom?

Modes of involvement of water and air (and others)

These vignettes permit grasping different modes of involvement with and in water during the construction of Yellow Dust, and the unexpected consequences of considering water 'just' a resource to remediate the air. Water, in different physical states, involves other people, things and technologies, among others. A symbolic involvement turned an element into an ecology and as a consequence re-enacted sociopolitical struggles; a passive involvement of human bodies brought in climate-scale temporalities; a technical involvement enrolled other unexpected humans and techniques.

The air remediation trough Yellow Dust involved water as a common good. Even at the Biennale, where water came from and how it arrived at the site could not be taken for granted. Access to water had to be negotiated among aesthetic and functional criteria and our own anxieties. It also demonstrated how elements, compounds and ecologies sometimes confront each other. The same can be said for their infrastructures. Even if they are built for the common good, they are valued differently.

Water also showed us how it is not only the liquid that comes out of the tap. The water-air of the monsoon involved water as an environmental condition: it involved our bodies without consent, disrupting the assembly of Yellow Dust. At the same time, this unexpected and strong water-air created some sort of intimacy with our own bodies. We had to care for them in order to care for the air, against the water's strength, and against our own perspired water, as a form of intimacy that emerged in the labour of research (Fraser and Puwar, 2008). Forms of intimacy also emerged with unexpected others. For instance, the manager of the very cheap hostel where I stayed, with whom I could not communicate because of language barriers, started leaving in my minuscule room a plate with a fresh piece of fruit and a frozen beer every evening (Figure 8.4). I will never know if he did so because I was the only woman in the hostel, or because he could

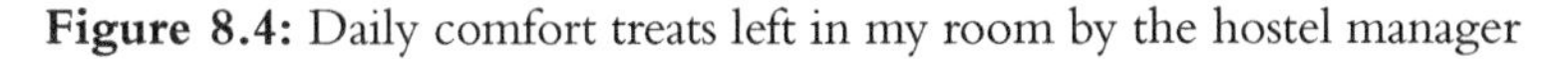

Figure 8.4: Daily comfort treats left in my room by the hostel manager

see my soaked face of exhaustion at my arrival every day. In any case, these refreshments nurtured my body and felt warm and tender.

Before we could start remediating the air, water involved more – and unexpected – people and technologies that needed to be sorted out – or repaired. When it seemed that we could just plug the hose, in a nearby toilet, water took over again. The first builder made a professional fixture, but despite his care, the water started leaking. He called another colleague to try a different fix, but with no success. Others came and went, trying strength, beautiful joint techniques with tape and cloths, changing the connector, interlacing elements instead of collapsing them and so on. But the problem was not the joint, it was the lack of pressure of the pump. I do not know if the builders cared for, suffered or knew much about air pollution. But water enrolled them in it, and they came back to visit the installation.

These involvements with the air through water brought to the fore the complex web of situated interactions and interdependencies that were part of a small airborne particulate matter experimental remediation project in Seoul. They also show how water cannot be just a resource, because its physical state (vapour, liquid or gas), where it comes from (from a hose or from the sky), who provides it (the atmosphere or a Biennale builder), and who it affects (farmers, our bodies or passers-by lungs) and how (as flesh, breathing, pain or joy), matter.

Involvement as ethics

We have seen how involvement is a useful analytical tool to unpack the interdependencies between elements, compounds and ecologies, and the multiple and specific ways in which air and water, among others, involve each other. In addition, I suggest that involvement is also a form of ethics for repair/action, because it puts at the centre the ethical decisions involved in every intervention with/in ecologies. For instance, which ecology is put to work for another, which element is extracted from where, and who is excluded or who gains power? In the case of the air, it draws attention to what it takes to repair and who or what benefits when remediating. It demands the evaluation of which toxicity becomes relevant, and where and for whom. It contributes, in sum, to bringing an ecological approach to environmental pollution practices and ethics.

Involvement as an ethics also expands the modes of engagement in environmental issues. At an individual level, it does not require will or care, as we are always involved in some way or another within ecologies. This involvement demonstrates that, if we look at the details of where, for whom or when, everything and everyone is involved; although, of course, some more than others, as the responsibility of destruction – of humans and ecologies – and its burden are unequally distributed. At a collective level, involvement does not require consensus or agreement among the people involved, it works towards the best possible form of co-habitation. Through small-scale testing, we managed to open on time, and only one male toilet was sacrificed for the evening – as well as our nerves for half a day. With this, I am not suggesting a de-politization of ecological matters. On the contrary. I am trying to re-politize by adding more forms of doing politics by making it collective (Calvillo, 2018). Many of these arguments resonate with Maria Puig de la Bellacasa's ethical dimensions of care (2011, 2017). However, the concept of care is loaded with 'good will', in particular within environmental issues – and Puig de la Bellacasa's complex proposal of care is often simplified or misread – at least in architecture and design. I have tried to think through involvement as ethics to complicate our love and care for non-humans, to highlight that every intervention is a messy and uncertain process that does good for some, but, inevitably, involves many others whose livelihoods might be damaged.

An ethics of involvement then recognizes and acts upon the messy and situated web of interdependencies that articulate ecological relations, and puts at the centre the ethics and politics that are part of any intervention in and with ecologies. It makes visible the moral dilemmas (what to prioritize, what can be let go) so as to keep in mind that there is no 'solution' to ecological issues. Therefore, improvements can only be achieved by taking the ethics seriously. What, how, when, with whom, for whom – these are not just technical questions. And they are fundamental to aim for more just socio-ecologies.

Notes

1. www.theguardian.com/environment/2021/jul/08/high-impact-wildlife-projects-aim-to-restore-habitats-across-england
2. I borrow the concept from the panel 'Involving Compounds' chaired by Maria Puig de la Bellacasa, Dimitris Papadopoulos and Manuel Tironi at the conference EASST2018 in Lancaster, where I presented the first draft of this chapter.
3. www.etymonline.com/search?q=%5Cinvolve
4. Authors: C+arquitectos/In The Air (Nerea Calvillo with Raúl Nieves, Pep Tornabell, Yee Thong Chai, Emma Garnett, Marina Fernandez). Collaborators: Victor Viña. Yellow Dust was commissioned by the Seoul Biennale of Architecture and Urbanism 2017, with the support of Acción Cultural Española and an impact ESRC IAA grant from University of Warwick and the Economic and Social Research Council (ESRC). http://yellowdust.intheair.es/

Acknowledgements

I would like to thank my Yellow Dust colleagues and the early In The Air collaborators for making the project happen, in particular to Raúl Nieves, Pep Tornabell and Thong Chai for giving it all in Seoul. My gratitude to the editors of this volume and Uriel Fogué for their feedback on early drafts of this text, and to Tim Choy for our 'patchy' and inspiring revision sessions.

References

Buck, H.J. (2019) *After Geoengineering: Climate Tragedy, Repair and Restoration*, New York: Verso.

Büscher, B. and Fletcher, R. (2020) *The Conservation Revolution: Radical Ideas for Saving Nature Beyond the Anthropocene*, New York: Verso.

Callén Moreu, B. and López Gómez, D. (2019) 'Intimate with your junk! A waste management experiment for a material world', *The Sociological Review*, 67(2): 318–39.

Calvillo, N. (2018) 'Political airs: From monitoring to attuned sensing air pollution', *Social Studies of Science*, 48(3): 372–88.

Calvillo, N. and Garnett, E. (2019) 'Data intimacies: Building infrastructures for intensified embodied encounters with air pollution', *The Sociological Review Monographs*, 67(2): 340–56.

Coccia, E. (2018) *The Life of Plants: A Metaphysics of Mixture*, London: Polity Press.

Fraser, M.M. and Puwar, N. (2008) 'Introduction: Intimacy in research – Mariam Fraser, Nirmal Puwar, 2008', *History of the Human Sciences*, 21(4): 1–16.

Jackson, S.J. (2014) 'Rethinking repair', in T. Gillespie, P.J. Boczkowski and K.A. Foot (eds) *Media Technologies*, MIT Press, pp 221–40.

Levidow, L. and Raman, S. (2020) 'Sociotechnical imaginaries of low-carbon waste-energy futures: UK techno-market fixes displacing public accountability', *Social Studies of Science*, 50(4): 609–41.

Manufacturer (nd) *Regenerative Thermal Oxidizer, Anguil Environmental Systems, Inc.* Available from: https://anguil.com/air-pollution-control-solutions/regenerative-thermal-oxidizer-rto/ [Accessed 15 September 2021].

Moore, K.D. and Moore, J.W. (2013) 'Ecological restoration and enabling behavior: A new metaphorical lens? Ecological restoration and enabling behavior', *Conservation Letters*, 6(1): 1–5.

Nunez, C. (2016) 'How one country is making rocks out of air pollution', *National Geographic*. Available from: www.nationalgeographic.com/science/article/iceland-turns-carbon-dioxide-into-rock-capture-sequestration [Last accessed 15 September 2021].

Palmer, J. (2011) '"Smog-eating" material breaking into the big time', *BBC News*, 12 November. Available from: www.bbc.com/news/science-environment-15694973 [Last accessed 15 September 2021].

Patel, R. and Moore, J.W. (2017) *A History of the World in Seven Cheap Things: A Guide to Capitalism, Nature and the Future of the Planet*, Berkeley: University of California Press.

Puig de la Bellacasa, M. (2011) 'Matters of care in technoscience: Assembling neglected things', *Social Studies of Science*, 41(1): 85–106.
Puig de la Bellacasa, M. (2017) *Matters of Care: Speculative Ethics in More Than Human Worlds*, Minneapolis, MN: University of Minnesota Press.
Suri, M. (2017) 'Delhi tests water cannons to combat deadly air pollution', CNN. Available from: https://edition.cnn.com/2017/12/22/health/india-new-delhi-water-cannon-pollution-intl/index.html [Last accessed 15 September 2021].
Ureta, S. (2014) 'Normalizing Transantiago: On the challenges (and limits) of repairing infrastructures', *Social Studies of Science*, 44(3): 368–92.
Zaera-Polo, A. and Pai, H. (2019) *Imminent Commons: Commoning Cities*, Barcelona: Actar & Seoul Biennale of Architecture and Urbanism.

9

From Museum to MoB

Timothy Choy

AT THE EDGE OF THE FOREST YOU'LL FIND IT.
THE MUSEUM OF BREATHERS.
I LIVE HERE.
ITS COLLECTION IS GATHERED BY A SIMPLE LOGIC: ANYTHING WHO BREATHES, IN ITS BREATHING CONDITION. WITH THAT CURATORIAL CUT, A WORLD OF COMPARISONS AND DIFFERENCES EMERGES.
I BUILT THIS PLACE. IT IS SLOW WORK, NOTHING REALLY HAPPENS.
UNTIL IT DOES
Target acquired.

SEARCH AND GATHER
SUBJECT 4504
65, 42, 35
INTERESTING. IT'S CLOSE. I CAN MAKE IT IN 2 DAYS.
I'LL WAIT UNTIL DARK.
YOU'RE FREE! FOLLOW ME!
!

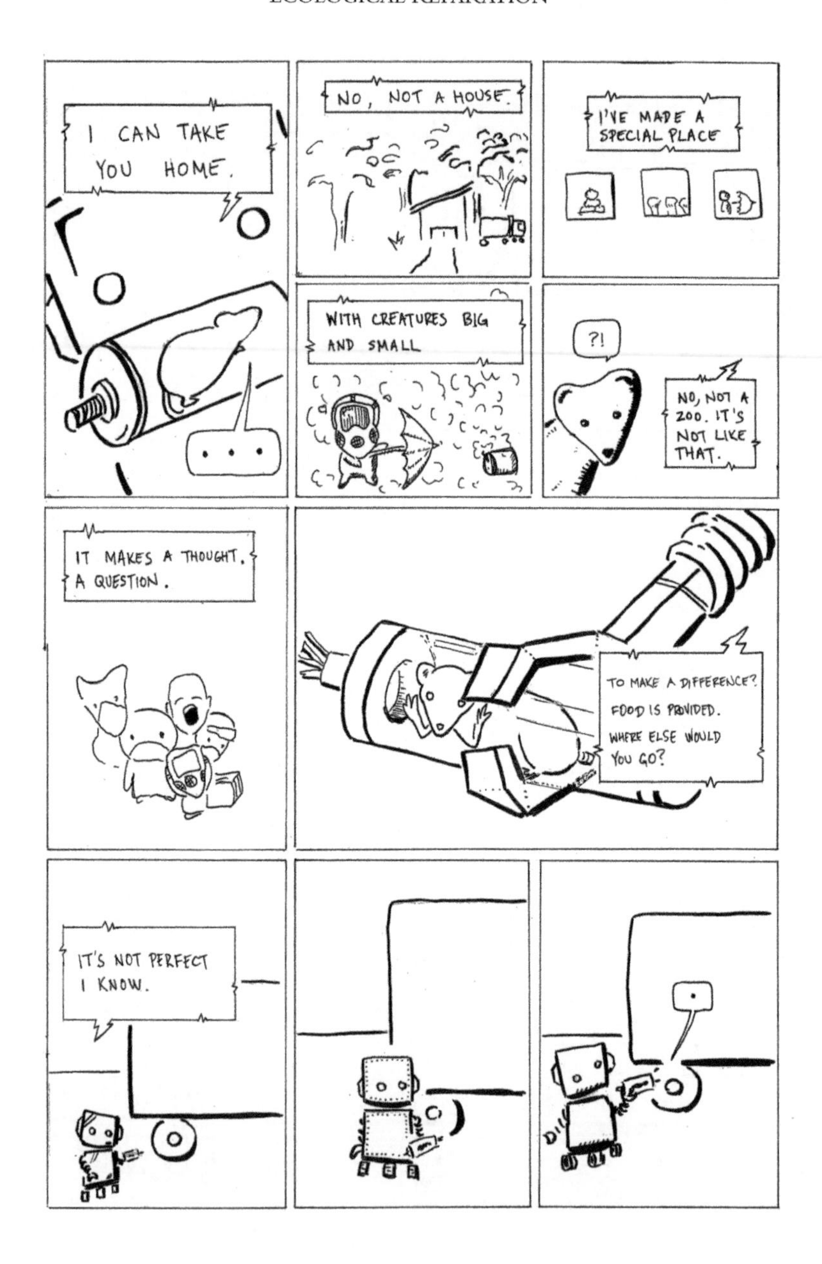
I CAN TAKE YOU HOME.
...
NO, NOT A HOUSE.
I'VE MADE A SPECIAL PLACE
WITH CREATURES BIG AND SMALL
?!
NO, NOT A ZOO. IT'S NOT LIKE THAT.
IT MAKES A THOUGHT. A QUESTION.
TO MAKE A DIFFERENCE?
FOOD IS PROVIDED.
WHERE ELSE WOULD YOU GO?
IT'S NOT PERFECT I KNOW.
.

YOU'RE PART OF SOMETHING BIGGER NOW.
YOU'LL MEET YOUR NEIGHBORS.
WHAT DO YOU MEAN, "WHAT NEIGHBORS?"
OH NO.
WHERE ARE THEY?

PART IV

Enclosing<>Reclaiming Land

10

Land in Our Names: Building an Anti-Racist Food Movement

Sam Siva

> If you have come here to help me, you are wasting your time. But if you have come because your liberation is bound up with mine, then let us work together.
>
> Aboriginal activists group, Queensland, 1970s

Land in Our Names (LION)[1] began in 2020 as a young collective with big plans for the year.[2] At the Oxford Real Farming Conference in the UK, in January 2020, we facilitated a workshop with Leah Penniman of Soul Fire Farm[3] on the topic 'Farming So White: Land, Ownership, Race and Racism in Britain' and held the first ever Caucus for Black and People of Colour (BPOC) growers, land workers, environmentalists and food justice organizers at Willowbrook Farm. It was an emotional experience for everyone who attended. We all remarked that we had never been in the countryside in Britain surrounded by People of Colour before. The care, excitement, community and safety that we shared with each other was energizing and made us even more dedicated. We want to speak to BPOC growers, land workers, land and food justice organizers.

Then our plans for the year took a dramatic turn with the pandemic and the Black Lives Matter demonstrations in the US, Britain and elsewhere. We have been humbled by the amount of people who have reached out to us asking for our thoughts and feelings. The truth is that we carry each life lost to racism with us always. It is a grief that we must manage every day. This is one of the reasons why we founded Land in Our Names – we want to create an anti-racist and inclusive land and food justice movement that speaks to Black and People of Colour. Racism is a structural and systemic problem that needs reparative justice in order to stop the unnecessary violence that People of Colour experience. That is why we see land reparations as crucial towards building resilient and sustainable anti-racist communities. We want to continue nurturing our fledgling network of BPOC growers, land workers, organizers, educators and enthusiasts with empowering and inspiring events. We want to create resources that look at food and land justice through an intersectional lens. We want to support BPOC growers' access to land and work.

Racial justice is the heart of LIONs' mission. Our work is inspired by Black and Brown revolutionaries who shed light on land inequalities as a hidden driver of much racial injustice. Unequal access to land is particularly stark in Britain, where land ownership is often inherited, and concentrated into the hands of a few wealthy (White) individuals and families. Little research exists on the whiteness of land ownership in Britain. But we do know the agricultural and environmental sectors are not diverse, representative or welcoming sectors employing BPOC; farming is the least diverse profession in Britain – 98.6 per cent of farmers are White British. The environment sector is the second least diverse at 96.9 per cent employees being White British. This is not just an employment issue but also one of access to space. A study carried out by Natural England in 2013–15 reported that only 56 per cent of under 16s from 'BAME' – British and minority ethnic – households visited the natural environment.[4] BAME communities are 60 per cent less likely to be able to access green space and natural environments than their White counterparts. Often Black and other People of Colour experience different forms of racism and othering when we do manage to be in the countryside. LIONs is working to address the barriers preventing us from dwelling and belonging in both urban and rural green spaces.

Despite the grief and rage that we feel at this time, we are inspired and energized. The future we are working towards is one where Black and People of Colour in Britain are able to access nutritious food, green spaces and nature, and grow food in both rural and urban spaces. To make this a reality we need more than your outrage or sympathy. We don't need your help, we need allies and accomplices who can recognize that systemic racism is tied to climate change, misogyny and the other oppressive structures that perpetuate widespread violence and destruction. This means we need you to recognize how our struggles are tied together, our liberation is bound together, and that dismantling racism is not just the work of Black people or something that is only remembered when one of our deaths is recorded and broadcast. It is work that we are doing every day and it is work that we all need to do if we want real change.

Reparations is about recognizing that the wealth and privilege of White people came from the exploitation of Black and Indigenous people and taking the necessary steps to heal that trauma. That means taking time to understand how you benefit in whatever way from White supremacy. It means using what resources and platforms you have access to and giving them to People of Colour. It means supporting BPOC organizations and institutions. It means dedicating time to learning more about the histories and cultures of racialized people. It means listening to and learning from Black and People of Colour. As the Nigerian journalist and poet, Professor Chinweizu said,

> Reparation is not just about money. It is not even mostly about money; in fact, money is not even one percent of what reparation is about. Reparation is mostly about making repairs, self-made repairs, on ourselves: mental repairs, psychological repairs, cultural repairs, organizational repairs, social repairs, institutional repairs, technological repairs, economic repairs, political repairs, educational repairs, repairs of every type that we need in order to recreate and sustain black societies.[5]

★

If you would like to support Land in Our Names, you can do so here: https://landinournames.community/support-our-work

Notes

1 https://landinournames.community/

2 This is a slightly edited version of a text originally published in Resilience the 24 June 2020. www.resilience.org/stories/2020-06-24/building-an-anti-racist-food-movement. It is republished here with permission from the author.

3 www.soulfirefarm.org

4 Hunt, A., Burt, J. and Stewart, D. 2015. 'Monitor of Engagement with the Natural Environment: a pilot for an indicator of visits to the natural environment by children – interim findings from Year 1 (March 2013 to February 2014).' Natural England Commissioned Reports, Number 166.

5 Chinweizu, Ibekwe. 1993. 'Reparations and A New Global Order: A Comparative Overview.' A paper read at the second Plenary Session of the First Pan-African Conference on Reparations, Abuja, Nigeria, 27 April 1993.

11

Land Reparations and Ecological Justice: An Interview with Sam Siva

Sam Siva, Dimitris Papadopoulos and Maria Puig de la Bellacasa

Dimitris and Maria: We would like to thank you again, Sam, for contributing your text 'Land in Our Names: Building an Anti-racist Food Movement' to this book and accepting to follow up with an interview. We reached out to you for an intervention because we felt your voice and the work of LION are absolutely necessary for this book. We are immensely grateful for your time as we know how precious and precarious activist time is. To start maybe we would like to ask a bit more about LION about the organization, how you work, and about your plans.[1]

Sam: So Land in Our Names is a collective of BPOC. We are an antiracist, land justice collective using a reparative justice framework to achieve land reparations, and increased access and support for BPOC to land and farming and nature more generally. An example of what we have been doing recently is a micro-grant programme (up to £1000) providing support to BPOC who want to get into growing or already do it as a hobby, or who want to follow related courses or have projects, who want to set up projects and so on. We also organize events about gardening or growing, skill sharing or just healing spaces where BPOC come together, connect with each other to learn in London, the city where most of us are based. We have also done a lot of research around access to food growing as a livelihood.[2] Our eventual goal is to get some land to set up a BPOC land cooperative, a land cooperative, where we're growing food, creating a space for people to rest and heal, that's in the works. We're quite young, we've only been around for two years and we're doing quite a lot of stuff all the time, and have changed formations a lot, but, yeah, it's great.

Dimitris and Maria: We have been following LION's work and different interventions about your vision and have felt inspired and challenged to think and act. We strongly felt we couldn't speak about ecological reparation in Britain today without including the voices, ideas and struggles you stand for. So maybe right away we'd like to ask if, and how, the notion of "ecological reparation" that we are trying to explore in this project resonates with you and your work.

Sam: I think for me reparations are connected to everything. Ecological repair and reparations are intrinsically tied. With *Land in Our Names*, we have a working definition of reparations very much linked to an idea of repair and that is also about healing. It is about addressing the harm resulting of colonialism, extractivism and exploitation of resources – whether these are human beings seen as property, or things, like minerals. And so, when I think of reparations, it's intrinsically tied to ecological action, whether that's about rebalancing ecosystems or about the fact that the wealthiest countries in this world are historically the biggest emitters of carbon emissions, the ones using the most resources, or energy, and consuming the most fossil fuels, while the people who are on the front lines of climate catastrophe, the people who are being harmed, are also the poorest or live in countries that were colonized and were victims of imperialism. Ecological reparations therefore go hand in hand with the idea of reparations for countries like Haiti or other countries in which populations were enslaved, exploited as workers and their natural environments exploited to generate products for the consuming home countries who in tun produced the most emissions and contributed more to climate change. At the COP26 march, I was representing *Land in Our Names* in the reparations block, constituted by groups such as *Tipping Point* and *Wretched of the Earth*, and we were saying that we need climate reparations for countries that are facing extreme floods, or more earthquakes, hurricanes, droughts. Countries that were exploited and deprived through colonization don't have the same financial resources as do colonizers who built a lot of their wealth on imperial exploitation and extraction of resources, and the corelated transatlantic slavery, even though right now there's a lot of historical amnesia about this connection. And while these former colonizer countries might be facing effects of climate change (flooding and so on), they have the infrastructure to deal with them while most formerly colonized countries do not.

Dimitris and Maria: We are very interested in what you say, that ecological reparation combines the work of repairing and healing damaged ecologies with reparations for damage done – by certain countries, corporations, governments. Could you tell us a little bit more about this side, which is not abstract but very concrete and specific and involves also compensations?

Sam: Yeah, definitely I really agree with you, it not just an abstract concept. This is something I see happening too with the term decolonization right now as (White) institutions are trying to claim decolonization while taking away from the concept the actual act of decolonization by which colonized people take back their country for reshaping and reforming post colonialism. It's not like your EDI (equality, diversity and inclusion) programme. The same with reparations, it's about that concrete compensation.

There are many aspects of reparations – redistributing money, compensation, investing into healthcare services, learning and education – and those are all tied to our work, but our main focus is about land specifically because of how so much is rooted back to that. Britain's wealth came from its empire, the exploitation of India, the Caribbean, the African continent, and other places including Australia. The extraction of natural resources in these places allowed the creation of goods and funded the economic system within Britain. *Land in Our Names* focuses on land reparations specifically understanding that land hoarding in Britain is a particular extension of colonialism. The elite, the people, who own land, benefited from colonialism and the slave trade, and the power and control and wealth gained then has been passed on through inheritance, trusts, foundations. Land ownership is somehow still feudal in the sense that 50 per cent of the land in England is owned by 1 per cent of the population, and a third of that is by the aristocracy, the same people who have been in power and have retained this wealth for hundreds of years, going back some time as far as William the Conqueror. You mentioned corporations too, for instance the Duke of Westminster has recently formed the Grosvenor group, a corporation owning an enormous amount of land and property in London and across the world.

The control of land is linked to the control of the food system, of housing, all these things that have implications on our wellbeing. Very few people actually can own land and even if you own land, it's very difficult to get the planning permission to build somewhere to live on that land. It is just so important to redistribute land. Most Black and People of Colour, people who are descended from colonized people from countries whose resources were extracted to build wealth in this country, do not have access to land and nature. This is why *Land in Our Names* is working towards land reparations. Redistributing land is redistributing resources. If you have land, you can build a school on it, for example. If you have land, you can build facilities to support people and also, most importantly for our work specifically, to grow food.

Many of the problems that BPOC face in Britain come from not having access to stable housing, to quality air, to quality nutritious and culturally relevant food, as well as to nature and greenspaces which is good for your mental health and your physical health. The food system within Britain is very complicated in that a lot of the food is imported, food items are very cheap but then not very good. So, the global economic food system is not only really exploitative to the poorer countries but also the people who cannot access quality food in Britain. Malnutrition is said to have risen massively under the Tories,[3] people aren't getting the quality of food that they need, and often the poorest people in this country are Black and People of Colour. Adding to this, with the pandemic and Brexit, certain food shortages have made many people more aware how these things are precarious and how we end up losing out through not knowing what type of foods can grow here locally.

So much of what Land in Our Names is looking for is to empower Black and People of Colour to get on the land to have access to nature, to get into growing because we need to be able to grow our own food for our communities, to change the narrative that quality food such as organic or agro-ecological food is only for White and middle-class people, that it is a luxury rather than a political issue. If you can eat well and can have access to these spaces, if you can feel like you have enough space to breathe and process, you can live well.

Many kids grew up in cities like London and have never seen a cow in their life. I mean, they probably know what a cow looks like but feel no connection to the land or have awareness of that sort of relationship, of how food it is connected to climate change and ecology. This is my experience of growing up in a city. And it was only through learning about food and growing that I learned a lot more about all the important aspects of the ecosystem that make growing food possible. About the tiny microbes and fungi and arthropods and so on that are an essential part of soil life, about how vegetables and other plants need these in order to grow and flourish, and how different insects and pollinators, as well as honeybees and bumblebees and hoverflies and all the way up to predators that keep animals like badgers or deer populations balanced … About how all that's so essential, how it's all interconnected. That's something I only was able to really access through learning how to grow food myself and being put in those situations. So, if we're able to redistribute land and create more spaces where people can be in nature, and be on the land, to feel how healing that is, how that helps you get out of your head … and to recognize how this is constant, this is ancient, this is everywhere and it doesn't belong to anyone, it belongs to us all, and we belong to the land. Compensating and redistributing land is also about the opportunity it creates for people, and how that can change you.

Dimitris and Maria: LION is promoting reparations of land and access for BPOC to the countryside in an ecological way, not only just "food", but good food, which means food that takes care of the environment too. You are clearly promoting what the mainstream would call an environmentally sustainable way, although there are many other ways of defining these new paths of relating to nature and non-humans.

Sam: Yes, this is very much the case with agro-ecological farming – which is how the Landworkers' Alliance for example defines it, and also *La Via Campesina* and organizations, unions and cooperatives of peasant farmers working together to promote other ways of farming. These ways are actually a lot older than what we call conventional farming – that is farming which relies on chemical fertilizers and pesticides that can first bring big harvest, but in the long run create dependency on more fertilizer more pesticides, hybrid seeds and so on, and benefits people who have more land and more money to afford them. Agro-ecological or just ecological methods of farming were used by our ancestors but also many people today. They are very much more about learning and listening to nature, learning *from* nature. About paying attention to what is around you. Take for instance mulch, using dead leaves or decaying matter to cover soils. That's what forests do. Trees shed leaves in autumn, and this protects soil life in winter by keeping roots and soil warm so soil life doesn't die out, and then decays and feeds the roots so that the next year more leaves and fruit can grow. Paying attention, looking around you, is an ecological aspect of what reparation is, learning to listen, not just thinking that humans know best and that we just can learn from our unique innovative abilities. So much of the designs and sciences and technologies that we have are probably better as they go on because we are mimicking what we are seeing in nature and also recognizing that we're a part of nature. So to work in, and to live in, an ecological way is beneficial for us.

As is being on the land. There are all these scientific reports coming out saying 'Oh, being near the sea is good for you, and being in nature is good'. And it's like, oh, yeah, who knew that being around green and blue would be beneficial to our mental health rather than being around grey concrete all the time? These things are so obvious, but we act as if it they aren't, maybe because we haven't done enough scientific studies. I was reading this article in the *Evening Standard* in the Tube and it was talking about how humans have brutalized nature, and I'm like, wait a second, I don't think that's true. That is the case for *some* people, because of specific dominant cultures. But a majority of Indigenous and Black peoples and societies had a different relationship to nature before colonialism that was not brutalizing, a relation much more ecological and balanced. There is an understanding of seeing yourself as an animal rather than as separate from nature and land. And I feel like that's what we want to return to. Because it's exhausting to have this narcissistic world view that we are somehow superior or better or separate from the environment. Our ancestors, cultures from which maybe we've become so severed from because of colonialism had other views. If we look at Yoruba or other Indigenous or traditional African spiritualities ... as a Black person, for me it's like a lot of what I'm learning, is about seeing yourself as part of nature and seeing yourself as an animal, and seeing yourself as part of the ecology and that you have an impact on your environment. So much of the healing is to heal from this mind-set that somehow you are not affected or you cannot have an effect on the environment, or that somehow your role is to always have power over those, not knowing that this animal, or this river, can know more than you do, or can have more of an important role than you do. It is getting out of this westernized mind-set that we are somehow separate and superior to non-human beings, and I feel like there's a lot that we need to learn in this way by moving towards ecological ways of growing and being.

It's not as if this is something new. It's something that's always existed but we've been told to forget or to not take it seriously or to dismiss it, and maybe there's a way that we can draw on certain aspects of modern Western science as well as indigenous practices and knowledge systems and tie them together, to draw the aspects of both that are beneficial. I'm not saying to replace cancer treatments with herbalism, it's more to see that it's not one or the other but a synthesis of both. That's what maybe a lot of people from my generation, who come from places like Jamaica or other colonized places feel, places where a lot of the traditional cultures, our ancestral cultures, have been erased by the colonizers' cultures. We feel we have to create something new which is about learning from everything that we have available to us and looking back at what techniques were available. It's not like, say, permaculture, for instance, would be the 'new' step, this is rather a specific branded synthesis of lots of indigenous ecological farming practices or design practices. We can take from all these different spaces, but also give back, through practicing these ecological ways of growing and designing and living.

Dimitris and Maria: I wanted to come back to something related to what you mentioned before about access to land and the right to roam that I [Dimitris] have felt many times when I'm in the countryside, both in Britain and when I lived in Germany. It took me many years to find the confidence to go for a walk in the countryside and then I came to realize that as a migrant, my accent and the different way of moving, I don't know what it is, it's probably many things, you feel uncomfortable because it is less diverse than the urban centres and you feel that you don't belong there. I want to ask you about your experience and also about the broader issue of racialized exclusion of black and people of colour from the countryside in Britain, about the lack of safety and how fear becomes embodied.

Sam: Yeah, yeah, totally. I have a lot of different points to comment on, and hopefully they all make sense. About Brexit for instance, and links of racism and xenophobia in Britain. … Until the '80s or '90s, Black people in Britain experienced a lot of racism before the concept of Black Britishness existed, in ways that told them 'you're not from here' – those were the terms in which the English Defence League worked. Then now it's very much rooted in an idea of White identity that is still also extremely xenophobic.

A lot of British people don't think of people who are from Spain or Greece or Romania or Bulgaria as White, even if that person considers themselves White. The idea of whiteness connected to this country is very fallible and determined by an idea of the other, and I definitely see the solidarity.

I was born in Jamaica and I came to London when I was seven, and my whole life is always having people asking me where I'm from, because I have an accent which I don't know where it is from, it's sort of like amalgamation of different things. And, yeah, growing up in London, I didn't have the same access to the countryside and when I did occasionally go on trips with my family, I never really felt safe and welcome. A lot of the times I felt a lot of hostility. I've experienced racism and people being quite horrible to me. That's just the experience for a lot of Black and People of Colour, especially when you grow up in the countryside. And that that's a whole other thing, some people experience some really horrible, violent behaviour, related to how threatening your identity is perceived to some people, and that's really sad but I think we can change that.

Paraphrasing Leah Penniman,[4] the violence may have happened here, but it wasn't the land … the land didn't harm us, the violence might have occurred on the land, especially in the US, in the South. It may have happened there, but the land wasn't the one who did it, the land didn't hurt us, it was the people, it was the system. Whenever I'm in nature, and even if there's some landowner or a farmer who's maybe eyeing me up, I know that this tree or this field or these butterflies or, I don't know, these cows, they're not the ones directing their hostility towards me. I can't control how people will react towards me, I can't keep modifying myself to make myself more palatable because it doesn't matter what I do, I could be the most inoffensive Black person ever and in the end someone who wants to hate me will always find a reason to hate me, but the land won't.

And that's how I have been healing in my body, being able to be with friends who don't police the way that I act when I'm in nature, when we can dance and play and sing, and run and enjoy it, learn about different trees, plants, or fungi. … These are the things that help heal me, it's about learning about nature, learning from nature and non-human beings, without getting distracted by how people treat me. I've accepted that I'm going to experience racism anywhere I go. I experienced it in London, I experienced it in the countryside. In the countryside I think British people have like this myth that is linked to the romantic movement I guess, an idea incredibly tied to racist concepts, an image of all White people just frolicking and smiling in the country, before the industrial revolution, before all this migrant intermixing happened. But then that never existed. There's always been a migrant and Black presence within rural spaces in Britain, but that history has been forgotten and hasn't really been investigated because it isn't valued, that's a violence of the archive in a lot of ways. And when it is archived, it is in line with the existing power schemes of the time.

Migrants, Black and People of Colour, are as deserving to be in these spaces as anyone else, and as much as ruling classes want to make the British countryside into this holiday resort where they can escape from the brutalities of the city and exclude other people, we also need to remind them but also to remind ourselves: we can take this, we are deserving to go there. Remind ourselves that me accessing this countryside is actually beneficial to non-human beings too, to biodiversity, to ecological repair. Learning all the new things that they try and tell us. Remember that saying, nature is first, that we need to protect nature before allowing people to access it, is forgetting these things go hand in hand.

It has taken me a long time. It has taken me having friends who already have that relationship because of their own privileges as being White and English, that have held my hand and looked after me and made me feel right when I felt angry because someone has made me feel uncomfortable. And then also connecting with other Black and People of Colour who loved being in nature and loved learning about plants, want to grow food, want to live on the land … but at the moment the only place that they can live is London, one of the biggest cities in the world, because there's not that many places where they feel safe, not that even living in London promises safety.

But it's just that we've been told you don't belong here and it's like: actually what if we just decide to go and live in there? What will happen then? And what if we do it in a way in which we build up community, and transform the space so that it is revitalized with culture and care, transformed into something exciting and not just towns becoming tourist traps. I think that's what a lot of people are excited about. It can help us heal. Nature doesn't belong to anyone as much as they want that exclusivity, to feel like they can exclude people. But if we keep just saying no, we're going to be here, and we're going to take that space, and we're going to make you uncomfortable, and we're going to keep talking about it until your ears bleed.

Dimitris and Maria: Picturing the image that you are giving us. A joyfully subversive story of what the countryside could be. It just feels like the way to go. We have discussed many times the possibility of living in the countryside but every time we imagine ourselves in an English village, we feel uneasy. Living in a very diverse town is what we have loved most in Britain. We really want to live closer to nature, but we keep wondering if we could move close to a community somewhere doing things differently … Imagine what would it be if we could make a third of each village reoccupied by different people! Somebody should write a speculative utopian fiction about this!

Sam: Yeah, this culture. It is something I've been thinking about because I also work with the *Right to Roam Campaign*,[5] and realizing, saying, that there is so much of this culture that is unnamed and not highlighted. The culture that is expected in the countryside, that is also rooted in the heritage of colonialism and enduring racism and exclusion. What if we recreate a new culture in which most people don't have to live in cities, and people can live in all different types of places in different ways, and living in the countryside doesn't have to mean living in this very particular way that feels quite suffocating if you don't fit or tick certain boxes. What if we were able to revitalize these spaces to make them a safe space for different types of people whether they're queer or racialized, or disabled. There are all these things that are culturally expected if living in the countryside, you're White, cisgender heteronormative, and able-bodied. What if we could create a space that is actually a lot more exciting that makes like people want to live on the land. But people who own most land don't want to redistribute it, and are afraid of people, are afraid of diversity.

Dimitris and Maria: It feels from reading your text that in the first calls for BPOC growers to come together, LION and others were surprised by the response, by the number of people who came to join. It seems that there is a critical mass out there and a desire that has been made invisible. Can you tell us about this?

Sam: Yeah, I remember the Caucus, I think you're talking about the BPOC Caucus after the Oxford Real Farming conference at the beginning of 2020, before the pandemic. This was before I was involved in *Land in Our Names*, and it really was so healing and so transformative to be in the countryside surrounded by Black and People of Colour and to just be, wow, there's so many of us here and we all want to do similar things!

And a lot of us were from cities. And shared the problems: you can want to go into the countryside, but you can't access land that is close enough to a town centre where you can sell that produce that you're growing; you want to be a farmer, but you don't want to feel isolated where you're just like on a farm in the middle of nowhere and then there's no communities near you. But the fact that there were so many people who were interested or doing work to do with these concerns, whether they were herbalists or land workers or environmentalists, it felt like really, oh, we exist! And we can connect, and we can build relationships. And from then, people who went to the Caucus or who have connected to Land in Our Names have gone into pursuing growing, whether that's doing training, or setting up community gardens, and there is such a strong desire of so many people to get into growing.

I got into growing because I've tried lots of different types of work and I've hated most of them but then I was like, oh, yeah, if I get into growing, like gardening and growing, that's supposed to be really good for your mental health, so if I'm around plants and my hands are in the soil and I'll feel good.

… And I did, and I love it. And this is shared by a lot of people, it's like one of the oldest things that we've done, it's in so many cultures in the world, it is about food, about just this ancient thing that is so grounding. Sometimes you wonder what's the point of what I'm doing, does this mean anything? But then you're like, oh well, I'm growing food, that's the point, I'm growing food so I could eat, and other people could eat. And so, I think a lot of people feel very connected to that and the biggest issue remains how can they access land, how can they access space to grow. Because the other thing with growing food is that the benefits of your work, especially ecologically grown food, you don't even see until after a few years because you've been investing in that land and space, you've been nurturing it. And so, the fact that in London, whether it is you cannot afford to buy land, or if you're renting land, that's still precarious. That is sad. That's why we need to have land distribution, land reparation, because so many people want to change the way that we grow food and access spaces, and engage with land, but then we don't have that power, we don't have that ability to just grow and be on the land.

But it is really exciting because that first event really made me aware that we're not alone, that this is a space that it is not only for White people. Obviously around the world there's lots of different people farming, but especially in Britain, farming being one of the whitest professions, it's just really exciting to know that so many people want to change that and connect.

Dimitris and Maria: We have two last questions. If you could say a bit more about food insecurity and food inequalities. Food apartheid is a very strong way to control people and marginalize people, and ecological reparation is also about fighting these food inequalities that are so deeply ingrained in social injustice.

Sam: Regarding food apartheid and food sovereignty, we think waiting for people who have power and privilege to make things better for you just doesn't work. Like *La Vía Campesina* and the Landworkers' Alliance and other affiliated groups, we're fighting for food sovereignty, the right for communities and for everyday people to grow their own food, to share seeds, to distribute that food and have a livelihood. To sustain themselves from good growing is everyone's right. There are less people working in farming, and people who have land, people who can grow food agro-ecologically in this country because of the enclosure of lands and because of how expensive it is and how little money you get for selling food in this country as well. This right is also constantly under the threat of land grabbing and agrochemical companies imposing their practices onto small-scale farmers. Even though we know that small-scale farming is better for the environment and better for communities and everyone's wellbeing, since the Green Revolution's obsession with caloric intake, many still think we're solving malnutrition and world hunger by just making people grow monoculture crops like wheats and corn. But malnutrition has increased since the 1960s and more farmers have died by suicide because of not being able to keep up with having to buy seeds or chemical input. And the only ones who have benefited are large landowners and big companies. The fight for food sovereignty includes recognizing that even though we have enough for everyone to be well fed in this world, there's still so much world hunger, and that even though Britain is an incredibly rich country, so many people are reliant on food banks. Relying on charity doesn't work: what we need is to empower people to have the skills to grow food for themselves, for their community, and also to know how to grow nutritional, culturally relevant, delicious food, not just boring, bland food that you have no connection with. We need to have connections to food, not just a foodie culture only accessible to middle-class people, or organic food only accessible to the middle classes. It needs to be accessible to everyone if we want to change and move away from chemical-reliant food growing. Food justice is not a consumer choice but about people being marginalized from having access to nutritional and diverse food. A lot of the people who end up living in food apartheids are those living in extremely exploitative conditions, who need to give most of their income towards ridiculously high living costs. Many families living in cramped conditions where people are having to work so much that they don't have the time and energy to put into making a delicious meal. Something that we shouldn't have to choose between is whether to eat well, or rest, or be able to pay for your rent. So much about food sovereignty is also about making food, returning food to the centre of our lives, and that includes people being able to farm and grow their own food.

Notes

1. This is an edited version of an online interview conducted on 11 November 2021. The video recording of the interview is featured in the *Ecological Reparation* You Tube Channel www.youtube.com/c/EcologicalReparation
2. See for instance the *Rootz into Food Growing. Knowledge and experiences of social enterprise food growers from Black/communities of colour, March 2021*, report based on research seeking to understand and reduce barriers to entry into the social enterprise growing system: https://landinournames.community/projects/rootz-into-food-growing
3. Several reports on this issue based on NHS figures analysed by the Labour Party were published in the Press in 2019: For example, 'Malnutrition cases treated by the NHS have almost trebled under Tory government', *The Mirror*, 24 August 2019, www.mirror.co.uk/news/uk-news/malnutrition-cases-treated-nhs-almost-18955553 [Accessed 10 January 2022]; 'Malnourishing Britain', *Tribune Magazine*, 26 November 2021, https://tribunemag.co.uk/2021/11/poverty-malnutrition-food-insecurity-conservative-party-universal-credit-jamie-oliver [Accessed 10 January 2022].
4. Penniman, Leah (2018) *Farming While Black: Soul Fire Farm's Practical Guide to Liberation on the Land*, Vermont: Chelsea Green Publishing.
5. www.righttoroam.org.uk/

12

Waste, Improvement and Repair on Ireland's Peat Bogs

Patrick Bresnihan and Patrick Brodie

Introduction

About 20 per cent of the surface of Ireland is covered in bogs, including the blanket bogs along the western coast and the raised bogs in the midlands. Long treated as 'wastelands' by colonial and postcolonial policy, it is estimated that there has been a 99 per cent loss of the original area of actively growing raised bogs in Ireland (EPA, 2011). This happens when bogs are drained – without water, the bogs cease to be bogs. The rural midlands, since the 1950s, have been the site of large-scale industrial turf-cutting by the semi-state company Bord na Móna (BnM). These cut peatlands, open mines across the island's boggy central regions, are now post-industrial landscapes or 'brownfields', as official policy refers to them. While the remnants of living bog provide habitats to many rare species of plant and animals and sequester carbon from the atmosphere, most drained and cut bogs today act as wastelands of modernity, as after their transformation for industrial activity, these largely barren landscapes continue actively emitting carbon which for centuries has been locked underground.

In 2018, BnM announced plans to phase out industrial extraction of peat by 2030, joining the Irish state's efforts to de-carbonize. As one of the country's largest landholders, with a landbank of around 200,000 acres (about 1 per cent of Ireland's total landmass), these 'ruined' environments may be central to the country's 'green' and 'smart' futures as sites of de-carbonization and remediation, carbon sequestration and information services. The company must balance its obligation to deliver a 'just transition' for the hundreds of industrial workers and communities organized around peat-cutting with contributing to the state's commitment to transition to 70 per cent renewable energy in the country by 2030.

Current discourses and associations of peat bogs and waste resonate with discourses from the 18th and 19th centuries, when efforts to drain and reclaim these waterlogged 'wastelands' by British colonial authorities were justified by moral and economic studies and mapping projects. The colonial division of the world into a patchwork of territories for extraction has frequently stressed treatment of the 'wild' or the 'terra nullius' in colonial imaginaries, concepts which allowed administrators and capitalists to make landscapes legible, populations manageable, and territories inert for extraction (see Yusoff, 2018). Ireland occupies a peculiar place within these narratives, as a European nation that was colonized until relatively recently, and where legacies of colonialism continue to pervade ways of governing rural landscapes.

In this chapter, we aim to disrupt linear accounts of bogland progress and ruination that rely on binaries of waste/value, to surface 'submerged' (Gómez-Barris, 2017) bog relations that pre-date and endure in spite of their designation as somehow 'un-valuable' within colonial and post-colonial value metrics, including contemporary efforts to render the bogs productive within emerging energy, data and carbon economies. These less visible bog relations include forms of cultural attachment, commons-based energy cultures, ecological care and inventive practices of 'repair' that, in different ways, suggest forms of situated, more-than-human life beyond the rationalities of bog productivity.

The temporality of the current 'green transition' in the midlands of Ireland locates the beginning and end of bog development as unfolding across the 70 years since the establishment of BnM and the state-led effort to put the bogs to use. The temporality of the bogs that is centred by BnM and state modernization concerns a particular phase of development, and a particular understanding of beginning and end widespread across 'Anthropocene' discourse. But, as we deploy throughout this chapter, the linear progression of 'modernity' is disrupted in Ireland by the coexistence of various kinds of life, governance and practice that persist on the bogs throughout history. Ireland's modernization was stunted and bisected by two distinct periods, marked by revolutionary rupture, and can be subdivided into its colonial period (~1600s–1920), centred on the enrolment of Irish space into the physical and epistemological project of waste/value necessary for colonial capitalism, and its postcolonial period as a developmental state (~1920–1960), inheriting many of these value systems as a way towards self-sufficiency and economic autonomy. However, within Ireland itself, modernity unfolded unevenly across the country's urban and rural regions, and thus today the anachronistic features of rural Ireland inform our approach to this coexistence – where the ongoing universalist, scientific mastery required by previous forms of development sits side-by-side with more situated practices of community, environmental care and conservation.

We are interested in how these practices endure in the midlands of Ireland 'post-extraction'. Perhaps, more generally, we are resisting discourses of 'breakdown', 'crisis' and 'ends' because they are in fact teleological, not paying attention to those living *with(in)* multiple temporalities, and are instrumental to capital's quest for new frontiers. People have been living in these 'wasted' spaces and their temporalities for a long time. Decolonial and indigenous scholarship has argued that for colonized peoples across the world, the 'catastrophe of the Anthropocene' was experienced 500 years ago, and thus the 'aftermath' becomes something quite different (Whyte, 2020). Elizabeth Povinelli (2011), for example, has argued that the 'endurance' of particular ways of indigenous life – and relations with land (2016) – refuse the 'eventfulness' required by late liberal forms of reform and recognition, which are themselves methods and technologies of 'abandonment'.

Beyond merely tracing the histories and political economies that have laid waste to particular landscapes, infrastructures and communities, this literature also has in common a tentative hopefulness around situated practices of repair and care (see Lepawsky et al, 2017). The temporality of repair is continuity and endurance; maintaining things against breakdown 'allow[s] what already exists to continue or persevere, to carry on being', as Lisa Baraitser (2017) writes. Our contribution is that 'repair', rooted in maintenance, endurance and liveability, while not spectacular, resists the temporality of boom/bust, progress/backward, even if remaining ambivalent in political expression. We outline how practices of repair in relation to deeper bog ecologies offer viable and already-existing, if ambivalent, alternatives to the future-driven enterprises proposed within the just transition.

From mine to sink: a brief history of making bogs productive

After the Cromwellian Wars in 1641 and the Act of Settlement, Irish Catholics who had conspired against the Protestant faith and new English parliament had their lands confiscated. Morally justified as punishment for their subversive actions and barbarous religion, but economically justified as a necessary means of paying soldiers in arrears and debtors who had financed the war of conquest, this settlement required unprecedented land surveys of Ireland intended to determine the value of the lands to be distributed to those with a claim. The most comprehensive of these surveys was the Down Survey, conducted between 1654–1658 by William Petty, physician-general of the English Army in Ireland. Undertaken by soldiers under Petty's supervision, the survey was at the time the most ambitious and detailed mapped survey of any country and set the standard for future colonial surveys.

While not specifically documenting the bogs, the Down Survey illustrated the extent of boglands on the island of Ireland, and was among the first of many colonial projects that sought to make legible the boggy Irish landscape for profit. It was not long before new landowners and colonial administrators seeking to tame and benefit from Irish territories targeted the drainage and reclamation of boglands for agriculture. From 1716 onwards a series of Acts were passed by the Irish parliament to encourage peatland reclamation. One such Act in 1731 set out to 'encourage the Improvement of Barren and waste Land, and Boggs' (Atkins, 2020). In 1776, Arthur Young, social commentator and future establishing member of the British Board of Agriculture, undertook a travelogue of Ireland focusing on the 'improvement' of what he saw as wasteland landscapes and impoverished ways of living on them. As historian Esa Ruuskanen articulates, 'Young envisioned how a country so widely covered with "wastelands" could possess a lush, cultivated countryside and wealthier population and serve as the granary of the industrialising England' (2018, 22).

Ruuskanen (2018) traces how science and technology were mobilized and circulated to otherwise distant and distinct environments in a shared effort from the 18th to 20th centuries to transform bogs, fens, marshes and mires from wastelands into valuable territories via maps, drainage and botany, specifically for agricultural production, as well as canal transport and silviculture. European, colonial modernity's vastly transformative scale dictated how practices and knowledge deployed in one place could be extended to other places, meaning that the calculative rationalities applied to bog improvement could be (and have been) repeated and re-applied to landscape and territorial governance elsewhere in the colonies (see Bhandar, 2020). European modernity's universalizing scientific knowledge was enacted and often violently enforced through techniques, methods, classificatory systems and technologies. Within landscapes and communities affected, especially in a place as loaded with historical memory as Ireland (see Beiner, 2018), these legacies are not easily forgotten.

Ireland emerged from British colonial rule after a long independence movement in the early twentieth century, establishing as a Republic in 1937. Nonetheless, the postcolonial institutions formed to govern and manage the bogs inherited British 'imperial eyes' (Pratt, 1992) in terms of how the bogs were viewed and how efforts were mobilized to reclaim and 'improve' them. English Bog Commission maps, drawn up between 1809–1814 and seeking new suitable lands for agriculture during the Napoleonic Wars, were used in the 1930s by the Turf Development Board, which include details on the ideal geographical locations of drainage ditches and soil utilization (Feehan et al, 2008). These maps not only relayed topographic or general environmental information, but informed the fledgling Irish state's steadfast strategies of bog drainage and reclamation for capitalist activity.

Bord na Móna (BnM), the present-day semi-state body governing the bogs, was established out of the Turf Development Board in 1946. In the

1940s, the company industrialized the process of peat extraction and in the 1950s–1960s built several peat-fired power plants across the country's rural midlands and west in partnership with the Electricity Supply Board (ESB). These were related to the postcolonial state's efforts to build institutions of progress, improvement and modernization, which often clashed with rural ways of life (see Quigley, 2003) but also fuelled narratives of national progress via the 'quiet revolution' of rural electrification (Shiel, 1984). Through transformative, industrial-scale activities on the bogs, BnM was key to the generation of a 'modern' form of social and cultural life in the midlands. The post-WWII developmental state, responding to changes in the world economy and still striving to establish national autonomy and self-sufficiency in the postcolonial era, invested in an energy industry (extraction and generation) and associated infrastructures for workers, including housing and schools, that actually drew populations into previously 'underdeveloped' rural regions. Thus, within any 'just transition' out of peat extraction and its carbon-intensive economies and infrastructures, there is not only the 'traditional' ways of relating to the bogs at stake like subsistence turf-cutting and the energy cultures that it supports, but also substituting the modern promise of bogland industrialization – and its provision of jobs and prosperity – with something different, providing for the workers and their families who face new unemployment.

The current dominant solutions and most visible projects for the just transition are associated with 'green' energy and the tech economy. BnM plans to place itself at the centre of these future renewable strategies, capitalizing on both new energy and data infrastructural investment. As the goal of their just transition strategy within the future-driven Ireland 2040 National Planning Framework, BnM argues that they 'seek to generate commercial, environmental and social value from all our lands through a mix of appropriate new uses' (Bord na Móna, 2017, 2), which are overwhelmingly based around attracting new investment and capitalizing on market-driven value. The most prevalent projects in this regard, not yet developed, are potential data/energy 'hubs' that will pair wind, solar, biogas, biomass, and/or battery storage facilities with data infrastructures and services, part of an emerging closed circuit between multinational technology companies and public energy infrastructure, where the Irish state instrumentalizes its 'green' environmental image and infrastructural conditions for transnational capital (Bresnihan and Brodie, 2020, Brodie, 2020). The post-extractive landbanks in the midlands are especially strategic, as part of a transport and infrastructural corridor between Dublin and the south and west of Ireland – including large cities like Cork, Galway and Limerick.

A second, equally speculative future for the cutaway, as the industrially drained and mined midland bogs are termed, is carbon sequestration enabled through bog rehabilitation. Even as the peat cutting machines and peat powered power stations are brought to a halt, the tens of thousands of acres of cutaway continue to emit carbon into the atmosphere. BnM is required under its EPA-granted license to extract peat, to 'rehabilitate' the bogs after operations have ceased. What this means is less clear. In late 2020, the government announced an investment of €108 million (along with €18 million from BnM) to 'rehabilitate' 33,000 hectares of degraded bogs for climate and biodiversity benefits (BnM, nd). These projects will require large-scale re-wetting and aim to provide up to 300 jobs for existing BnM workers. Re-wetting reverses the four-hundred-year focus on bog 'improvement' as drainage, and the announcement drew widespread praise from scientific and environmental communities. However, scepticism has already surfaced from these same communities (Woodworth, 2021). Claims by the state and BnM that these rehabilitation projects would sequester millions of tonnes of CO_2 either suggests a failure to fully understand the complex interactions between bog ecosystems, hydrology and atmospheric carbon, or a more opportunistic attempt on the part of the state to appear to be taking action on climate change. Tech companies are also getting involved in these projects, demonstrating how strategic these sites are becoming for potential carbon off-setting – Microsoft has partnered with researchers at the National Science Foundation and Maynooth University on a carbon mapping project called Terrain AI, and Intel is re-wetting bogs in the Wicklow Mountains as part of a larger water conservation strategy. Whatever the project, re-wetting bogs may reduce the amount of CO_2 that is emitted from the cutaway (regardless of whether or not peat is extracted and burnt), but carbon is only *sequestered and stored* by *living* plant ecologies, particularly species of Sphagnum moss, which are not so easily corralled into large-scale climate offsetting projects.

If the data/energy projects as mentioned envisage the wastelands of industrial peat extraction as potential sites for large-scale storage and transmission infrastructures, then plans to transform these same landscapes into carbon sinks present 'rehabilitated' bogs as literal green storage infrastructures – a sink for carbon emissions released elsewhere. Both sets of ideas propose to enrol bog landscapes into a wider green economy based on carbon reduction and carbon accounting *at scale*. What is consistent in this accounting and logistics logic is one of substitution, externalizing and offsetting, the fixing of bog wastelands as new frontiers of value through their capacity to store what is not of use, to hold data, energy and carbon that is extracted and used elsewhere.

Environmental humanities and political ecology scholars have extensively unpacked the relations between capitalist waste/value and its mediation through colonial knowledges and practices of science and technology (Gidwani, 2013; Goldstein, 2013). Jennifer Wenzel (2020) discusses John Locke's seventeenth-century treatises on wastelands and value in the 'frontiers' of the Americas, which promised abundance from enclosure and improvement as a moral mission as much as economic interest. Deciding not to make these 'virgin' lands productive would not just be foolish, but profligate, a *waste*. The product of this modern improving enterprise has been wasted lives and places, as well as vast quantities of toxic material waste that continue to amass today in living bodies and landscapes. This stands in stark contrast to Locke's original idea of wastelands of abundance. What does not change, however, is the ideological motivation to transform these anthropogenic, industrial wastelands into new landscapes of value, now shrouded in a 'green' mission of moral necessity. In the case of the bogs, twentieth-century modernization and development were predicated on *extraction* – drainage, mining and burning of solid fossil fuels. In a strange inversion of this, efforts to render these industrial 'wastelands' productive today emphasize *storage* – the building of infrastructures and the modifying of landscapes to store energy, data and carbon emissions: a transition from mine to sink.

In the case of these data/energy/carbon hubs, it appears as though we are already seeing the intensification of extractive processes within more networked or 'circuited' formations (Arboleda, 2019). 'Bringing "waste" land into capitalist production', writes Jennifer Wenzel, 'means excluding people from their livelihoods and rendering them surplus, their previous lifeways and very presence often criminalised' (Wenzel, 2020: 142). The visions of post-industrial bog life – automated data centres drawing energy from the wind through algorithmically controlled power grids; re-wetted cutaway sequestering and measuring carbon through a network of environmental sensors and instruments – do not appear to require human labour or presence, as after construction, most of these projects will be managed remotely, with only small maintenance crews required to sustain them. Against this vision, we focus attention on forms of bog life that invite different futures based on situated relations of more-than-human care and repair that might mend collectives of humans, plants, soil, animals and machines.

Temporalities of repair: bog commons, ecologies and modernity

Bogs have long been sites of quotidian, and sacred, relations with those living on and around them. As Ray Stapleton, a manager of Lullymore Heritage and Discovery Park, a bog heritage tourism facility in Kildare, told William Atkins, "[t]he bogs have always been wild" sites of uncertain and competing sovereignties across clans (2020). Like other wetlands, for example swamps, bayous and mangroves in North America (Ogden, 2011; Sayers, 2014), these were places that colonial powers and the postcolonial state found very difficult to colonize, develop, make profitable, and only were able to do so with considerable and expensive effort (Loftus and Laffey 2015). Ireland's boglands have also historically provided refuge for people and communities fleeing colonial domination, as their inaccessibility allowed people to make lives and communities at arm's length from colonial and capitalist discipline, policing and surveillance (Gladwin, 2016).

Following Gomez-Barris' (2017) call to 'consider realms of differently organized reality that are linked to, yet move outside of, colonial boundaries ... submerged perspectives that perceive local terrains as sources of knowledge, vitality, and liveability' (2017: 1), the rest of this chapter considers some forms of bog life that continue to occur in the face of and in spite of systems of colonial and postcolonial domination, '*remaindered* within the enlarged production time of capital' (Tadiar 2009: 20). What follows is an attempt to trace out some forms of bog life that unsettle the boom/bust temporality of improvement which may re-orientate alternative practical imaginaries of and for the bogs. *Bog commons*, *bog ecologies*, and *bog modernity* each correspond to different communities of work, practice and culture that connect bogs to people in particular, socio-material and technical ways. These three categories arose out of our understanding that within the bogs many different histories and forms of life co-exist and play out simultaneously, especially as they come under threat by economic development. These three forms of bog life are connected through their familiar positioning and depiction as *remnants* of the past, somehow enduring in the present but experienced as not of the present.

Bog commons

> The Representatives of Bellanagare bog confirmed that there were multiple cutters on the bog in question, which was not accurately reflected on the title. They expressed concern for those not expressly on the title, as they were part of their community. Further, they were finding it extremely difficult to ascertain their identity – i.e. gave an example of an 85 year lady who had approached them after mass (the priest mentioned the matter) confirming her family had been in the bog for generations and this was the first she had heard of it. She was visibly upset at being told she could no longer cut for fuel.
>
> Quirke Report, 2012: 28

In early 2012, the Peatlands Forum was held in a hotel in Athlone, Co. Westmeath, in the heart of the midlands. The Forum was convened to hear the views of turf-cutting communities who were set to have their rights to cut turf taken away by the State in the interests of habitat conservation. The background to this was Ireland's failure to implement the EU Habitats

Directive of 1992, which sought to establish a network of protected sites and a strict system of species protection across member states. Ireland's failure to designate or protect any sites provoked the European Commission to bring its first infringement proceedings against Ireland as early as 1999. It began new proceedings in January 2011, with the threat of legal action and potentially serious fines in the EU Court of Justice. It was in this context that 53 raised bogs were designated as Special Areas of Conservation, with the State taking a strong line with prohibitions on turf-cutting on these sites.

The commons are about subsistence – ways of meeting needs and wants beyond the sphere of waged labour and the state. The bogs have historically been direct sources of food, medicine, shelter and land for grazing animals (Feehan et al, 2008). They have also, and continue to be, important sources of domestic fuel in the form of turf. Turbary rights grant individual rights of access to households living near or adjacent to specific bogs, but the 'winning' of turf via hand or machine, the drying of turf and its final transfer to houses, often involves collective labour and shared resources. Turf-cutting is not just about *social* relationships (organizing, working, sharing), but *ecological* relationships, knowing the material properties and qualities of the bog and the weather. Turf-cutting in Ireland today traverses commercial and subsistence economies, manual labour and mechanization. But, as the quote at the beginning of this section attests, historical attachment, customary use rights and energy subsistence remain key parts of what we are calling the *bog commons*.

Bog commons also allows us to distinguish the more localized, collective and largely subsistence turf economies that continue to exist, from the large-scale, state-led industrial peat extraction and associated cultures of energy that developed from the mid-twentieth century. The everyday cultures of energy that have sustained ways of life for centuries – such as gathering around the fire or cooking on a stove – are often impossible to account for within these larger, transformative processes. In Henry Glassie's famous ethnography *Passing the Time in Bellymenone*, his stories constantly mention turf smell, sitting in front of a purring turf fire, interspersed with tales of the wet, muddy bogginess of his walks between houses and interlocutors (1982). The intimacies of these 'energy cultures' (Szeman and Boyer 2017) that are domestic and communitarian, based on non- or semi-commercial social relations and material affects rather than large-scale extraction or transformation, are part of social reproduction grounded in the reparative practices of how communities form, relate to one another and 'pass the time'.

In Ireland, by the beginning of the eighteenth century most bogs belonged in law to individual landowners, with many of the large bogs incorporated within great estates. But longstanding customary rights were able to survive, in part because these rights of turbary can be invoked by anybody coming within the custom of the locality, even if they have no other land rights

(Feehan et al, 2008). This is what makes the governance of such rights, particularly via the state, so contested. The conflict between 'official' landscapes of state administrators and the 'vernacular' landscapes of those who continue to live on and by them is captured in present-day disputes over boundaries, as mapping EU directives onto the messy, informal and often commons-based energy cultures continues to arouse tensions.

The vilification of turf-cutters shares something with the vilification of other subsistence energy cultures that rely on wood or charcoal in a time of climate change and biodiversity loss, what Larry Lohmann calls small 'e' energy (2016). Small 'e' energy cultures embed social activities within particular ecologies – turf-cutting and bogs, for example – with 'all the limits to interchangeability and accumulation that the existence of such disparate meshes of relationships imply' (2016). It is exactly these sorts of more-than-human relations that ideas of 'terra nullius' and wilderness remove, leave out, and/or abandon to history in the creation of frontier landscapes. This rubric, applied *with* renewables, continues and intensifies in the present, under the guise of a sustainable 'future'.

Bog ecologies

> Best of all with each step we took, we made squelchy watery sounds. This bog is alive – snipe rose from the ground and flew away from us, and we came across signs of the Irish Hare (droppings). But most of all the colours of the sphagnums were simply a feast for the eyes.
>
> Creative Rathangan Meitheal, Facebook post, nd

The preservation of the remnants of raised bog in Ireland has an interesting history that parallels developments in global conservation science and activism since the 1970s. In 1974, the government of the Netherlands established the first Peatland Conservation Plan to inventory, purchase and restore hydrological function, and manage peatlands for the long term. This was the background for research being conducted in Ireland in 1978 by Matthijs Schouten. There are interesting connections to be made between the Dutch 'discovery' of Ireland's bogs, and the 'edenic narratives' discussed by Slater (1996) and others in relation to the 'discovery' of the Amazon by European scientists in the 1980s. Comparisons between Ireland's bogs and the Amazon have become more explicit through global carbon accounting and the promise of carbon sequestration and sinks. In both cases, efforts to protect unique species and environmental sites become tied into global discourses and comparisons by environmentalists and their cultures (Choy, 2011).

These different fields of expertise – climate science and peat ecology – emphasize different elements of the bog, activating different ways of knowing and engaging with the bogs. Communities of scientists, environmentalists, artists, bird enthusiasts and local residents have been instrumental in protecting bogs in Ireland in large part because of their wonder and fascination with the life that exists in and through them. This intimate and ground-focused attention to observing and documenting the strange life flourishing in the bogs underpins many community conservation efforts. The education work of Creative Rathangan Meitheal, for example, which involves guided, creative walks with local school children, led by local artists, ecologists and historians.

There are at least 30 species of Sphagnum mosses within Ireland. They are what produce the distinctive hummocks on the raised bogs; they are what literally makes the bog through their unique composition and ability to shape the environment in which they live. Most of the plant is dead. These dead cells can't photosynthesize or reproduce and yet they are integral to the success of the plant: they allow *Sphagnum* to absorb 20 times its body weight (Kimmerer, 2003). This water is what creates the anaerobic conditions of the bog, which prevents some species of plant growing (including trees that would overtake the *Sphagnum*), and enables others – carnivorous plants like Sundew, for example. Sphagnum thus creates the conditions which enables it to thrive. These are also the conditions which slow the decomposition of organic matter, storing carbon and slowing climate change.

Sphagnum moss is depicted by Monica de Bath, a midlands-based artist, in her paintings as uncanny, supernatural and monstrous, outflanking industrial infrastructures and buildings (de Bath, nd). This has the effect of turning a relatively brief industrial history into something minor when compared to the deep time of moss. Questions of ecological temporality have come more to the fore in the last few years as peatland science, global climate science and state policy have converged around the promises of carbon sequestration. Rather than foregrounding the kinds of intimate knowledge that undergird community conservation practices of raised bogs, emphasis has shifted to geo-chemistry and pollution sinks, where seductive logics of offsetting play a part in shaping the 'restoration' and 'remediation' of the cutaway bog.

Maria Puig de la Bellacasa's work on soil and 'care time' is instructive here, looping ideas of maintenance and continuity with posthumanist ontologies and technoscientific practices (2015). Against the productivist, anthropocentric temporality of progress, a 'restless futurity', that insists on a state of permanent soil fertility, the kinds of soil care that Puig de la Bellacasa draws attention to engage soil as an extensive, relational ecology within which multiple temporalities exist. 'Making time for soil' thus involves different technoscientific practice (not their abandonment) that object to the short-term, productivist temporalities of soil management, expanding possibilities for relations of care to thicken in the present and thus open up alternate futures. Bogs are also soil ecologies, with *Sphagnum moss* as one, central, plant relation. How does bog repair look if you begin with this plant and the relations, contexts and timespans it requires to flourish? Carbon can be part of this slowed-down relational view because it too finds a place in the watery, acidic ecology produced by *Sphagnum*. The tension arises when spectacular claims by the state to store 100 m of carbon overtakes the slow, careful work required to muster living, bog ecologies from the ruins of industrial peat extraction.

Bogs are highly dynamic, diverse and complex environments that are being enrolled to fit within the projects of Big Data, Big Energy and Big Climate (carbon sinks). The comparison to the Amazon is instructive here, where efforts to instrumentalize (and financialize) the carbon sequestration services of the forests have stumbled in the face of social, political and ecological obstacles once they jump from the spreadsheet to the ground. We see similar tensions playing out between the kinds of situated, embodied, careful knowledge of community conservation groups and networks, attuned to the temporal scales and registers of bog repair, and efforts to draw cutaway into the more large-scale, state-supported projects of carbon accounting and storage. The tension, or potential conflict, is best expressed in a contrast: between seeing a bog as a unique ecology of plants, animals, chemicals, humans, geology, and archaeological remains that involves ongoing ethical and political consideration and care; and seeing a bog as a space to store anthropogenic carbon pollution.

Bog modernity

It was not until the mid-twentieth century that the midlands of Ireland underwent significant social and economic transformation under the coordination of BnM and the still newly independent Irish state. The forms of life that BnM's economic development generated in the midlands – planned towns, migrant communities, new social infrastructures – and the robust social, cultural and affective attachments this architecture continues to produce is what we call *bog modernity*. Cutaway landscapes are not just examples of the catastrophic extraction of resources, they also contain material and affective presences (and absences) characteristic of older industrial ways of life that seem to have no place in a 'greener' future. Like post-industrial mining and steel towns in the US and elsewhere, midlands communities are crafting futures that must navigate within and around the landscape of de-industrialization.

For the midlands, the development of BnM and ESB was like something out of science fiction. Up until this point, the midlands were places of outward migration, small-scale farming and subsistence economies. The sudden arrival of waged jobs, of modern technology and expertise from the Soviet Union, and inward flows of people keen to find work, was transformative. As it became clear that there were energy resources and state support for power stations, BnM invested more significantly and permanently in the midlands region, including carefully designed and built housing estates for workers (O'Riordan, 2014). The modernity that shaped the midlands in the 1950s and 1960s was a mash of Soviet peat mining expertise, German engineering, Irish urban design and materials and a workforce and community that brought these elements together and combined them through their own skill, invention and cultural activity.

A documentary film produced for the Irish language television station, TG4, in 2019 depicts the demise of the peat industry in the midlands. The film follows several work colleagues and friends walking through an abandoned machinery depot or repair shop where they apparently once worked. As they touch the dusty, broken machines they are reminded of what it felt like to work there together, anecdotes are told, and what can only be described as a pleasure they derived from their work and the skills and social relations that infused that work. The identity of workers and communities attached to the industrial heyday of BnM oscillates between a tragic depiction of redundancy and abandonment, or the (paper thin) promise of jobs attached to foreign direct investment (FDI)-led projects that tend to be based on automation and technology, not labour-intensive activities. What comes next for these workers, but also for the thousands of kilometres of industrial railway networks, power stations, briquette factories, machine sheds and machinery, much of which was designed and maintained by the workers themselves as they refined their ability to extract and process peat more efficiently? This is a question Leila Dawney (2019) poses in the context of Visaginas, a former nuclear town in Lithuania. Decommissioned in the late 1990s as Lithuania gained entry to the EU, the town and its inhabitants now live in the wake of the Soviet nuclear dream, which, like towns in the midlands, is not just an object of nostalgia or derision but a material context and structure of feeling that continues even after the power plant is closed (Berlant, 2011). Dawney shows how life continues in the quotidian practices of the people who remain. Importantly, Dawney highlights how inherited material, institutional and cultural affordances constrain and enable the conditions of possibility for the kinds of living that persist in Visaginas – including affordability of housing, lack of regulation of space and residual Soviet political subjectivities.

Lough Boora is former cutaway that has been transformed into a Sculpture Park and public amenity with walkways and cycleways. As one of the first former industrial peat cutting sites to undergo this kind of transition, the project has received local, national and EU funding since the late 1990s. Initially, the focus was on re-wetting several hundred acres of cutaway for habitat and natural fen-regeneration to occur but Tom Egan, land project manager with BnM and a key force behind the project, identified '[b]rown carpet open spaces' where there was room for experimentation (Collins and Goto, 2020). This was the context that would enable the development of the Sculpture Park, a radical departure from previous approaches to the cutaway that had sought to reclaim the land as agricultural or silviculture land. The Sculpture Park project developed over a period of ten years, with a key feature of the experiment being annual collaborations between artists (Irish and international) and former BnM workers – both those who had moved into restoration efforts, and those who were former machinists with knowledge of the materials and former machinery that were used by the artists, recuperated and transformed into public art works. The first sculpture symposium, held in 2002, attracted a mix of Irish and international artists to make large works of art arising out of the history, materials and setting of the cutaway boglands (Tipton, 2010). The model was repeated regularly over the next eight to ten years as the curator, Kevin O'Dwyer, refined and developed partnerships to produce world-renowned landscape art in the midlands of Ireland.

Ecotourism projects, archival and heritage projects and community transport links are all ideas that get raised and circulated in the context of community development in the midlands in the wake of the peat industry. Such projects will never be motivated or mobilized by FDI and speculative financial capital – it takes too long, and produces value that is not easily monetized or measured. But, as with the *bog commons* and *bog ecologies*, this is not just a question of funding or state support. Ecological reparation in the midlands requires the active involvement of former workers and communities who may still be attached to (soon to be) historic forms of work, organization and energy. The everyday politics of repair and maintenance involves making spaces and environments where these attachments can loosen, where alternative futures become more desirable, more liveable, 'a landscape for a specific form of commoning' (Millington 2019). This process is not simply about restoration of place, because this is simply not possible, but about something different coming in its place. Repair as practice is important not only as a way of warding off breakdown, but as a way of inflecting the present, of creating something new, and ultimately, maybe, undermining the given-ness of dominant regimes of improvement and accumulation. In the post-industrial midlands, this might involve living with and around, and carefully re-shaping, the often ugly residues of the extractive past, without resorting to a more aesthetically 'cleaner' – but still extractive – future.

Conclusion

If one of the aims of critical, engaged theory in science and technology studies and energy humanities is to disrupt the temporalities of projects like data centres and their associated imaginaries of green progress in a time of climate breakdown, then we have sought to do this by drawing attention to alternative temporalities of place, culture, ecology and knowledge in the bogs of Ireland. In doing this, we hope to contribute to a broader project of dismantling colonial epistemic practices and relations to matter, while searching for alternative forms of experimental practice (Papadopoulos, 2018; Papadopoulos et al, 2021). This is necessary when the catastrophic history of the past 500 years continues to be cast as a long series of unintended side effects of Western progress. Integral to this positive green agenda is a belief in 'like for like' substitution in the progressive transition from 'dirty' to 'clean' economies. Materials, energy sources, labour, technologies, even behaviours, can be swapped in and out as though globally composed, place-specific ecologies were a single, vast accounting ledger. Ongoing efforts to transform industrial, 'wasted' boglands into landscapes for a new 'green' economy is one example of this. Rather than an innocent move towards a cleaner, more sustainable future, however, this process can be understood as a continuation of the extractive logics of the past that have not served the majority.

Against this, we can see these concepts and practices of repair as subversive, as operating within the 'ruined' landscapes of post-industrial extraction (Tsing, 2015) in spite of the ongoing incursions of capitalist value. If capital is continually incorporating new frontiers and forms of practice within its operations, perhaps we can brush against the grain of capital's dominant expansionary pathways by paying closer attention to places where its logics of value are being materially disrupted, however imperfectly and small scale. Within the human and more-than-human relations sustained on the bogs through colonial modernity and post-colonial developmentalism, we see vital and active ways of living and being against the grain of capitalist value, more attuned to the unique rhythms and timescales of the bogs. These forms of life occupy different histories, present to their own pasts.

In this chapter we have described these as *bog commons*, *bog ecologies* and *bog modernity*. These labels are not exhaustive, nor do they reflect the ways these different forms of life overlap and intersect. They are intended as invitation, not closure. All three connect with the idea of ecological reparation. First, they help foreground the more-than-human relationships, attachments and labours that exist on and through the bogs – the *ecological* – which have historically been ignored in process of making bogs productive. Second, they disrupt the neat progress temporalities of colonialism and capitalism by refusing to disappear, slowing down the advance of new 'green' enclosures and offering hope that it may not be too late to make amends for an extractive past while building a more liveable present – *reparations* attuned to the ecologies of life and practice that make up the bogs today.

References

Arboleda, M. (2019) 'From spaces to circuits of extraction: Value in process and the mine/city nexus', *Capitalism Nature Socialism*, 31(3): 114–33.

Atkins, W. (2020) 'Letter from Ireland: Boglands', *Harper's Magazine*, July. Available from: https://harpers.org/archive/2020/07/bogland-bog-of-ireland-peat-bog-bord-na-mona/.

Baraitser, L. (2017) *Enduring Time*, London: Bloomsbury Publishing.

Beckett, C. and Keeling, A. (2019) 'Rethinking remediation: Mine reclamation, environmental justice, and relations of care', *Local Environment*, 24(3): 216–30.

Beiner, G. (2018) *Forgetful Remembrance: Social Forgetting and Vernacular Historiography of a Rebellion in Ulster*, Oxford: Oxford University Press.

Berlant, L. (2011) *Cruel Optimism*, Durham, NC: Duke University Press.

Bhandar, B. (2020) Lost property: the continuing violence of improvement. *The Architectural Review*. 8 October. Available from: https://www.architectural-review.com/essays/lost-property-the-continuing-violence-of-improvement. [Last accessed 27 September 2022].

Bord na Móna (2017) Submission to the Department of Housing, Planning, Community and local Government on the 'Ireland 2040 Our Plan – Issues and Choices' paper for the Proposed National Framework Plan. 31 March. Available from: https://npf.ie/wp-content/uploads/2017/09/0457-Bord-na-Mona.compressed.pdf

Bord na Móna (nd) '100 million tonne carbon store secured under major Peatland Restoration Plan'. Available from: www.bordnamona.ie/100-million-tonne-carbon-store-secured-under-major-peatland-restoration-plan/ [Last accessed 14 September 2021].

Bresnihan, P. and Brodie, P. (2020) 'New extractive frontiers in Ireland and the moebius strip of wind/data', *Environment and Planning E: Nature and Space*. https://doi.org/10.1177/2514848620970121

Brodie, P. (2020) 'Climate extraction and supply chains of data', *Media, Culture and Society*, 42(7–8): 1095–114.

Choy, T. (2011) *Ecologies of Comparison: An Ethnography of Endangerment in Hong Kong*, Durham, NC: Duke University Press.

Collins, T. and Goto, R. (2020) *Deep Mapping Lough Boora Sculpture Park*, Offaly: Offaly County Council.

Creative Rathangan Meitheal (nd) Available from: www.facebook.com/creativerathangan/ [Last accessed 21 December 2020].

Dawney, L. (2019) 'Decommissioned places: Ruins, endurance and care at the end of the first nuclear age', *Transactions of the Institute of British Geographers*, 45(1): 33–49.

De Bath, M. (nd) Available from: https://monicadebath.wixsite.com/monicadebath [Last accessed: 14 September 2021].

EPA (2011) *BOGLAND: Sustainable Management of Peatlands in Ireland-Final Report*. Dublin: Environmental Protection Agency.

Feehan, J., O'Donovan, G., Renou-Wilson, F., and Wilson, D. (2008). *The Bogs of Ireland: An Introduction to the Natural, Cultural and Industrial Heritage of Irish Peatlands*, Dublin: University College Dublin, Environmental Institute.

Gidwani, V. (2013) 'Six theses on waste, value, and commons', *Social & Cultural Geography*, 14(7), 773–783.

Gladwin, D. (2016) *Contentious Terrains: Boglands in the Irish Postcolonial Gothic*, Cork: Cork University Press.

Glassie, H. (1982) *Passing the Time in Ballymenone: Culture and History of an Ulster Community*, Philadelphia, PA: University of Pennsylvania Press.

Goldstein, J. (2013) 'Terra economica: Waste and the production of enclosed nature', *Antipode*, 45(2), 357–375.

Gómez-Barris, M. (2017) *The Extractive Zone: Social Ecologies and Decolonial Perspectives*, Durham, NC: Duke University Press.

Hennessy, M. (2017) '"This is about heritage and tradition": Turf cutters say the new bogs strategy won't solve the problem', *Journal.ie*, 22 December.

Kimmerer, R.W. (2003). *Gathering Moss: A Natural and Cultural History of Mosses*, Corvallis, OR: Oregon State University Press.

Lally, D. (2019) 'The sacred fire of a data centre', *Strelka Mag*, 2 October.

Lepawsky, J., Liboiron, M., Keeling, A., and Mather, C. (2017) 'Repair-scapes', *Continent*, 6(1): 56–61.

Linebaugh, P. (2008). *The Magna Carta Manifesto*. Berkeley: University of California Press.

Living Bog – Raised Bog Restoration Project (nd) Available from: http://raisedbogs.ie/the-dutch-connection/ [Accessed 21 December 2020].

Loftus, C. and Laffey, J. (2015) *Powering the West: A History of Bord Na Móna and ESB in North Mayo*, Crossmolina-Bord na Móna/ESB Commemoration Committee.

Lohmann, L. (2016) 'What is the "green" in "green growth"', in G. Dale, M.V. Mathai and J.A.P. de Oliveira (eds) *Green Growth: Ideology, Political Economy and the Alternatives*, London: Zed Books, 42–71.

Millington, N. (2019) 'Critical spatial practices of repair', *Society and Space*, 26 August. Available from: https://www.societyandspace.org/articles/critical-spatial-practices-of-repair [Last accessed 27 September 2022].

National Parks and Wildlife Service (2012) The Quirke Report. Available from: www.npws.ie/sites/default/files/publications/pdf/QUIRKE%20REPORT.pdf [Last accessed 21 December 2020].

O'Riordan, C. (2014) 'An experiment in sustainable architecture. Irish Architectural Archive'. Available from: https://iarc.ie/an-experiment-in-sustainable-architecture/ [Last accessed 21 December 2020].

Ogden, L.A. (2011) *Swamplife: People, Gators, and Mangroves Entangled in the Everglades*, Minneapolis, MN: University of Minnesota Press.

Papadopoulos, D. (2018). *Experimental Practice: Technoscience, Alterontologies, and More-than-social Movements*, Durham, NC: Duke University Press.

Papadopoulos, D., Puig de la Bellacasa, M. and Myers, N. (eds) (2021) *Reactivating Elements: Substance, Actuality and Practice from Chemistry to Cosmology*, Durham, NC: Duke University Press.

Povinelli, E. (2011) *Economies of Abandonment: Social Belonging and Endurance in Late Liberalism*, Durham, NC: Duke University Press.

Povinelli, E. (2016) *Geontologies: A Requiem to Late Liberalism*, Durham, NC: Duke University Press.

Pratt, M.L. (1992) *Imperial Eyes: Travel Writing and Transculturation*, London: Routledge.

Puig de la Bellacasa, M. (2015) 'Making time for soil: Technoscientific futurity and the pace of care', *Social Studies of Science*, 45(5): 691–716.

Quigley, M. (2003) 'Modernity's Edge: Speaking Silence on the Blasket Islands', *Interventions: The International Journal of Postcolonial Studies*, 3(5): 382–406.

RTE (2018) '€150m investment to build two power plants in Offaly', 23 April. Available from: www.rte.ie/news/leinster/2018/0423/956516-offaly/ [Last accessed 21 December 2020].

Ruuskanen, E. (2018) 'Encroaching Irish bogland frontiers: science, policy and aspirations from the 1770s to the 1840s', in J. Agar and Jacob Ward (eds) *Histories of Technology, the Environment, and Modern Britain*, London: UCL Press, pp 22–40.

Sayers, D. (2014) *A Desolate Place for a Defiant People: The Archaeology of Maroons, Indigenous Americans, and Enslaved Laborers in the Great Dismal Swamp*, Gainesville, FL: University Press of Florida.

SEI and EPA (2008) 'Assessment of the Potential for Geological Storage of CO2 for the Island of Ireland', Sustainable Energy Ireland and Environmental Protection Agency. Available from: www.seai.ie/publications/Assessment-of-the-Potential-for-Geological-Storage-of-CO2-for-the-Island-of-Ireland.pdf

Shiel, M.J. (1984) *The Quiet Revolution: The Electrification of Rural Ireland, 1946–1976*, Dublin: O'Brien Press.

Slater, C. (1996) 'Amazonia as edenic narrative', in W. Cronon (ed) *Uncommon Ground: Rethinking the Human Place in Nature*, London: WW Norton & Company, pp 114–31.

Star, S.L. (1999) 'The ethnography of infrastructure', American Behavioral Scientist, 43(3): 377–91.

Storm, A. (2014) *Post-Industrial Landscape Scars*, New York: Palgrave Macmillan.

Szeman, I. and Boyer, D. (eds) (2017) *Energy Humanities: An Anthology*, Baltimore, MD: JHU Press.

Tipton, G. (2010) 'Culture in the cutaway', *The Irish Times*, 3 April. Available from: www.irishtimes.com/culture/art-and-design/culture-in-the-cutaway-1.647575 [Last accessed 21 December 2020].

Tsing, A. (2009) 'Supply chains and the human condition', *Rethinking Marxism: A Journal of Economics, Culture & Society*, 21(2): 148–76.

Tsing, A. (2015) *Mushroom at the End of the World: On the Possibility of Life in Capitalist Ruins*, Princeton: Princeton University Press.

Wenzel, J. (2020) *The Disposition of Nature: Environmental Crisis and World Literature*, New York: Fordham University Press.

Whyte, K.P. (2020) 'Too late for indigenous climate justice: Ecological and relational tipping points', *WIREs Clim Change*, 11: e603.

Woodworth, P. (2021) 'Bord na Móna's big shift: How the peat giant pivoted to bog restoration', *The Irish Times*, 7 January 2021. Available from: www.irishtimes.com/news/science/bord-na-m%C3%B3na-s-big-shift-how-the-peat-giant-pivoted-to-bog-restoration-1.4438767 [Last accessed 14 September 2021].

Yusoff, K. (2018) *A Billion Black Anthropocenes or None*, Minneapolis, MN: Minnesota University Press.

13

New Peasantries in Italy: Eco-Commons, Agroecology and Food Communities

Andrea Ghelfi

The agroecology of Genuino Clandestino

Starting from the end of the 1960s the so called 'Green Revolution' significantly transformed the ways through which agriculture has been developed on a global scale (Shiva, 2008; Altieri and Toledo, 2011; Rosset and Altieri, 2017; Altieri, 2018). The central role of mechanization, the adoption of new technologies, the selection of high yielding varieties of cereals and the extensive use of chemical fertilizers and agrochemicals are the main features of current 'industrial' agriculture. These technologies of food production have wide-ranging eco-social implications on biodiversity and climate change, and they entail a relationship of strong dependency between farmers and the world's largest chemical producers. Agroecology (Altieri, 2018) appears as one of the main alternatives for overcoming the shortcomings and damages that the 'Green Revolution' has caused. Agroecology is a response to the question of how to transform and repair our food system and rural life, starting from the ecological practices of peasants and farmers, artisanal fishers, pastoralists, indigenous cultivation methods, urban food producers and so on.[1] In this sense, food sovereignty movements and agroecological farming are creating alternative politics of matter (Papadopoulos, 2018) and by seeking different material circulations and channels of involvement, they enact different possibilities of human-soil-food relations.

Permaculture, organic, bio-dynamic, regenerative agriculture, alternative food distribution:[2] these are some of the names given to practices by which movements of ecological agri-food transition are converging today in emphasizing a need to attend to the health of the soil and the broader ecologies in which we grow food (Altieri, 2018). Food sovereignty campaigns entail the simultaneous responsibility of participants to be food growers and consumers, which means being involved in the processes of food production and distribution by inventing alternatives to the large supply chains that currently dominate the existing agri-food system. However, food sovereignty is something more than the consumers and growers' right to choose what to consume and what to grow and how. Agroecology and food sovereignty are, first of all, about creating alternative ways to deal with the ecological interactions and interdependencies involved in the processes of farming: the collective enterprise of creating an alternative lifeworld within the interactive dynamics that inhabit the soil and its inhabitants.[3]

Following the example of an Italian network of farmers, called *Genuino Clandestino* (Genuine Clandestine in English), we can see agroecology as a set of practices that are transforming the everyday doing of farming, as a movement that is redefining the economic and juridical space of action of organic food producers and as a process of reinvention of rural forms of life. The network was born in 2010 to defend the existence of a multiplicity of alternative rural forms of living and promoting agroecological knowledges and practices. Genuino Clandestino's practices include grassroots markets that promote food sovereignty, innovative forms of trust between producers and consumers through a self-organized process called the Participatory Guarantee System (PGS), civic use and collective care of land as commons and strong links between the movement and scientific research on soil ecology, seeds biodiversity and food sustainability. Genuino Clandestino can be seen as an example of a novel movement that fuses traditional environmental movement campaigns – for example against the use of pesticides in agriculture – with the experimentation and the building of food communities and alternative forms of living.

Reclaiming rural forms of life, though, is not a way for restoring some form of premodern vision of social conditions. In the politics of Genuino Clandestino the farmers and activists who define themselves as 'contadini' (peasants in English) reactivate the capacity to invent other spaces and times of existence. The peasants of Genuino Clandestino reclaim an alternative use of the land and the right to make their own food in polyculture farms. Here farming is not an entrepreneurial activity but a diverse way for cultivating a 'practicality' of life within the cycles of the land. How to become a companion of the Earth by taking part in more than human communities of food? This is the open question that accompanies the making of alternative forms of living in the Genuino Clandestino network. This is the open question that forces this more-than-social movement (Ghelfi and Papadopoulos, forthcoming) to invent – from the farm to the table – alternative networks of eco-social (re)production.

In this chapter, I explore three key dimensions of these new ways of becoming a peasant: farming ecological practices, making food communities that links countryside and urban spaces, civic and collective use of land. By exploring the peasant return and its culture of eco-sharing, the development of self-organized peasant markets in Bologna (*Campi Aperti*) and the community of *Mondeggi Bene Comune, Fattoria Senza Padroni* (Mondeggi Commons, Farm Without Owners in English), we will scout the multiple practices of mending and ecological reparation that animate the everyday politics of Genuino Clandestino.

Returns

In the Italian context the expression 'new peasantries' has a double meaning. In an epoch deeply marked by multiple ecological crisis, the first meaning of this expression refers to the fact that the experience of 'becoming a peasant' is rooted in an ecological approach to farming. Even if there are significant historical legacies with previous cycles of 'back to the land' movements and rural experiences developed from the 1970s, the forms of activism that characterize Genuino Clandestino situate the struggles for food sovereignty and agroecology inside a historical context deeply marked by the 'ecological condition': sixth mass extinction, climate crisis, soil depletion, ocean acidification, human displacement, forest destruction. The traces of global ecological crisis are everywhere. The unpredictable consequences of the ongoing modifications of the chemical, biological and geophysical composition of the Earth are perceived by a significant portion of civil society as ungovernable. Even if these global issues often create a feeling of impotence, the typical 'too big to face' situation, multiple forms of organization – for example farmers markets, solidarity purchase groups, community support agriculture, ethical consumerism promotion of recent rural, ecological movements – have been capable of offering modest and situated ways to encounter a few of these issues by proposing a pragmatic alternative starting from the food that we eat or grow. Furthermore, an ethical-political approach concerning food and agriculture is deeply present in the networks of a solidarity economy, within many of the environmental mobilizations that crisscrossed Italy in the last ten years, in large strata of the 'generational' climate activism, in numerous self-managed social centres, in significant sectors of the cultural and social left and catholic associationism. To put it briefly, the ecological question has a political centrality that may be impossible to imagine in previous decades, and the campaigns for food sovereignty represent a situated way for enacting practices of ecological reparation starting from a modification of everyday behaviour.

The second meaning of 'new peasantries' speaks about the biographies of peasants. And the adjective 'new' refers to the simple fact that in the last six years, since I started my engagement as an activist and a practitioner within Genuino Clandestino, I rarely met a peasant who has been a peasant for her/his entire working life. They are back to the landers, even if in most cases they inhabit territories they never met before. The new peasants did not grow up on a farm. Rural life lives in the memory transmitted by previous generations, it lives in the storytelling of grandparents much more than in that of parents. The word peasant has started to circulate again in recent years after disappearing in Italy in and around the 1950s and 1960s: the return to it requires a reinvention. A word that traditionally is synonymous with poverty, social subordination and economic dependency – as sharecropper or as a salaried agricultural worker – nowadays is used again in order to evoke independence and alternative forms of rural existence. In the context of Genuino Clandestino people call themselves peasants and not farming entrepreneurs simply because there is a refusal to qualify their daily activity within the logic of economic, social, cultural and material productivism. More than a job, the word peasant here evocates a form of life. Becoming a peasant is a transition process marked by an everyday relation with a territory which demands an alternative use of the land and simultaneously an alternative use of the self.

In the context of Genuino Clandestino, becoming a peasant, as they call themselves, is an existential and ethical-political transition. A transition to a 'practicability of life' (Bertell, 2016), to a form of living in which self-subsistence and ecological care are inextricably intertwined, starting from the reinvention of daily practices of existential regeneration and socio-ecological repair. The desire of an embodied, everyday, dirty, material relationship with the land: this is the peasant return. Permaculture, organic, bio-dynamic, regenerative agriculture,[4] the return brings with itself a multitude of practices of care (Puig de la Bellacasa, 2017) in which material engagement meets an obligation to make an ecology a liveable place for all the participants in it. But this is a more than a human affair because material regeneration is a practice of interrelation: I can't prune an olive tree without trying to understand where it wants to be cut, without acting with it, without situating myself within the web of ecological interdependencies that sustain simultaneously its life and my harvest.

In permaculture practices include the making of an embodied ecology of proximity with certain entities that populate an ecology, a capacity to feel and act with significant others, with human and more than human elements and forces. Permaculture's essential kit of knowledge politics consists in cultivating a capacity to feel and act with such forces: composing your design with the activities of these entities, calculating interdependencies and in the ability to act inside an 'ecology of canals' (Ghelfi, 2015) in which commensality (Papadopoulos, 2018) defines the ontology of more than human life. In permaculture human activities meet interspecies engagement: making food means acting within an ecological food web in which different species exploit the same class of environmental resources. In commensal coexistence actors just leave stuff around, and other actors use them.

This is an ontology of coexistence. To foster coexistence, actors must reduce their presence, their subjectivities, and leave space for other actors to exist. Commensality presupposes the careful retreat of the self: de-individualization, interdependence, contingency, involvement. In permaculture design a farm is crisscrossed by a multitude of ecologies of canals which work via transforming the surpluses of one entity into the substance of the other. In permaculture nothing can be done alone, no one thing has the power to take the place of the universe and nothing can be done without participating in the more than human practices of eco-sharing that constitute and maintain the commons of the Earth.

The main difference between 'back to the land' practices and Latour's recent 'back to Earth' (Latour, 2018) proposal consists in taking seriously, or not, the transformative role of material practices. Latour reminds us that a back to earthly basics politics needs to take into account the many actors and the traffic of relations that make a territory habitable. The necessity of recognizing the fabric of multispecies interdependencies enabling the generation of terrestrial life is a significant contribution for thinking ecological reparation. At the same time 'back to the land' practices show us that acknowledging the human-nonhuman traffic of relations is not enough for making a politics of material regeneration. It matters how practices generate, or not, other entanglements and alternative politics of matter. The transformative power of back to land movements consists in their capacity to set up alternative material configurations and everyday practices that aim to materialize justice in the human-nonhuman everyday continuum. By engaging with a form of transformative politics in which commoning is primarily about experimenting with material practices in which self-subsistence meets ecological care, Genuino Clandestino is a lure for thinking the role of everyday, more-than-human, practices of material regeneration in a politics of ecological reparation.

From the farm to the city

The ecology of proximity is a way for accounting how biodiversity works inside an agroecological farm. Genuino Clandestino extends this more than human sense of community from the farm to the city, generating material proximity and trust throughout economic, cultural, material and political canals that extend into urban spaces. In this section we will see how the canals of agroecology come to the city. The experiments of peasant markets organized by Campi Aperti (Open Fields) allow us to think food communities within and beyond the daily experience of relationship with the land. Food is the material mediator, the key actor that allows the construction of new alliances and transversalities among producers, consumers and social centres, because here food brings with itself novel practices of care and commensality, solidarity and cooperation between the city and the countryside.

The story of Campi Aperti started almost 20 years ago, with a stall of vegetables in a squatted social centre called XM 24 in Bologna. In the beginning, May 2002, several farmers brought their vegetables and cheese from the Samoggia Valley once a week to the social centre and organized a small clandestine market together with an urban collective interested in food sovereignty politics. From the very first moment a mixed assembly, composed of producers and buyers, organized the weekly market, public discussions and convivial events as a form of political action. Around 2006 the Thursday evening peasant market of XM 24 became a reference point for students, residents and workers of the working-class district of Bolognina. It is a key principle of Campi Aperti that each producer must only bring to the market their own products. Clearly, sellers reduce the variety of their supply, and the market increases the overall variety by expanding the numbers of producers. In the same years weekly markets were inaugurated in other social centres: the traffic of genuine food, artistic and cultural initiatives transformed the forms of sociality and the spatial organization of social centres. In 2018 a formal association – Campi Aperti, Associazione per la Sovranità Alimentare (The Open Fields Association for Food Sovereignty) – was built with the aim of obtaining from the municipality the use of public squares to multiply the weekly markets. Organizing a clandestine market in an autonomous, independent zone where police and health officials do not enter, such as a self-organized social centre in Italy, is totally different to organizing a legal market that hosts clandestine farmers and clandestine products. The use of public squares has been a turning point in the history of Campi Aperti, a local network composed today of more than 160 farmers and thousands of customers.

This turning point, which has been underpinned by the strategy of making the presence of markets in Bologna irreversible, coincides with the opening of a national campaign of civil disobedience[5] to challenge the social and health regulations governing food production. These regulations make it particularly difficult for small farms to store and process seasonal surpluses. A communication campaign has launched around a logo attached to the products on sale. Around the organization of this campaign that reclaim the legitimacy of a series of peasant practices, such as the exchange and distribution of genuine agricultural products, mutual work aid and the reproduction and exchange of seeds, a national peasant network is formed. The name of the label is Genuino Clandestino. Usually, to obtain the label 'organic' and thereby enter the organic trading circuit, farmers must pay a private agency to come and check their land and take soil samples for laboratory testing. This certification process has become an expensive procedure for small farmers (costing hundreds if not thousands of euros). Underground markets selling genuine products are made up of a mix of producers who own legal and formally recognized enterprises and those who call themselves peasants regardless of having a legal title allowing them to produce and sell.

When ecological practices come before state regulations, the process of self-organization makes its own rules and boundaries. The same could be said for the products that circulate in the market. The everyday practice of agroecology implies a disconnection from the standards of food production and circulation (especially in relation to processed food such as cheese, marmalade, wine, tomato sauce) simply because these standards are thought to be in relation to the infrastructures of industrial production. The Campi Aperti network of farmers and consumers has developed a space outside of state-regulated organic certification: a participatory guarantee system through which producers and consumers (called co-producers in Genuino Clandestino) decide together prices, organize visits to the farms in which they check farming conditions (the type of fodder used, the living conditions of animals, the revenue and working practices of farmers and their co-workers, and so on), make public reports on the strengths and limits of each farm and set up self-education workshops on agroecological knowledge.

Self-organization and trust are key features of Campi Aperti's modes of organizing. Thanks to this capacity of assembling sensibilities coming from heterogeneous socio-political and cultural backgrounds around concrete practices of material regeneration, in recent years the projects of food sovereignty have multiplied in Bologna: a complementary currency – called Il Grano, Wheat in English – which connects the activities of farmers' markets with other practices of solidarity and transformative economies in the city; a self-organized community emporium –Camilla Cooperative – whose clients are also members, and each member works one afternoon every four weeks in the food emporium; the birth of a Community Support Agriculture – called Arvaia – which involved 500 members who are supporting the cooperative production and distribution of organic food in the countryside of Bologna; participatory scientific practices around the genetic improvement of seeds with pioneer researchers on seed biodiversity for organic farming; peer to peer tutorials and shadowing activities for sustaining new members' transition to agro-ecological practices; the making of 'mutual aid agreements' in which the whole community of Campi Aperti becomes a guarantor for farmer investments developed in relation with cooperatives of mutual aid and solidarity finance.

While I am writing this text on ecological reparation, I look at the motto printed on the 2021 membership card of the association: 'Campi Aperti is community, trust, autonomy, politics'. Social movements are constituent of social power by inventing and creating alternative institutions whose aims consist in transforming the governance of social life. Food communities do the same through the co-emergence of politics and matter which gives birth to alternative spaces of existence and alternative material circulations. The autonomous infrastructures of food communities make agroecology durable, generate 'generous' encounters and dislocate 'politics' within everyday practices. The food community of Campi Aperti reinvents canals of cooperation between the countryside and the city, creating autonomous infrastructures capable of rearticulating the food web within and beyond the farm. Ecological reparation brings with itself novel forms of social and more than social cooperation, in which the making of a food movement is not separable from the invention of mundane alternative of eco-social (re)production. By politicizing every aspect connected with the supply chain and inventing pragmatic alternatives from seeds to kitchen, Campi Aperti offers a significant example of eco-social reparation in which there is no split between the material and the political: soil ecological reparation and struggles against agribusiness reinforce each other, daily access to genuine food and the empowerment of food sovereignty are walking together. Eco-material regeneration *and* the construction of autonomous infrastructures: food communities achieve their political autonomy, their capacity to act and repair economies, ecologies and social relations, through the making of alternative infrastructures.

Thinking ecological reparation through these infrastructures allows us to acknowledge the central role that every day, more-than-human, practices of material reparation play in crafting viable alternatives in food systems. The activists, peasants, and co-producers of Campi Aperti are dealing with a form of transformative politics in which commoning is primarily about experimenting with pragmatic alternatives in the ways in which food and ecological care relate to each other. To make it possible, they are dealing with a plurality of practices and processes of political organizing that are redefining the social, cultural, material, political, geographical and technological dimensions of food networks. Material regeneration animates an ecology of reparation in which agroecology, transformative economies, self-organization, infrastructures, participatory guarantee systems and knowledge production are deeply intertwined. An ecology of proximity that is rearticulating the traffic of relations between urban and rural life.

The eco-commons of Mondeggi

Eco-material regeneration and the construction of political autonomy meet again in another key experiment of Genuino Clandestino: the eco-commons of Mondeggi. The estate is located in the municipality of Bagno a Ripoli, about 12 kilometres from the centre of Florence. It covers about 200 hectares and includes about ten hectares of vineyards, 10,000 olive trees, about ten hectares of arable land, a fourteenth-century Medicean villa and eight farmhouses. It is a complex of medieval origin, passed over the years into the hands of various families of the Florentine nobility, who used it as an agricultural estate and hunting ground. In the 1960s of the last century, the villa and the entire estate passed into the hands of the Province of Florence, becoming a public property. With this transfer of ownership, the Mondeggi estate became the headquarters of the Mondeggi Lappeggi S.r.l. agricultural company, of which the Province of Florence (a public institution) was the unique shareholder with 100 per cent ownership of the capital, and it became a conventional, intensive, monocultural and mechanized farm: olive trees, vineyards and other crops such as wheat and sunflower have been planted in accordance with this idea of farming. In the last years of its existence, the company accumulated a debt of over one million euros, leading to the bankruptcy of the company and its subsequent liquidation in 2009. Since then, the entire complex has been progressively abandoned, leading to the deterioration of buildings and the degradation of crops, including vineyards and olive groves. Since 2011, the Province of Florence has been trying to sell the estate through auctions, none of which have been successful to date.

In June 2014 the estate was occupied by a group of activists and became the largest land occupation in Italy. The process of collective appropriation, custody and civic use of Mondeggi's land, called *Mondeggi Bene Comune, Fattoria Senza Padroni* (Mondeggi Commons, Farm without Bosses), is the most significant experiment of the *Terra Bene Comune* (Land as commons) campaign promoted in 2013 by the Genuino Clandestino network against the alienation of publicly owned land. The aim of the campaign, animated by a heterogeneous group of activists, producers, farmers and students, was not only to oppose the sale of many publicly owned lands but also to claim 'a management of public land by local communities, according to forms decided in autonomy at territorial level and the distribution of public land for rurality projects' (Genuino Clandestino, 2013).

This occupation, which combines self-organized management of commons and agroecology, started locally in 2013 thanks to the encounter among the campaign *Terra Bene Comune* and various groups including the student collective of the Faculty of Agrarian Studies of University of Florence, the network of GAS (Solidarity Purchase Groups) of the Florentine area, activists from social centres and inhabitants of the territory of Bagno a Ripoli. In January 2014, the Charter of Principles and Intents was approved, in which the values, ideas and key points of what is collectively imagined and desired for Mondeggi are clearly expressed:

> Promoting the management of Mondeggi as commons and preventing its privatisation; creating experimental pathways for the custody of the

> commons by communities of people; maintaining a strong relationship with the territorial community; generating widespread wealth (social, environmental, relational) by building a local economy; promoting peasant agriculture, food self-determination and agroecology in full respect of the environment, living beings and human dignity. (Mondeggi Bene Comune, 2014)

In the last days of June 2014, a three-day event was organized in the Mondeggi Farm, supported by an appeal signed by hundreds of citizens, associations, movements, academics and jurists in support of the civic use of Mondeggi. At the end of the event, the Committee for Mondeggi Commons declared the birth of a permanent peasant residency at the Mondeggi estate: within the collective appropriation of land and farmhouses starts the story of Mondeggi as 'emergent commons'.

At present, the peasant garrison is made up by more than 20 people living and working in Mondeggi. The farm produces, following the principles of agroecology, bread, honey, beer, wine, oil, herbal medicine products (herbal teas, natural cosmetics, tinctures, etc.), vegetables and saffron. These products are sold on the farm, at peasants' markets in Florence and in the self-organized circuits of Solidarity Purchase Groups. A significant part of the land is available to anyone who wants to join the MO.T.A. (Mondeggi self-organized terrains) and MO.V.A. (Mondeggi self-organized vineyards) projects: the olive grove, the vineyard and the terrains are divided into parcels (in the case of the olive grove composed of about 30–35 olive trees, calculated to guarantee the self-sufficiency of olive oil for a nucleus of about 3–4 people, and the same criteria lead the distribution of terrains and the vineyard). These projects of civic use and food self-production involve more than 300 people ensuring the care of the olive trees, the vineyard and giving birth to about 50 vegetable gardens. Moreover, the community of Mondeggi organizes with experts, practitioners and natural scientists a peasant school – *La Scuola Contadina*: courses and workshops on agroecology involving hundreds of people well beyond the regional territory. Participation is free of charge and includes modules on brewing, beekeeping, bread-making, natural horticulture, olive growing, viticulture and herbal medicine.

The commons of Mondeggi are inseparable from the making of the non-proprietary and non-enclosed territory. But the eco-commons of Mondeggi are more than that. Here commons refer not only to the invention of practices of collective management, even if access to land and its civic use are key elements in the politics of Genuino Clandestino. From its inception, the people involved in the project started to call Mondeggi a territory of 'emergent commons', which means that the commons emerge inside an activity of commoning which involves a daily cohabitation with other people and with animals, plants and the soil. This is not only the social commons but the worldly commons, ecological commons that emerge out of the process of commoning matter. The eco-commons of Mondeggi are inseparable from agroecology, from material reparation, from a reinvention of rural forms of living, from the desire of cultivating an everyday relation with the land. The 'emergent commons' comes within the experimentation of territories that are at the same time existential, relational and material, they come with the experimentation of daily practices of regeneration and socio-ecological reparation. As Derek Wall (2014: 127) wrote, 'different forms of commonality give rise to different sustainable environments'. But the multiple temporalities of care and ecological reparation are complex, and we can think to what extent a monoculture of vine and an olive grove made by ten thousand trees can suit an ecological understanding of agriculture in which the webs of biodiversity and commensality make a farm sustainable. In ecological reparation we rarely start from scratch: the commoners of Mondeggi are struggling to inherit a territory marked by 'industrial' agriculture, to take part in a complex ecological transition process by planting a forest garden, introducing animals to the farm, changing olive trees pruning techniques, fostering the increase of organic matter in soils, recycling rainwater, and so on.

As in the case of Mondeggi, ecological reparation asks as much as possible to distance ourselves from two trends that in numerous ways criss-cross debates on the relationship between ecology and society: the prophecies of catastrophe (Beuret, 2016) and the teleologies of salvation (Castillo, 2019). Both these approaches to ecological conflict suggest that we will eventually have to abandon this world to start over from zero, be it after the catastrophe or to initiate a new world. Rather than denying the world, the experiments of ecological reparation represent a way for us to learn to inherit what has been damaged without denying it (Tsing, 2015). The task of ecological reparation practices is to stay with the problem of damaged worlds by making politics alive. To make politics alive involves not only moving beyond cynicism and despair, but it involves also populating our politics with nonhuman actors that were traditionally excluded by the eternal deferrals of a modernist idea of a rational order to come. Ecological reparation takes seriously the grounded problematics that demand us to think and act now, the actual occasions that we face, the present continuous of mending.

New peasantries, agroecology, food communities, eco-commons. When movements encounter matter as a strategic field of action for experimenting generative practices of justice, a new idea of autonomy emerges. Autonomy historically refers to the idea that social mobilizations and social conflicts drive social transformation instead of being a mere response to social and economic power. The key strategy of Genuino Clandestino consists in something less and something more than contesting and addressing existent political institutions. Ecological reparation requires material interconnectedness, practical organizing, everyday coexistence. And these are always more than human, more-than-social. They entail interactions, ways of knowing, forms of practice that involve the material world, plants and the soil, infrastructures and energies, other groups of humans and their surroundings, and other species and machines. Ecological reparation politics rework and expand autonomy to engage with questions of justice in more than human worlds by highlighting the relevance of creating alternative everyday politics of matter, transformative economies, rural-urban interdependencies, ecological commons. Here autonomy is a call for direct action, for material recombination, for practical reparative justice.

Notes

1. For different approaches and cases see García López et al (2019), Lanka, Khadaroo and Böhm (2017), Rosset et al (2019), Altieri and Toledo (2011).
2. See, among others, Holmgren (2002) and Mollison and Holmgren (1978) for permaculture, Paull (2006) and Pfeiffer (2004) for bio-dynamic agriculture, Masters (2019) and Chase (2014) for regenerative agriculture.
3. See Bertoni (2013), Krzywoszynska (2020), Puig de la Bellacasa (2014).
4. In Genuino Clandestino the notion of agroecology takes in account the proliferation of these different farming techniques.
5. Further details on this campaign are available here: www.vag61.info/vag61/articles/art_5435.html and www.campiaperti.org/

References

Altieri, M.A. (2018) *Agroecology. The Science of Sustainable Agriculture*, Boca Raton, FL: CRC Press.

Altieri, M.A. and Toledo, V.M. (2011) 'The agroecological revolution in Latin America: rescuing nature, ensuring food sovereignty and empowering peasants', *Journal of Peasant Studies*, 38: 587–612.

Bertell, L. (2016) *Lavoro Ecoautonomo*, Milano: Elèuthera.

Bertoni, F. (2013) 'Soil and worm: On eating as relating', *Science as Culture* 22: 61–85.

Beuret, N. (2016) 'Containing climate change: The new governmental strategies of catastrophic environments', *Environment and Planning E: Nature and Space*, 4(3): 818–37.

Castillo, D.P. (2019) *An Ecological Theology of Liberation. Salvation and Political Ecology*, Maryknoll, NY: Orbis Books.

Chase, L. (2014) *Food, Farms and Community: Exploring Food Systems*, New Hampshire: University of New Hampshire Press.

Garcìa Lopez, V., Giraldo, O.F., Morales, H., Rosset, P. and Duarte, J.M. (2019) 'Seed sovereignty and agroecological scaling: Two cases of seed recovery, conservation, and defense in Colombia', *Agroecology and Sustainable Food Systems* 43: 827–47.

Genuino Clandestino. (2013) *Terra Bene Comune*. Available from: https://genuinoclandestino.noblogs.org/files/2013/09/MANIFESTO-CAMPAGNA-NAZIONALE-TERRA-BENE-COMUNE.pdf [Last accessed 7 May 2020].

Ghelfi, A. (2015) *Worlding Politics: Justice, Commons and Technoscience*. PhD dissertation, University of Leicester.

Ghelfi, A. and Papadopoulos, D. (forthcoming) 'More-than-social movements: Politics of matter, autonomy, alterontologies', in L. Pellizzoni, E. Leonardi and V. Asara (eds) *Handbook of Critical Environmental Politics*, Cheltenham: Edward Elgar.

Holmgren, D. (2002) *Permaculture: Principles and Pathways Beyond Sustainability*, Victoria: Holmgren Design Services.

Krzywoszynska, A. (2020) 'Nonhuman labor and the making of resources: Making soils a resource through microbial labor', *Environmental Humanities* 12: 227–49.

Lanka, S.V., Khadaroo, I. and Böhm, S. (2017) 'Agroecology accounting: Biodiversity and sustainable livelihoods from the margins', *Accounting, Auditing & Accountability Journal* 30: 1592–613.

Latour, B. (2018) *Down to Earth: Politics in the New Climatic Regime*, Cambridge: Polity Press.

Masters, N. (2019) *For the Love of Soil. Strategies to Regenerate Our Food Production System*, New Zealand: Printable Reality.

Mollison, B. and Holmgren, D. (1978) *Permaculture One: A Perennial Agriculture for Human Settlements*, Sisters Creek: Tagari Publications.

Mondeggi Bene Comune. (2014) *Carta Principi e Intenti*. Available from: https://tbcfirenzemondeggi.noblogs.org/carta-dei-principi-e-degli-intenti/ [Last accessed 4 May 2021].

Papadopoulos, D. (2018) *Experimental Practice. Technoscience, Alterontologies, and More Than Social Movements*, Durham, NC: Duke University Press.

Paull, J. (2006) 'The farm as organism: The foundational idea of organic agriculture', *Journal of Bio-Dynamics Tasmania*, 80: 14–18.

Pfeiffer, E. (2004) *Soil Fertility: Renewal and Preservation: Bio-Dynamic Farming and Gardening*, East Grinstead: Lanthorn Press.

Puig de la Bellacasa, M. (2014) 'Encountering bioinfrastructure: Ecological struggles and the sciences of soil', *Social Epistemology*, 28: 26–40.

Puig de la Bellacasa, M. (2017) *Matters of Care. Speculative Ethics in More Than Human Worlds*, Minneapolis, MN: University of Minnesota Press.

Rosset, P.A.A., Val, V., Barbosa, L.P. and Mccune, N. (2019) 'Agroecology and La Via Campesina II: Peasant agroecology schools and the formation of a sociohistorical and political subject', *Agroecology and Sustainable Food Systems* 43: 895–914.

Rosset, P.A.A. and Altieri, M.A. (2017) *Agroecology: Science and Politics*, Black Point, Nova Scotia: Femwood Publishing.

Shiva, V. (2008) *Soil, Not Oil: Climate Change, Peak Oil and Food Insecurity*, London: Zed Books.
Tsing, A.L. (2015) *The Mushroom at the End of the World: On the Possibility of Life in Capitalist Ruins*, Princeton, NJ: Princeton University Press.
Wall, D. (2014) *The Commons in History: Culture, Conflict and Ecology*, Cambridge, MA: MIT.

14

'Obedecer a la Vida': Environmental Citizenship Otherwise?

Juan Camilo Cajigas

Obedecer a la Vida[1] (Obey Life, Abide by Life) was the expression Roberto Saenz used when referring to what made him feel and think. This is the way he expressed his commitments to the Tenasuca Lagoon (located in the Tequendama Andean Cloud Forest of Colombia); to his colleague Arnovis, who worked every day in agroecology;[2] and to his private nature reserve and small organic market.[3] In addition, he added, "I do not practice civil resistance. I obey life." Roberto's statement struck me deeply and led me to think about the different ways of considering emergent practices for repairing a certain place. These practices respond to what the situation demands in the face of changing ecological and social conflicts. However, in my interpretation, Roberto's practice doesn't follow the same framework as the nation-state's liberal environmental policies – at least, in Colombia. What kind of environmental-oriented action are we talking about? How does the statement 'obey life' relate to concrete practices in the field of agriculture? And then, how does this practice differ from other ways of practicing a policy that addresses nature conservation? These are some questions I will unravel in this chapter. My aim is to elucidate whether liberal citizenship can promote and support ecological reparation. More broadly, I wonder if the liberal framework of political thought is the only way in which the current environmental crisis might be addressed. Colombian nature conservation policy has centred on the production of discourses, knowledge and subjectivities that aim at, on the one hand, overcoming the negative effects of development, and on the other, securing strategic ecosystems in terms of biodiversity. However relevant these efforts are, biodiverse ecosystems are still valued under a utilitarian logic based on capitalist profit-making. In this sense, sustainable development discourses are articulated within liberal societies' economic structure.

To this end, I will narrate the story of agroecological practitioners in the Tequendama region of Colombia, aiming to elucidate how their practice explains what 'obey life' means. If the liberal framework is based on the notion of rights and obligations (and its cognate notions, namely, individuality, universal thinking, self-interest and antagonism among political actors), 'obeying life' seems to follow another framework. Such a framework features obligation as a relation that is established by a place's dynamic unfolding, foregrounding practices of ecological reparation that both diffract liberal environmental citizenship and open up the possibility for an ethic of abundance.

Roberto and Arnovis' life trajectory

Roberto Saenz belongs to the Saenz-Santa María family, an affluent family who once owned the majority of the south side of the Cerro Manjui, in the Tequendama valley of the central Andes of Colombia.[4] At the beginning of the twentieth century, they owned more than 700 hectares. There was a long military tradition in his family, and his ancestors got this significant amount of land as a war reward. Roberto is the 'black sheep' of the family, although a loved one. He studied computer sciences, and he worked for several years in the Colombian health system, helping to develop software to control tropical diseases. Fifteen years ago, he bought his aunt's farm, located near the Pedro Palo Lagoon (well-known for the 'rolos' – people who live in Bogotá), and decided to create a 'natural reserve'.

He was quite emphatic when talking about the difference between a reserve and a farm: 'reserves' were established under the principles of conservation and sustainable production; 'farms' focus on productivity no matter the cost. For instance, Roberto's cousins own several farms nearby. They relentlessly look for 'productive' activities – especially cattle grazing – in order to increase their income. As a result, farms often lead to environmental depletion. Being aware of this regional situation, Roberto decided to create a reserve, and named it 'Tenasuca' (Figure 14.1). He has encouraged his family members to create reserves around the lagoon to preserve it. However, for his family members, reserves are just a waste of time and money because those are 'unproductive lands', as he confessed to me.

The very name of the nature reserve and the lagoon 'tenasuca' acknowledge pre-colonial Indigenous habitation. Muisca people inhabited this region of Colombia between 800 A.D. and 1600 A.D. The word 'tenasuca' in muisca language means 'the eyes that witnessed the first sunrise' or 'behind the deepness'. For Roberto and Arnovis, his main partner at the nature reserve, this place-lagoon is alive, as they said: 'it is a very special *ser vivo* [living being]'.

Figure 14.1: Tenasuca Lagoon, 2015

This lagoon is frequently visited by contemporary Colombian Indigenous people (such as Kogui and Guambianos) for it belongs to a network of 'black water lakes' that originated before the origin of this time: *antes de la Creación* (before the beginning of this creation). Indeed, for them, there is a 'sacred' geography that overlaps official territorialities. Particularly, this place-lagoon marks a cosmological beginning, and as a ritual device enhances people's life rebirth. Tenasuca Lagoon has a special character, an inherent charm that they think must be honoured and preserved. For Roberto, this is not a matter of exotic belief, rather it is something he practices daily as a way of reconnecting to place. Given that Roberto owns the land nearby the lagoon, the aforementioned group of Indigenous people invited him to become the protector of this lagoon, offering *pagamentos* (ceremony) with tobacco for the spirits of the forest and the lagoon. Roberto took this proposal as a mandate that obligated him to this place.

In 1998 the regional environmental state institution Corporación Regional de Cundinamarca (CAR) intended to develop a *Plan de Ordenamiento Territorial* (Rural Territorial Management Plan), focused on tourism. As the area of the lagoon belongs to the municipality, the major at that time designated it as a tourist destination, requiring the building of restaurants, shopping areas, sports such as kayak and water-bicycles. In short, it was to be a whole touristic complex that would capture thousands of visitors and *capitalinos* (people from the capital city of Bogotá). The lagoon was well known by urbanites as a camping spot, but also as a space for relaxing and 'being in touch with nature'. For Roberto, it was unbelievable that this would be the conservation objective of the supposedly regional environmental institution. At that point Roberto realized that he was more a conservationist than the environmental institution in charge. In fact, developing Tenasuca Lagoon in this way would destroy it ecologically. In the past, Roberto told me, there were many controversial actions linked to tourism such as illegal fishing, fires produced by irresponsible camping, even people drowning while swimming drunk. Eventually this touristic plan was aborted because of successive failures in administrative procedures, combined with Roberto's relentless efforts to intervene in the messy local politics. The touristic project failure provided an incredible opportunity to continue Roberto's conservation efforts.

Currently, Roberto organized his reserve into three sections: a forest area, an organic crop field and a nursery of native species. This reserve was originally created as a *Reserva de la Sociedad Civil* (Civil Society Natural Reserve), a Colombian government environmental policy which Roberto thinks insufficiently protects the forest from the threats of mining or irresponsible tourism. The Environmental Ministry created such policy to designate complementary natural preserved areas within the *Sistema Nacional de Areas Protegidas – SINA* (National System of Protected Areas). This institutional state designation is nominal, that is, it gives official recognition to conservation spaces as a private initiative, yet there is neither financial nor technical support by the government. Precisely, this is the reason why Roberto decided to affiliate with a grass roots initiative: *Red de Reservas de la Sociedad Civil Resnatur* (Civil Society Reserves Network) a communally-oriented independent organization which gathers citizens that voluntarily aim to preserve fragments of forest on their property, just because, in their words 'they feel love and responsibility for a piece of this Earth'.[5] From the mid-1990s, the Civil Society Natural Reserve Network linked informally hundreds of individual and communally-oriented conservation initiatives all over the Colombian territory. This network created a *mancha social* (social patch) of various conservation strategies, which has been a socio-ecological experiment embodied by peasant and middle-upper-class individuals that fought through legal and institutional lobbying for a different legal understanding of the relationship between private property and conservation.

Arnovis technically manages the organic crop field. Arnovis is Roberto's agriculture teacher – truly a very patient one. When I first meet Arnovis, he was cooking vegetables and chicken. He was taking a break after having worked all morning in the field. He explained to me what I was supposed to do as a volunteer: picking up cow faeces for hours, and then help him to prepare *abono* (organic compost).

Arnovis is from Risaralda, a region well known for intense agro-industrial coffee production. On his family farm – a considerably big peasant natural

Figure 14.2: Tenasuca Natural Reserve logo, 2016

reserve called *Miraflores* – they used to grow black beans as an agro-industrial crop. At that time, Arnovis and his family sold their products to the juice factory 'Tutifruti', which was owned by Postobon (a huge national beverage industry). After having some business problems with them the family decided to quit, because, as Arnovis said, *el pez grande se come al chiquito* (the big fish eats the small one), meaning, they, as fruit producers, received much less than what was legally and typically paid for that crop. As a consequence, they quit this business and followed their neighbour's advice, and turned to what in the region is called *cultivos limpios* (clean crops), crops that do not have chemical fertilizers.

They started a process to convert to agroecology by participating in several workshops with well-known Colombian agroecology leaders, Ricardo Ramirez and Mario Mejia. They later visited them for a short period in the Miraflores natural reserve. At first, the production of their farm failed, and they had to invest money to maintain their crops. Later on, they diversified their organic field with new products such as lettuce, potato and vegetables, and became independent. They sold their products at a local farmer's market managed by small peasants in Pereira City where market prices were low and affordable for the low and middle classes. Market prices are another reason why Arnovis is so sceptical about what he terms 'hippie-chic' organic markets in Bogota the capital city of Colombia. There, sellers duplicate the crops' original cost. In Bogotá, he asserts, "this is another candy for rich people". In Pereira, instead, he witnessed how clean food (organic food) could be sold at affordable prices.

Arnovis decided to move out from the region called Risaralda with his son, and learned about Roberto's project. They met and agreed to work

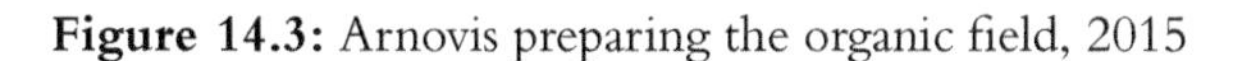

Figure 14.3: Arnovis preparing the organic field, 2015

together regardless of the difficult economic situation of Roberto as an organic entrepreneur (by the time of my fieldwork Roberto's debt was around $10,000, USD). The organic field was relatively small, organized in the traditional structure of *eras* (fields). They grew different varieties of lettuce, potato, tamarillo, squash and tomato fruit.

Arnovis is quite emphatic when he explains how to prepare a field. First, you should 'clean' the field of weeds, which extracts soil force; second, you should open small holes, as shown in Figure 14.3, about 3 or 4 inches each; third, the compost, made from a mix of cow dung and chicken bones, should be placed in the holes; fourth, each hole should be watered until the soil is wet enough; finally, the seedlings of different vegetables are planted. Roberto was not just an observer; he worked hand in hand with Arnovis in the whole process. At the end of the week, on Thursday or Friday, they harvest the vegetables that are ripe.

Roberto calls himself a *neo-campesino* (neo-peasant). Fifteen years ago, he was employed in the government as a computer scientist and not that much into 'getting dirty'. He never explained to me why he changed; perhaps because of being forewarned against social accusations of his family and friends. Certainly, he has a deep desire for protecting and living with this forest-lagoon expressing how he feels and thinks with the statement: 'I am an environmentalist 24 hours per day.' Roberto unlearns certain urban habits and learns new ones. I affirm he *unbecomes* by withdrawing himself from his mainstream society environment, and learns to be affected by the Lagoon-Arnovis partnership – the cloudy sky and foggy forest, that is, this rural assemblage.

Roberto is not the only one. An article published by the national newspaper *El Tiempo* on the 'neo-peasant phenomenon'[6] reported people are moving from the city to the countryside looking for a rural or natural way of life. Although there are no official statistics, this is undoubtedly an emergent diasporic phenomenon within nation-states, and transnational if one considers the international volunteer network of eco-villages and organic farms (Nates, 2008; Mendez, 2012; Litfin, 2014). Some integrate an economy based on urban and rural activities, whereas others dedicate their efforts to exclusively rural ones. The aforementioned article highlights how escaping city life to pursue a more ecologically oriented way of life is without a doubt one of the reasons behind the neo-peasant phenomenon. Among the ecologically oriented activities, neo-peasants practice agroecology by banking seeds, composting, microorganism conservation and restoration.

Reconfiguring environmental citizenship

On one of our trips Roberto and I ended up having a very interesting conversation with one of the agriculture departments' officials working for the municipality, who is in charge of defining the spatially-explicit policy for protecting the environment in the region, *Plan de Ordenamiento Territorial* (Environmental Zoning). In this conversation, Roberto explained to her the activities he pursued on his private nature reserve, mainly focusing on organic food production and growing native trees. I recognized a tone of urgency but also his veiled interest. Certainly, he knew how to move skilfully within the local bureaucratic machine. As I mentioned earlier, he asserted with conviction: 'I am an environmentalist 24 hours per day.' This assertion was interestingly further elaborated when he claimed that environmental civil society was defined by the statement: '*no hacemos resistencia civil, sino que obedecemos a la vida*' (We do not practice civil resistance, rather we obey life).

Obeying life seems to complicate how environmental claims have to be resolved within the liberal environmental citizenship framework. The latter introduces physical environments as an object in relation to liberal national citizens who have rights and obligations (Bell, 2004; Smith and Pangsapa, 2008; Dobson, 2015).[7] For instance, in the Colombian constitution the environment appears as the physical conditions that guarantee citizens' good health. In this sense, Colombian citizens have the right to clean water or unpolluted air as long as their health as human beings is concerned. At the same time, citizens have obligations to their physical environment as an expression of their personal preference and, more broadly, the moral duty of environmental protection. In this sense, Colombian liberal citizens have rights and obligations with respect to the environment. So, it is by extending liberal rights that environmental protection is secured. However, Roberto and Arnovis' life trajectory is centred around a very concrete form of obligation to the Tenasuca Lagoon. Their everyday struggle for an alternative agricultural practice (which is, at the same time, a conservation strategy) has eventually crystallized into a form of political engagement, which does not claim the extension of their rights as citizens, but, more deeply, 'life obedience'. Indeed, their practice responds to the call of this place-lagoon, as there is a feeling of urgency about what is disappearing due to ongoing regional ecological depletion. In short, there is a call of urgency that demands an act of obedience. Yet still, in theory they belong to a local liberal political community. In this context, one might ask various questions: what kind of life-politics are emerging in these ecological reparation efforts? Do these life-politics – rooted in a sense of obligation – go beyond the notion of liberal citizenship? How does the modern political demos change? What drives the emergence of new peasantries?

In Roberto's case, I suggest, the mandate to obey life is not performed as an application of an abstract moral rule nor as a locationless liberal-citizen obligation. Quite the opposite, it is lived as a specific obligation to the Tenauasa Lagoon. It is lived as an effort to take risks, to begin an adventure only partially knowing what the next step will be, which is a normal situation, for a small entrepreneur in the emergent organic farmers' market in Bogotá (with all the social and economic contradictions this could imply). His obligation to this lagoon leads Roberto to participate in the regional strike against the state's huge electrical infrastructure project. It is also lived as an effort to be open to sensing things differently, to open his body to new connections, to re-learn and un-become. Roberto learns from the Kogui and Guambiano Indigenous people how to properly smoke tobacco and ask permission from the spirits of the forest as part of a practice of what he terms *manejo integral del territorio* (integral land-based governance).

However, Roberto maintains his project through the legal guarantee offered by the nation-state's structure of 'private property'. This legal structure is also shared by all participants in the Civil Society Reserve Network, whether neo-peasants or small farmers. One might interpret this as an inherent contradiction in late capitalist formations, for such a structure could be used for both land over-exploitation (as in extractivist economies) and ecological reparation. Private property generates patterns of both social exclusion and inclusion. Normally it excludes and negates access to common resources. Paradoxically, private property, in this case, is the mechanism that allows the resurgence of forest life; Roberto's private nature reserve in fact preserves patches of the highland cloud forest. The formal and abstract structure of liberal ownership manifests in the form of substantial contents – in this case, conservation interests. The paradox emerges when a socio-ecological practice folds liberal ownership's universality. Indeed, this is the paradox of the concrete universal.

Moreover, thinking with Arnovis' practice and specific sense of obligation one may understand what a full engagement with this place would be. Arnovis' silent presence never leaves this lagoon and its surrounding mountains. His meticulous, daily caretaking of the fields maintains an unfolding relationship with soil, plants and foggy weather conditions. Arnovis cultivates an agricultural practice rooted in peasant knowledge and modern agroecology, which, for instance, allows time for the soil to recuperate and for paying attention to plants' sensitivities. Arnovis avoids employing intensive chemical pesticides because the soil could be *quemado* (burned), which means its vitality might be diminished. This kind of pesticide kills all sorts of micro-organism that maintain nutrient cycles. Soil vitality and abundance, for him, depends on this form of attention to the diversity of micro-organismic life. Arnovis pays attention as well to plant behaviour as they respond to people's intentions and emotions. He doesn't allow visitors to enter the fields, because several times crops were *burned*, meaning some of them died. People's mood affects plant development just as plants affect people's sensorial dispositions. '*Quemar*' (to burn) speaks about diminishing life forms by both contaminating and ignoring plant sensitivity. Arnovis' agricultural practice aims to compensate soil and plant life by avoiding harmful techniques of production that *queman* (burn) soils.[8] Yet, to practice such compensation, attention must be paid to the changing behaviour of plants and soil capacities; in short, to the qualitative unfolding dynamic of place. Arnovis' practice couldn't be done elsewhere, as he is not a locationless embodied subject but the very determination of this place-lagoon.

Roberto and Arnovis' reparative agricultural practice complicates assumed definitions of liberal environmental citizenship in manifold ways. Practices of ecological reparation could not be performed by locationless and disembodied subjects, as mainstream liberal citizenship would endorse, for repairing and compensating the non-human world is not done through an extension of moral obligation to an external physical environment. From the perspective of the liberal subject, to be concerned about the environment is a matter of personal choice, free will and, moreover, duty to obey environmental laws. Yet there is no particular attachment or connection to the physical environment (Bell, 2004). As shown by Roberto and Arnovis' agricultural practices, ecological repair follows the unfolding dynamic of place, so the latter could not be depicted as the motionless background of human drama; rather, place is the very active container of relations, which, at the same time, determines Roberto and Arnovis' sense of obligation.

More than a moral duty, obligation in this case points to what Puig de la Bellacasa refers as 'the ongoing reentering into co-transformation that further obliges the interdependent web' (Puig de la Bellacasa, 2017: 156). What needs to be done in order to maintain life is at stake, so we are obliged by the very ecology which we belong to. It points to a dense sense of reciprocity and care that implies our very existential-ecological conditions and not just our rational deliberation regarding an external environment. There is no external physical environment that needs to be managed or repaired; rather, what emerges is a form of reciprocal action without exteriority. My interpretation stresses the active emplacement of such an interdependent web, where obligation occurs as a mandate to follow place's determinations – namely, to obey life.

What reshaped their particular form of non-liberal environmental citizenship is their obligation to this place, which here I understood as the qualitative unfolding dynamic of this place-lagoon.[9] In this regard, an obligation is neither an abstract moral rule to be internalized nor a displaced material affective relation between two different entities. Instead, their obligation points to a form of co-constitution where people and place unfold as modes of the same intimate relationship. Such a relationship is the determination of Arnovis and Roberto's active emplacement. Theoretically speaking, bodies of different kinds relate to one another in virtue of their emplacement 'with-within-into' place (my fault! – aiming at a sort of speculative preposition). Their bodies are now seen as accumulations of energy related not in the place (the latter understood following the notion of independent Newtonian space), but as established by the place (the field which they belong to). In short, Arnovis and Roberto's bodies are the qualitative unfolding of this lagoon, and, at the same time, the lagoon is the result of their terraforming practices through agroecology.

The mandate of obeying life points to an obligation that sustains, for instance, Roberto and Arnovis' way of life; this obligation is the crystallization of the Tenasuca Lagoon's differential unfolding. Differential in this case points to minute and vibratory changes in a place's ecological rhythms. To obey life, then, is a concrete form of life-politics that emanates from the qualitative unfolding of place: micro-organism life, weather conditions, sounds of the lagoon, Arnovis' ability to sense plants, and Roberto and Arnovis' companionship. All these elements compose their daily practice, diffracting their state-based environmental citizenship. Active emplacement (place's dynamic unfolding) holds their obligation. As a consequence, they enact a collective *demos* in which social, organic and inorganic materials participate. This is a particular sense of obligation, for before knots of relation, there is active emplacement. In other words, relations are determined by the very place in which the relationships hold (Nishida, 2012).

That being said, one cannot refer to this kind of obligation employing the liberal framework of environmental citizenship. Arnovis and Roberto's obligation is based on active emplacement more than on individual responsibility and duty.

No end

Let me return to Arnovis and Roberto's story, as they were also drastically affected by drought conditions, alongside everyone who lives in the Tequendama region. At the end of one of our trips with Roberto, in which he bought organic fertilizer for his reserve, he expressed his distrust of public institutions, but also of social organizations.[10] In the latter, he was already worn out because he spent a lot of his time in discussions and efforts, often unsuccessfully. Nevertheless, with everything good and bad, these were the means to maintain his reserve project. Neither in the state, nor the social organizations, and at the same time, with the state and social organizations. Nothing is enough, and yet everything is. Precisely, the mandate to obey life contains multiple contradictions: scepticism regarding public institutions; doubts about the organizational capacities of civil society; confidence in the obligation that Roberto and Arnovis have with the lagoon; and hope in life's abundance. *With the state, against the state, inside the state, outside the state; with civil society, against civil society, inside civil society and outside civil society*. Fundamentally, obeying life points to that direction: a radical scepticism of the potential to resistance of civil society (environmental citizenship), given that both civil society and state-based programmes belong to a set of governmental technologies of control. In this sense, obligation is not pure: it dwells within the messy institutional and organizational politic of liberal societies in the Global South.

Obligation as a practice of ecological reparation performs as a device for re-enacting life's abundance. As a matter of care, there is a tighter relation between people and place, which articulates Roberto and Arnovis' obligation-life politics 'with-within-into' the lagoon. Such obligation makes them avoid living and working with an exhausted soil and a deforested forest. Life abundance is re-enacted as they follow the expansive movement of place unfolding dynamically (ecosystem regeneration). Restoring abundance points to a field of self-organizing life processes, within a circle of regeneration and re-birth. As Bautista points out (Bautista, 2017: 214), such movement is expansive as opposed to the concentric movement of capitalist accumulation (as practiced for instance by industrial fern agriculture in the Tequendama region). In this sense, there is no supplementary dimension that drains off the flow of matter-energy (soil nutrients, water sources, human labour) to a central point. Abundance is expansive and immanent, as opposed to concentric and transcendent. To repair by obeying life means to reproduce life itself, following an expansive movement of flourishing and multispecies wellbeing.

Notes

[1] The Spanish expression 'Obedecer a la vida' could be translated into English as: obey life, abide by life, or heed life. Despite the authoritarian connotations this expression has in English, it is not the case in Spanish. Rather, it points to an obediential form of power, which crystallizes a collective self across the human and non-human realm. The Zapatista movement in México popularized this form of power with the statement 'Mandar Obedeciendo' (To command by obeying).

[2] Agroecology is practiced in the Tequendama region as an alternative technique to the industrial monocultural model of agriculture; the latter is commonly employed for fern and ornamental flower crops production. Although this alternative technique is not massively used, it guarantees the conservation of fragments of cloud forest, as this form of agriculture is concerned with biodiversity enrichment and food sovereignty.

[3] This article presents some of the results of my dissertation research. With fieldwork in the Tequendama Cerro Manjui Cloud Forest in Colombia (carried out between 2016 and 2018), this research interrogates the practice of agroecology and conservation as performed by environmental citizens in private nature reserves and small peasant farms to address the local effects of deforestation, land degradation and global climate change. These groups are made up of small coffee farmers, middle- and upper-class internal migrants who moved from the city to the countryside (neo-peasants), and who individually and collectively constitute nature reserves. Their heterogeneous and experimental practices transform the cloud forest deforestation and land degradation patterns. I argue that a specific sense of obligation to a place articulates this form of environmental citizenship.

[4] I define a *life trajectory* as a mode of attention that ties together biography and place. A life trajectory is not linear, it is more like wayfinding. As a mode of attention, it points to performative relations with space-time. The perceptual capacities of humans develop as we walk (Ingold, 2011). Instead of thinking through the teleological implications of the word 'project' (objectives, methodologies and so on), I refer to trajectory highlighting; first of all, the practices which rise as the expression of connections in a given assemblage, and second, the nonlinearity of any lifeway's ongoingness. Hence, I take the statement of obeying life as an inspiration for thinking about life trajectory ongoingness, which may or may not disappear. On the other hand, a life trajectory breakthrough threshold, that is, *painful existential passages* largely conceptualized by Black and Indigenous abyssal thinking (Yountae, 2017). Abyssal thinking focuses on the memory of painful existential passages where human and non-human life is at the edge of extinction.

[5] See promotional video of RESNATUR: www.youtube.com/watch?v=TSq6P0iLYgc (retrieved 13 January 2019). Catalina Restrepo – a leading figure in this conservation initiative – stressed: 'we obtained the recognition that private property's social function might be framed as conservation … a piece of forest that a given family owned, inherited from their ancestors 100 years ago, had a social and ecological value for the nation as a whole. Conservation, from that moment on, no longer belonged uniquely to the nation-state' (Catalina Restrepo personal interview).

[6] Article from the newspaper *El tiempo*, 11 October 2014: 'Neocampesinos, gente de ciudad que se muda al agro' (Neo-peasants, city life people moving to the country-side). Increasingly, several studies demonstrate the role neo-peasant inhabitants play in local environmental governance. Neo-peasant inhabitants' everyday practices intervene in the dispute over the exploitation/preservation of different ecosystems across Latin America (Cortes-Vazquez, 2014). Particularly, from a critical perspective, sociological analysis in Colombia has called attention to the 'process of gentrification in rururban places' (Nates, 2008: 2). Gentrification occurs when urban actors displace rural populations because their presence increases the value of the land and commercial services. The gentrification

processes could crystallize in forms of exclusion and social differentiation. It is worth mentioning that in Colombia the expression 'rururban' points to the dilution of the boundaries between the city and the countryside.

7 In Colombia, liberal environmental citizenship signals a political strategy within contemporary liberal societies that addresses environmental claims. In this case, 'environmental' works as an adjective added to liberal citizenship used to speak about certain kinds of actions, such as public participation in environmental decision-making, and certain kinds of attitudes – notably, those that go beyond self-interest. Due to its articulation within the liberal thinking framework, it also defines the environment mainly as property and the provider of human needs. However, such a frame is insufficient when it comes to introducing the ecological politics of indigenous and rural movements. As different scholars point out, for the Latin American context, the link between the environment and citizenship needs to be conceived of more broadly, for it exceeds traditional debates on liberal normativity and questions of rights and obligations (Ulloa, 2013, Gudynas, 2009, Latta and Wittman, 2010). Furthermore, as I am arguing, there is little comprehension of political subjecthood in a wide variety of rural and indigenous movements as a locationless rational entity; on the contrary, without denying rationality, it is a persistent process of desubjectivization, as place embeddedness is central. This form of locationfullness does not exclude universality either; rather, it proposes a concrete form of universality.

8 Ecological reparation is done as a practice of restoring abundance (Fujikane, 2021), namely, producing life amplification and enhancement, as opposed to producing accumulation while diminishing life. Capitalist accumulation is exclusive and concentric, whereas abundance is open and expansive. Restoring abundance refers to flourishing and wellbeing across different species; in other words, multispecies wellbeing. This kind of flourishing is created by the collective intelligence that emerges from agroecological practices. So, there is a power of life (power from below) which generates flourishing. In short, abundance points to autonomy and also to fighting late capitalistic ideologies of scarcity. I define ideologies of scarcity as the economic practice that creates artificial conditions of scarcity. By privatizing natural resources (energy, water, air), capitalist relations of production end up creating conditions of scarcity for people who don't own those resources. The social production of scarcity is conditioned by capitalist relations (exchange value). For a Latin American theory of abundance see Rafael Bautista (Bautista, 2017: 215).

9 The flatness of environmental holism might be criticized/augmented in this respect, for what matters is the layering of relations; in other words, the factorialization of relationships: relationship of relationships. The dynamic unfolding of place speaks about a multi-layered and differential perspective that encompasses the flatness of relational organic connections. Relationships of life are subsumed under a different kind of relationship, located at an alternate and foundational phase. As Higaki stresses regarding this particular notion of place: 'It consists of topologically factorializing the relationships of life' (Higaki, 2020: 94). There are many examples of this active emplacement: fish become attuned to their river environment by changing skin colour, or individuals with similar DNA configurations manifest in a variety of forms, depending on the environment/place they are located in.

10 In Colombia, civil society operates as an abstract container that encompasses social and individual rights. Ideally, the state guarantees the protection of these rights, yet the internal armed conflict and the history of political violence turned the state into a war machine against people. In this sense, civil society as a liberal institution function contradictorily as a governmental technology of control and violence.

References

Bautista, R. (2017) *Del Mito del Desarrollo al Horizonte del Vivir Bien. ¿Por qué Fracasa el Socialismo en el Largo Siglo XX?* La Paz: yo soy si Tú eres ediciones.

Bell, D. (2004) 'Environmental refugees: What rights? Which duties?', *Res Publica*, 10(2): 135–52.

Cortes-Vazquez, J. (2014) 'A natural life: Neorurals and the power of everyday practices in protected areas', *Journal of Political Ecology*, 21: 493–515.

Dobson, A. (2015) *Environmental Politics: A Very Short Introduction*, Oxford: Oxford University Press.

Fujikane, C. (2021) *Mapping Abundance for a Planetary Future: Kanaka Maoli and Critical Settler Cartographies in Hawai'i*. Durham, NC: Duke University Press.

Gudynas, E. (2009) 'Ciudadanía ambiental y meta-ciudadanías ecológicas. Revisión y alternativas en America Latina', in J. Reyes Ruiz and E. Castro Rosales (eds) *Urgencia y Utopía frente a la Crisis de Civilización*, Guadalajara: Universidad de Guadalajara y Ayuntamiento de Zapopan, pp 58–101.

Higaki, T. (2020) *Nishida kitaro's Philosophy of life*, translated by Jimmy Aammes. Milan: Mimesis International.

Ingold, T. (2011) *Being Alive: Essays on Movement, Knowledge and Description*, New York: Routledge.

Latta, A. and Wittman, H. (2010) 'Environment and citizenship in Latin America: A new paradigm for theory and practice', *European Review of Latin America and Caribbean Studies*, 89(October): 87–96.

Litfin, K. (2014) *Ecovillages: Lessons for Sustainable Community*, Cambridge: Polity.

Mendez, M. (2012) 'El Neorruralimso como Práctica Configurante de Dinámicas Sociales Alternativas: Un Estudio de Caso', *Luna Azul*, 34: 113–30.

Nates, B. (2008) 'Procesos de Gentrificatión en Lugares Rururbanos: Presupuestos Conceptuales para su Estudio en Colombia', *Antropol.sociol.*, 10: 253–69.

Nishida, K. (2012) *Place and Dialectic*, translated by J. Krummel and S. Nagatomo. Oxford: Oxford University Press.

Puig de la Bella Casa, M. (2017) *Matters of Care: Speculative Ethics in More Than Human Worlds*, Minneapolis, MN: University of Minnesota Press.

Smith, M. and Pangsapa, P. (2008). *Environment and Citizenship. Integrating Justice, Responsibility and Civic Engagement*, London: Zed Books.

Ulloa, A. (2013) 'Controlando la Naturaleza: Ambientalismo Transnacional y Negociaciones Locales en torno al Cambio Climático en Territorios Indígenas en Colombia', *Iberoamericana*, 13, 49: 117–33.

Yountae, A. (2017) *The Decolonial Abyss: Mysticism and Cosmopolitics from the Ruins*, New York: Fordham University Press.

PART V

Loss<>Recollecting

15

Travelling Memories: Repairing the Past and Imagining the Future in Medium-Secure Forensic Psychiatric Care

Steven D. Brown, Paula Reavey, Donna Ciarlo and Abisola Balogun-Katung

Introduction

Memory is the connective tissue that makes lives meaningful. A connection to the past enables sense making in the present and renders possible futures as thinkable. In the case of traumatic or difficult pasts, this connection becomes intensely important. At personal, collective and national levels, past harms and injustices need to be made visible and subject to commemorative exploration in order for victims to 'go on' in the present. In this context, repair is usually considered to be a memorial work of putting the past in order to meet ongoing moral and epistemic demands (Margalit, 2002; Blustein, 2008; Campbell, 2014). Through this work it becomes possible to envisage a reconstruction or 'healing' of personal and social ecologies of thought and feeling.

This understanding of memorial work as repair is complicated by issues around mental health. For example, while some approaches to trauma (for example Johnstone and Boyle, 2018) emphasize the need to understand personal histories – 'what happened to you' – as a way of addressing current feelings and experiences – 'what's wrong with you' – there is also a counter-discourse around the inherently unrepresentable nature of traumatic pasts (Caruth, 1996). Pain and suffering incurred through extraordinary and horrific violations of social and personal relations may be simply incomprehensible and hence difficult to both recollect and to narrate.

Mental health issues may also call into question the reliability of memory. Victims – and in some cases perpetrators – may have their recollected experiences problematized or discounted (see Haaken and Reavey, 2010). They may also be accused of focusing unduly and unhelpfully upon the past rather than facing up to problems in the present. Here, repair can take the form of an injunction to disconnect from a difficult past in order to 'move on' with living.

In this chapter, we want to explore the tensions in memorial repair work around mental health. We will be concerned with the question of when and how the past comes to matter for persons managing severe and enduring mental health issues. Crucially, we look at the practices which are enacted to manage these tensions, and how they are collectively performed within an institutional setting. Our argument is informed by work we have conducted in a medium-secure forensic pathway in a large inpatient psychiatric unit. The mental health service users who participated in this study arrived at the unit in question through the criminal justice system, and some had been convicted of serious offences.[1] However, all of the participants had challenging life experiences, and many had also been victims of offences. As we will go on to discuss, memorial repair work is a complex matter for both service users and for the institution in which they are currently detained.

In conceptualizing how remembering is at stake within secure psychiatric care, we draw upon Astrid Erll's notion of 'travelling memory'. Erll uses the term to emphasize the fluid and mobile nature of memory as 'the incessant wandering of carriers, media, contents, forms and practices of memory, their continual "travels" and ongoing transformations through time and space, across social, linguistic and political borders' (2011: 11). The contrast here is with an approach that treats memory as stable units of meaning that are 'contained' within an individual, group or place (see Middleton and Brown, 2005). The relevance for us is that the service users in our study were typically detained for significant periods of time at a considerable distance from their usual place of residence in a specialist mental health care service. Memories of these geographically and temporally remote places and persons had 'travelled' with each service user. But at the same time, these travelling memories remained connected to these other spaces and times. As service users pass through the criminal justice and mental health care systems, these connections experienced through remembering became transformed.

We define our approach to memory as 'ecological' in the sense that we treat remembering as a practice that is shaped and enacted within the settings in which it takes place. There is, in fact, a long tradition of ecological thinking within Psychology, which dates from William James, but includes E.B. Holt, J.J. Gibson and Ulrich Neisser (see Heft, 2001). The potential of this work has often gone unrecognized when it has been read through the lens of the dominant Cartesian ontology of much twentieth-century Psychology.

The re-emergence of distributed cognitive, open systems and process thinking provides an opportunity to re-author psychological approaches to remembering, and much else besides in the discipline (see Brown and Reavey, 2015, 2020a). Remembering and imagining are central to how we engage with the organized settings in which we dwell, because they involve the mobilization of versions of the past in order to realize the potential futures that inhere in our fleeting and ongoing relationships. In this sense, the past is not an inert record of 'what happened' but is a living, active, dynamic presence at every moment, which is never settled or entirely put in its place.

We begin our argument by describing an approach to difficult pasts we have referred to elsewhere as 'vital memory' (Brown and Reavey, 2015). This approach seeks to expand the usual remit of memory beyond the individual, and indeed beyond the usual boundaries of what we take to constitute the person, while at the same time recognizing the unique trajectory that makes for personal remembering. We then offer some contextual details around forensic psychiatric care and discuss some of the clinical and social care practices which complicate memorial work in settings where this care is provided. On this basis, we then go on to describe some of the informal memorial repair work techniques that service users and staff have developed and the ways in which they impact upon capacities to imagine potential futures beyond present detention. We conclude by situating this very particular set of practices within the broader 'experience ecologies' of the contemporary mental healthcare system in the UK.

An expanded view of memory

Memory is often considered to be a personal property, something that is primarily 'held' by the person and which is organized subjectively as a function of their own thoughts and feelings. One of the principal concerns within the emerging interdisciplinary area of Memory Studies has been to counterbalance this with a notion of 'collective memory', originally derived from the work of Maurice Halbwachs. This is usually understood to refer to social and cultural processes occurring at group or community levels wherein versions of the past are commemorated and contested. From this perspective, while our own memories feel as though they belong uniquely to us, they emerge from the ecology of thought and feeling constituted by the memorial communities to which we belong. The 'collective frameworks' which Halbwachs (1980, 1992) theorized as at work here were posited as complex processes of translation and distribution of perspectives on the past – or 'vibration' between actors and the spaces they inhabit – which enabled memorial communities to make sense of their shared history. The term 'collective memory' is simply shorthand for this ongoing cognitive, social and material work.

While this approach arguably affords more elaborate and nuanced understanding of the work of memorial repair, it raises some difficult questions around personal participation. Does subjective repair involve a subsumption of personal memories to a collective framework? What is the nature of the agency that now intuits and 'holds' the tensions around trauma and reparation? We can begin to address this by working through two initial propositions. First, remembering depends upon others. The ways in which we speak of the past are shaped by the interlocutors with whom we tell our stories or provide accounts. Not only are there common frameworks and norms which structure how and when we are able to recount past events, but we rely upon others to assist in elaborating, completing and revising our accounts (Brown & Reavey, 2018a). Remembering personal experience is a thoroughly collaborative process (see Mead et al, 2018 for varied approaches to this). The interactional dynamics around *who* we address our recollections to, *what* is said and *how* it is received all shape the process of remembering. Moreover, even when those recollections occur privately, others are present both as actors within what is remembered and as anticipated interlocutors to whom we either desire or fear to share our accounts with. In this sense, while past experiences are 'ours', in that we are uniquely positioned with respect to them (that is, these are events *I witnessed*, these things happened *to me*), they are not our property in any clear sense, in that others are integral to how recollections are shaped and expressed (Brown & Reavey, 2018b).

Second, remembering is a situated, contextual activity. It occurs somewhere, in a definite time and place – *where* and *when* remembering happens shapes what can be told of the past. For example, recalling a past traumatic experience is very different in the context of therapy than it is in a court of law or a social-welfare interview. This does not solely concern the words that can be used, but rather more substantively the ways in which it is possible to make sense of what is being remembered and the meanings which will be accorded to it within the setting where recollection occurs. Settings – such as therapy rooms, legal chambers, administrative offices – provide normative constraints which serve to 'frame' how remembering can proceed within them. What counts as a 'good' or 'reliable' or 'acceptable' memory is thereby locally determined by the practices and properties of the setting. If we take it as a given that the fidelity or the truth of our recollections are always important to us, then the importance of these framings in shaping memory within the setting suggest that, in a strict sense, it is the setting-level interactional processes rather than the person who 'does' the remembering. Elsewhere we have referred to this as the 'setting-specificity' of memory (Brown and Reavey, 2020a).

The approach to remembering that begins to emerge from these considerations can be called an 'expanded view of memory'. In this view, memory is not strictly located within whatever is taken to be the boundaries of the person (for example the mind, the brain, the body), but rather extends out or 'travels' into the world in a diffuse way that moves across interactions and settings. Rather than treat memory as a property that is held by a given person or collective, we must look instead at the multitude of potential functions that remembering enacts in specific times and places, and how these are subsequently involved in reorganizing the ways each of us turns around on our past experiences, including the work of reparation. These functions cannot necessarily be established in advance. While some may reflect the action programmes of the settings in which they occur (for example the purpose of a trial is encode accounts of the past within legal claims), others emerge within interactions (for example managing one's status as a victim of a past offence). This can produce tensions within accounts, or enact contradictory effects simultaneously (for example an apparent disparity between the details and meaning of testimony given across different contexts). If memorial reparation is an activity that is distributed across multiple settings, then on this account it is unlikely to be enacted without conflicts, dilemmas and lacunas that arise from the incommensurability of setting-specificities.

We have previously drawn upon the work of Kurt Lewin (1936, 1997) to conceptualize this expanded approach to memory (Brown and Reavey, 2015). Lewin describes ongoing life experience as 'the totality of facts which determine the behaviour of the individual at a certain moment' (1936: 12). A 'fact' is to be understood here as an affective-behavioural potential that emerges in relation to others and the world. These relationships are structured topologically, rather than being clearly positioned in a Euclidean mapping of space or a strictly chronological ordering of time. Persons or places that are geographically remote from us can become relationally close at a particular moment, in the same way that the past can return to the present. Lewin defined psychological reality in terms of 'what has effects' (1936: 19), or that which provides the possibility of affording action in the present moment. Our psychological 'life space', as Lewin termed it, is a topologically organized field of relations that is continuously expanding and contracting as our actions realize possibilities for acting. Life space extends our experience beyond the confines of our immediate setting or the present moment. For instance, while service users may be detained within the relatively confined space of a locked ward for a prolonged period of time, their life space may expand beyond the boundaries of the ward through material practices of communication, imagination and remembering.

Lewin's notion of life space draws analytic attention to the relational structuring of experience. Our lives flow and travel through an ongoing weaving of relationships. For example, I hear a piece of music and the sounds open a point of contact to the places where I previously listened to it, the events which it has soundtracked, the people with whom I am emotionally connected through the music. My life expands beyond the here and now, this place where I am sat. I pick up a photograph of my mother and my father. I cannot bear, right now, to think of his voice and his behaviour in those last few months of his life. I put the photograph away. My life contracts and withdraws itself into the rhythm of my fingers as I type, the screen and the desk in front of me. Life space has a diastolic and systolic character. It pulses with the expansion and contraction of experience as relationships become affectively intimate and distant. Henri Bergson's (1991) notion of 'attention to life' as the mobile horizon of our ongoing concerns and experience describes this well. The boundaries of our experience, at any given moment, are not defined by the spatial limits on our movement, or the formal organization of the setting with which we are presently engaged, but rather by the shifting topological foldings of life space.

There is a risk here that life space might be conceptualized as sufficient unto itself rather than being culturally and historically grounded. Lewin (1997) introduced the somewhat confusing term 'psychological ecology' to refer to the 'boundary conditions' of life space. We prefer the term 'experience ecology' as shorthand for the distribution of ways of thinking and feeling which become available within a given setting and which are undergoing a constant process of transformation (Brown and Reavey, 2020a, 2020b). This is derived in part from previous work where we explored the ways in which secure psychiatric care promotes particular ways of understanding sexuality and sexual conduct (Brown et al, 2014). Service users are typically inculcated in a discourse that sexuality is a risk and a threat to their current wellbeing. This leaves service users feeling that sexuality is something that must be 'left behind' or 'amputated' until they are well enough to engage with it again. We argued that this experience of sexuality that had been 'learned' within the ecology of secure care travelled with service users as they transitioned back to the community, where it had demonstrably negative effects on their ability to engage in the kinds of relationships and forms of intimacy which many craved (and which would aid their mental health recovery). Our approach to memorial reparation then necessarily involves an understanding of what is afforded by differing experience ecologies and what happens to the ways persons relate to themselves as they move between them.

Forensic psychiatric care

The mental health care system in the UK has been progressively 'de-institutionalized' over the past century. Up until the 1950s, the system had been dominated by the network of large hospital institutions known as 'county asylums' where persons diagnosed with severe mental health conditions might be forcibly detained under the terms of a section of the Mental Health Act (that is, 'sectioned') for either a determinate or indeterminate amount of time (see Parr and Philo, 1996 for one local history). In the post-war era, social reform was coupled with economic reform to reduce the numbers and size of county asylums, with care being provided in community settings by teams of clinical and welfare professionals. However, unlike other European countries such as Italy which attempted to permanently move away from secure hospital care for mental health entirely, the UK retained a significant number of inpatient bed services (roughly eight beds per each 100,000 of population).

This is particularly noteworthy for forensic pathways through inpatient care. Persons detained within a forensic pathway are typically transferred from the prison estate to secure care (that is, they are detained on a locked hospital ward under sections 37 or 41 of the Mental Health Act). Forensic psychiatric care is provided with three large 'high secure' units, around 80 'medium secure' units, and through an extensive network of mostly community-based 'low secure' housing facilities. Section 37/41 detention does not have a specified time limit but is subject to review by clinical staff and yearly tribunals. The process of ultimately being discharged from inpatient care typically involves a gradual 'stepping down' through levels of secure care which can take several years and may involve the use of a 'Community Treatment Order' specifying provisions for living within the community, with an involuntary return to secure care mandated if they are breached. A primary concern within psychiatric care is managing the risks which a patient poses to 'themselves', 'others' and the 'environment' (Drennan and Alred, 2012). The dominant language currently used in this context is 'recovery', which is conceptualized as the process whereby a person gains awareness of their mental health conditions, establishes some form of control over their own thoughts and behaviour, and demonstrates adherence with their treatment regimen (usually taking psychiatric medication on a regular and permanent basis).

The site where our study was conducted was a large specialist provider of inpatient psychiatric services in England. This charity provider supplies mental health services on referral across the UK, meaning that a great many patients at the time the study was conducted (2018/19) had been referred at some distance from the hospital site. When we began collaborating with the institution, clinical staff there identified facilitating contact of patients with relatives and prior communities as a significant issue. They were aware that a noticeable number of patients were electing to remain in the local area following their discharge from the service and had hypothesized that this might be related to problems of maintaining contact with 'home' during their time in detention. Our study was designed to explore how practices of remembering might maintain attachment to 'home' for patients on the forensic pathway in the main hospital site.

The hospital site itself is a large green site with multiple units and wards offering a wide range of inpatient mental health services. Service users admitted to the forensic pathways are placed on gender and age specific wards, mostly within new-build specialist facilities. As with all secure psychiatric care in the UK, wards on these units have limited space, with patients having small, single-occupancy bedrooms, along with communal areas and some enclosed outdoor space. Some service users may have other conditions attached to the detention, which mean that they are effectively confined to the unit and may not be able to access the wider hospital grounds. Visits from relatives and friends are permitted, depending on current capacity of the patient, and are sometimes conducted through telephone and video calls. All visits are heavily supervised and physical contact is typically not permitted. Patients have limited access to other means of maintaining contact (for example internet, social media etc) outside of these times.

The number of possessions which service users bring with them on admission is highly limited and may constitute only a small number of clothes and personal possessions such as books or photographs. Sections are often initiated following police arrest or intervention by a local mental health crisis team. On these occasions there is limited scope for gathering the service user's possessions. Depending upon the nature of the offence (referred to as 'index offence'), it may not be possible for service users to acquire some of their possessions once they are detained, for instance, in cases where the index offence involves family members who may subsequently break contact. Detention may not only fracture the service user's relationship with 'home', it can also make it difficult to retain possession of the items which might make it possible to preserve a memorial link.

From previous work, we were aware that service users in medium-secure forensic care have often moved multiple times through different institutions, and in some cases have a long history of being placed in institutional care prior to incurring their index offence. Their memories have travelled some considerable distance with them throughout their journey, although service users themselves leave very little by way of traces of themselves and their time spent in the institution when they leave. In general, service users do not tend to build longstanding friendships during their time in detention. In part, this is because there is a reticence to discuss index offences with other service users, or to disclose personal information which might come to be used by others as a kind of social currency on the ward. Friendships are also typically not encouraged by ward staff, out of concerns that either controlling or intimate relationships might subsequently develop. It is not uncommon that patients to be moved in order to break up such developing relationships (Ravenhill et al, 2020).

The most common non-pharmacological therapeutic intervention in forensic care is the use of variants of Cognitive-Behavioural Therapy (CBT), in particular Dialectical Behaviour Therapy (DBT). This latter was developed for working with patients who are given the controversial and contested diagnosis of Borderline Personality Disorder (BPD), and its use is widespread within UK inpatient psychiatric care (Ivanoff and Marotta, 2018). Like all forms of CBT, DBT focuses on supporting the patient to manage their own current thoughts and emotions and to break free of existing behavioural patterns and the past environments which are supposed to have reinforced them. Dwelling on past events is antithetical to the approach. Moreover, DBT has a strong element of mindfulness, encouraging a focus on current experiences. The net effect of this is to continuously shift focus away from the past towards the here and now or 'present-ism'. As we have described elsewhere, this reinforces a concern with 'being well' in the present that can make forensic psychiatric care operate as a 'system of forgetting' (Brown and Reavey, 2016). In summary, while it might be assumed that sustaining a relationship to the past (that is, 'what happened to you') might be a highly valued aspect of in-patient psychiatric care, structurally and practically service users are strongly directed towards the present moment.

Memorial practices

There is not an abundance of activities to keep patients occupied during their time on a locked ward. While there may be some group sessions for service users to attend, aimed at either managing behaviour or dealing with strategies for improving recovery, hours are usually marked by mealtimes or by watching television. Detention under section 37/41 is indeterminate. Discharge dates are not fixed but are decided by clinicians with reference to indicators that gauge whether the service user is able to take charge of their own recovery in a community setting. The lack of fixed markers can mean that service users focus on what are actually arbitrary signs without any real clinical meaning – such as feeling that they are being treated as an object of reduced concern by ward staff. As is common in most secure psychiatric settings, patients are left for the most part to occupy themselves, with comparatively little intervention by staff, who tend to be engaged in either collecting routine wellbeing information about service users, or alternatively managing crisis events. Having long stretches of time to fill can leave patients with plentiful opportunities for chance distractions. Patients can find themselves 'triggered' to think about past events or persons through chance associations arising from music they hear or television programmes that have been selected by others on a communal screen. This gives rise to pleasant and threatening and/or disturbing memories in apparently equal measures.

A key issue here is that patients lack the ability to control or programme the sonic environment of the ward. While many are able to play their own music in their own bedrooms, patients are encouraged to spend time in communal areas (since this enables both better monitoring by staff and is driven by the belief that if patients spend time on their own, they will ruminate on their own mental health issues). But since the 'soundtrack' to the ward is a matter of, at best, collective negotiation, or, at worst, the wishes of the most persuasive patient, communal areas become filled with a wide range of unpredictable sonic cues for memory. Elsewhere, we have argued the relative lack of sonic agency has effects on recovery and undermines efforts to create therapeutic atmospheres on closed wards (Brown et al, 2020). But it also serves to mediate memorial agency. Sonic cues can expand life space, making otherwise distal relationships suddenly proximal, but do so in ways that do not allow patients the opportunity to engage in their own 'memory management'. For example, a music video that appears unexpectedly on the television can transport a patient back to the times when they first listened to it and make the people they were then close to suddenly reappear in their thoughts and feelings.

Modern mental health institutions have their roots in the former 'county asylum' system, where institutionalization took the form of a 'retreat' from the timings of everyday life in which daily markers of work or domestic labour were removed. Time can be experienced differently on a locked ward, particularly when embodied rhythms of sleep and wakefulness are further disturbed through psychiatric medication. The division of the day into meaningful units can be difficult to sustain. However, this does not mean that life is lived entirely without reference to calendrical and other temporal markers. Holidays and other seasonal events, such as Christmas and Easter, are celebrated on the ward. Other events may permeate the boundaries of the unit. For instance, the sounds of fireworks set off in celebration of 'Bonfire Night' on 5 November can act as memorial cues. The recurrence of specific dates, and the memorial associations they hold, creates a punctuated backdrop for everyday remembering, irrespective of whether or not the dates are specifically recognized by ward staff and the patient community.

Each patient has their own specific dates each year that represent challenges for memory management. These can include birthdays of family members, dates marking transition in and through the healthcare and prison system, and dates which are associated with an index offence. Staff are typically aware of these personal memorial cues and arrive at informally negotiated arrangements to help patients manage their remembering around these dates. These include practices such as temporarily taking custody of photographs, letters or cards such that patients will not feel drawn to ruminate on these objects at this time. Through providing managed access to tools through which remembering is accomplished, staff become custodians of memory. But the work they do here is informal – the institution does not see it as their role to share ethical responsibility over patient's memorial work. There is to some extent a tension here with the project of supporting recovery, which by definition includes repairing relations with the past in order to establish hope for a future life. While the institution promotes itself as 'recovery led' (in common with the dominant ethos in UK mental health care), the crucial temporal dimension of recovery as a form of repair between past, present and future is absent from formal care planning.

The design of the built environment in secure mental health units is often meticulous. A dominant concern is with managing patient safety through embedding technical security measures into the physical structure of the unit. For example, the furniture and fitting of patient bedrooms are designed primarily to remove ligature points from which patients might potentially self-harm. The hinges and mechanisms of doors function to prevent deliberate trapping of fingers in the closing mechanism or barricading of rooms. Every inch of ward space embeds a programme of risk management. As a consequence, it can be difficult for patients to personalize the space they occupy during their time in detention. It is as though the unit were designed to remove all traces of their presence as soon as they have departed (by contrast, it is not uncommon in much older units to see names and messages etched into external walls and other features of the environment – see Reavey et al, 2021). Patients do, nevertheless, find ways of marking out the space. One common practice is to fix photographs or cards to bedroom walls. This can be something of an achievement on some wards, where common fixing agents such as reusable adhesive putty (Blu Tack) are proscribed. Decorating walls in this way creates something of a 'world within a world'; it folds relations together to expand life space beyond the immediate boundaries of the unit. But it also makes these relations into live matters of concern for the patient, who may be continuously reminded of matters that need to be settled, issues that need to be addressed. Making the past both visible and intimately present – for instance, literally the first thing that is seen when waking – has its own particular risks for patients.

An alternative strategy is to have the means to access the past accessible but not necessarily available at any given moment. For example, some patients have a set of photographs or letters stored within a drawer in their bedroom or elsewhere on the ward. They may view these at particular moments when they 'need' or 'feel able' to do so. This is a strategy that arises in part through the experience of travelling through the mental health and criminal justice system. A service user must be able to pack and take with them all the personal possessions they need at relatively short notice. The practice may then be retained as a way of being able to unpack and repack a 'world within a world' depending on whether one seeks to expand or contract life space. In other studies, we have observed that female patients, for example, may show the contents of their handbags as a way of offering a narrative of their recent history. Memories travel not only in the sense of moving with persons as they engage in remembering across different times and places, but also literally in these material practices of temporarily unpacking the past for oneself and for others.

The 'unpacking' practice has its own particular risks. When the past is accessed through a specific set of images or artefacts (for example a letter, a signed card, a piece of clothing) it can become 'frozen in time' and unable to take on other forms. Some service users, for instance, have been distant from family and friends for some considerable period of time. During the course of this, they rehearse an account of 'home' that reflects a very specific moment, which is now long passed. Children, for example, may be remembered not as they are now, but as they were during the times when the service user was most involved in their lives. This can lead to an account of the past that appears to be idealized, along with the expression of a desire to return to a place that no longer really exists in the ways in which it has been remembered. We have termed this relationship to memory one of 'spectral agency', where a highly crystallized, frozen image of the past is invoked as a central facet of what one is seeking in the future (Reavey et al, 2019). This leads to what can appear to be highly unrealistic imaginings of life outside of secure care, since they involve an impossible return to a no longer existing state of affairs. By contrast, in other work, we have pointed to the importance for those who are living with a 'difficult past', to develop a view of their personal history and critical events as 'unfinished' or 'unsettled', and thus available for re-narration, and more important to be 'felt differently' at subsequent points in the future (Brown and Reavey, 2015).

Part of the difficulty here is the dominance of the idea around both personal and cultural memory that a coherent, well-structured narrative is necessary to come to terms with the past. On this basis, things such as gaps in memory, contradictory versions of events and ambiguity are seen as problematic and as obstacles to wellbeing. But in work we have done on memories of child sexual abuse and with adoptive parents, we have come to see what the value can be of maintaining a certain degree of ambivalence as a critical resource. The capacity to willingly hold onto differing potential readings of the same event can be a means to engage with tensions and contradictions that might otherwise be overwhelming. Coherency is not and should not be the absolute goal of remembering on all occasions (see Brown and Reavey, 2020b). As we see it, the problem on the unit is that the unpacking practice is not a focus for care planning. It could very well be taken as a starting point for a shared process of narrating and engaging with the past collectively (along the lines of the forms of reminiscence therapy common in dementia care). Or alternatively, it might be seen as a way into a collective 'memory work' practice, where persons might be encouraged to re-work and re-author their relationship to the past. Again, the broader issue here is that memorial work is not seen as integral to the way the institution supports recovery.

One outcome of the lack of attention to remembering is in the expressed future plans of patients around where they will live following discharge from secure care. Very few patients we spoke with had plans to return 'home'. Some patients imagined that the best option would be a 'fresh start' somewhere else, where they could seek new friendships and employment where they would be free of the past and those influences that had shaped their behaviours prior to secure care. This would both allow them to 'stay out of trouble' and not render them as a burden to their family. This imagined future seems to be of a piece with the personal change that is promoted by DBT, where it is necessary to break with past patterns of thinking, feeling and acting in order to be mindfully aware of one's present circumstances. But an equal number of patients imagined a future where they might be living independently close to 'home', but at a sufficient distance. This is a more demanding version of the future, which requires the person to maintain potentially contradictory versions of themselves as 'they were then', 'they are now' and 'they might become'. Such a version of self might potentially be challenged or rejected by others. In this instance, repair would involve a great deal of living with ambivalence and being able to interactionally manage the ambivalence of others, rather than closing down and moving on from the past.

What these memorial practices demonstrate is the difficulty service users face in mobilizing their travelling memories within the complex ecology of secure psychiatric care. We might say that the programme around which care is organized in this setting is 'recovery without reparation'. In other words, stabilizing the mental health of service users without opening up the question of how they are able to put the past in order and address the complex web of injustices in which they may be both victims and perpetrators. But for service users, it appears to be almost universally impossible to focus on their recovery without addressing these past relationships. The informal practices which have arisen within secure care then represent a delicate negotiation between staff and service users that goes on in the interstices of the setting. They keep memory and the possibility of reparation in play, but in ways which may ultimately not be sustainable as patients move towards discharge.

Memorial work as ecological repair

Memory is the connective tissue that makes lives meaningful. This truism is easily stated, but horrendously difficult to operationalize adequately. Our lives are not really 'ours', in the sense that the borderlines between the person and the collective, the past and the future, living and the ecologies that sustain it, are perpetually shifting and blurring. For each of us, our lived experience is most certainly a unique trajectory, but one that consists, that is constituted out of relations that are both distal and proximal. A '100%

relational' view of the psychological, to borrow a phrase from Eduardo Viveiros de Castro (2016), requires the invention of an array of concepts that do not really fit together well in analytic terms, and that perhaps expresses something of the ambivalence that is found in the object of enquiry. In our work, this is further complicated by the desire to have this make sense in very specific empirical settings, such as secure psychiatric units. We want a 'process-ecological' approach to remembering and personhood to be able to inform the practical care and design decisions made in these institutions (see Reavey and Harding, 2017). This would blend an understanding of the psychological as distributed and organized across ecological settings with a treatment of questions such as agency, subjectivity, health and relationships informed by process philosophy (see Brown and Reavey, 2020a).

For example, secure psychiatric care environments place considerable emphasis on boundaries, including: physical boundaries between the hospital and the community; symbolic boundaries between staff and patient and within patient groups; affective boundaries between mental health and other aspects of personhood; temporal boundaries between a difficult past and present thoughts and behaviours. From a process-ecological perspective, we can observe that experience cannot be contained within these boundaries – it expands and contracts through memorial work of engaging with photographs, listening to music, handling objects. Agency and recovery are not 'things' located within the person, but are, at any moment, the undulation of this movement between and across boundaries. It is possible to conceive of care practices and designs of the built environment that act to support rather than restrict this movement (see Reavey et al, 2021 for practical examples). In ecological terms, we can make judgements about the practices that enable reparative forms of living both within secure care and subsequently within the community.

Such ecologically grounded forms of repair may take many forms. In some instances, it may refer to the capacity to contract or narrow life space, such that the relationships which are in play become concentrated and focused on the present. When the past is literally unbearable, secure care can revert back to its roots in the notion of asylum, a refuge from everyday life. But repair must also involve the centrifugal movement of expanding life space, of going beyond the boundaries of the here and now. Memorial work can play a central role here in terms of mobilizing the past in order to invoke potential futures. But much turns on whether the past is treated as a living, dynamic presence rather than a frozen moment in time. As Michel Serres (2018) observes, the past is most valuable to us when its contingency can be reorganized in multiple different series – 'you will only discover it in its truth at the moment it will no longer do you any good'. Again, this is more easily stated than lived. Reparation, ultimately, is a not a putting of the past in order, but rather the capacity to 'bear' the past, to live and work with ambivalence. The memorial practices that 'work' in the ecology of secure care may not do so in the wider ecologies through which service users move. We all need multiple versions of ourselves and our pasts to get by. The experience ecology of secure care tends to encourage a thinning-out of persons such that they can reduced to their bare capacity to function – no longer a risk to self, others, environment. It could instead become a site for supporting the multiplying and expanding of personhood and, as a consequence, of the imagined futures that lie therein.

Note

[1] We use the term 'mental health service user' because it is the preferred non-medicalized descriptions of persons receiving mental health care. At points we use the term 'patient' when drawing attention to an aspect of inpatient or ward-based practice. While we are referring to same persons throughout, the distinction remains important.

Acknowledgements

The authors acknowledge the assistance of St Andrew's Healthcare in conducting this study.

References

Bergson, H. (1991) *Matter and Memory*, New York: Zone.

Blustein, J. (2008) *The Moral Demands of Memory*, Cambridge: Cambridge University Press.

Brown, S.D. and Reavey, P. (2015) *Vital Memory and Affect: Living with a Difficult Past*, London: Routledge.

Brown, S.D. and Reavey, P. (2016) 'Institutional forgetting/Forgetting institutions: Space and memory in secure forensic psychiatric care', in E. Weik and P. Walgenbach (eds) *Institutions Inc*, London: Palgrave.

Brown, S.D. and Reavey, P. (2018a) 'Rethinking function, self and culture in "difficult": Autobiographical memories', in B. Wagoner (ed) *Handbook of Culture and Memory*, Oxford: Oxford University Press, pp 159–81.

Brown, S.D. and Reavey, P. (2018b) 'Contextualising autobiographical remembering: An expanded view of memory', in M. Mead, A. Barnier, P. Van Bergen, C. Harris and J. Sutton (eds) *Collaborative Remembering: How Remembering with Others Influences Memory*, Oxford: Oxford University Press, pp 197–215.

Brown, S.D. and Reavey, P. (2020a) 'Memory in the wild: Life space, setting specificity and ecologies of experience', in B. Wagoner, I. Brescó de Luna and S. Zadeh (eds) *Memory in the Wild: Niels Bohr Professorship Lectures in Culture Psychology*, Charlotte, NC: Information Age.

Brown, S.D. and Reavey, P. (2020b) 'Remembering as difference', in B. Wagoner, I. Brescó de Luna and S. Zadeh (eds) *Memory in the Wild: Niels Bohr Professorship Lectures in Culture Psychology*, Charlotte, NC: Information Age.

Brown, S.D., Reavey, P., Kanyeredzi, A. and Batty, R. (2014) 'Transformations of self and sexuality: Psychologically modified experiences in the context of Forensic Mental health', *Health*, 18(3): 240–60.

Brown, S.D., Kanyeredzi, A., McGrath, L., Reavey, P. and Tucker, I. (2020) 'Organizing the sensory: Earwork, panauralism and sonic agency on a forensic psychiatric unit', *Human Relations*, 73(11): 1537–62.

Campbell, S. (2014) *Our Faithfulness to the Past: The Ethics and Politics of Memory*, Oxford: Oxford University Press.

Caruth, C. (1996) *Unclaimed Experience: Trauma, Narrative and History*, Baltimore: Johns Hopkins.

Drennan, G. and Alred, D. eds. (2012) *Secure Recovery: Approaches to Recovery in Forensic Mental Health Settings*, London: Routledge.

Erll, A. (2011) 'Travelling memory', *Parallax*, 17(4): 4–18.

Haaken, J. and Reavey, P. (eds) (2010) *Memory Matters: Contexts for Understanding Childhood Sexual Abuse*, London: Routledge.

Halbwachs, M. (1980) *The Collective Memory*, New York: Harper & Row.

Halbwachs, M. (1992) *On Collective Memory*, Chicago, IL: University of Chicago Press.

Heft, H. (2001) *Ecological Psychology in Context: James Gibson, Roger Barker, and the legacy of William James' Radical Empiricism*, New York: Psychology Press.

Ivanoff, A. and Marotta, A.L. (2018) 'DBT in forensic settings', in M.A. Swales (ed) *The Oxford Handbook of Dialetical Behaviour Therapy*, Oxford: Oxford University Press.

Johnstone, L. and Boyle, M., with Cromby, J., Dillon, J., Harper, D., Kinderman, P., Longden, E., Pilgrim, D. and Read, J. (2018). *The Power Threat Meaning Framework: Towards the Identification of Patterns in Emotional Distress, Unusual Experiences and Troubled or Troubling Behaviour, as an Alternative to Functional Psychiatric Diagnosis*, Leicester: British Psychological Society.

Lewin, K. (1936) *Principles of Topological Psychology*, New York: McGraw Hill.

Lewin, K. (1997) *Resolving Social Conflicts and Field Theory in Social Science*, Washington, DC: American Psychological Association.

Margalit, A. (2002) *The Ethics of Memory*, Cambridge, MA: Harvard University Press.

Mead, M., Barnier, A., Van Bergen, P., Harris C. and Sutton, J. (eds) (2018) *Collaborative Remembering: How Remembering with Others Influences Memory*, Oxford: Oxford University Press.

Middleton, D. and S.D. Brown (2005) *The Social Psychology of Experience: Studies in Remembering and Forgetting*, London: Sage.

Parr, H. and Philo, C. 1996. *'A forbidding fortress of locks, bars and padded cells': The locational history of mental health care in Nottingham*, London: HGRG.

Ravenhill, J., Poole, J., Brown, S.D. and Reavey, P. (2020) 'Sexuality, risk, and organisational misbehaviour in a secure mental healthcare facility in England', *Culture, Health & Sexuality*, 22(12): 1382–97.

Reavey, P. and Harding, K. (2017) *Design with People in Mind*, London: Design in Mental Health Network.

Reavey, P., Brown, S.D., Kanyeredzi, A., McGrath, L. and Tucker, I. (2019) 'Agents and spectres: Life space on a medium secure forensic psychiatric unit', *Social Science and Medicine*, 220: 273–82.

Reavey, P., Brown S.D., Ciarlo, D. and Lazenby, K. (2021) *Design with People in Mind: Borders and Boundaries*, London: Design in Mental Health Network.

Serres, M. (2018) *The Incandescent*, London: Bloomsbury.

Vireiros de Castro, E. (2016) *The Relative Native: Essays on Indigenous Conceptual Worlds*, Cambridge: HAU.

16

Conversations on Benches

Leila Dawney and Linda Brothwell

Introduction

The conversations that follow took place while walking through the green spaces of Bristol, in South-West England, in preparation for Linda's latest 'act of care': a commissioned artwork that performs a detailed, delicate repair on an object in public space. The act of care will take as its object a park bench, and on this occasion we're on a tour of benches in Bristol's urban parks, looking at their designs and materials, at how, when and where they were made, and where they have been adorned, damaged and repaired.[1]

Public benches hold us; they support us when we need to rest, or talk, or perhaps sleep. They are sites of breakups, of impromptu get-togethers, of wistful daydreams and of drinking and drug taking. They are truly public amenities: an often-overlooked aspect of everyday infrastructure that tends to our bodily comfort, and our need to pause.[2] Our focus on benches helps us to think about how we care for and relate to public objects, and how they in turn care for us. Here, we consider benches as sites for reparation, both against the neglect of public objects and spaces that has been an ongoing feature of the UK under neoliberalism, and against our alienation and separation from each other and from our worlds.

This chapter discusses our relationship to these everyday material infrastructures in the context of neoliberalizing forces that erode the idea of the 'public'. Drawing on the work of political theorist Bonnie Honig and psychologist Donald Winnicott, it argues for the importance of public objects: shared material forms that physically hold us, and provide material resources for the production of public life. Through a series of conversations, we explore what it means when these objects are invested in, valued and cared for, and what we can do when it is unlikely that this will happen. Linda's work has, for many years, involved care and repair of public objects. Her acts of care – towards public toilets, streets and benches – are a provocation to rethink how we might care for objects and things. They model an ethos of care towards the world which is not reciprocal yet embeds ways of relating to objects into wider networks of support and worlding. Small acts of care and repair for public objects acknowledge the stuff of collective life that has been lost; they reclaim things in the name of a public and actively remake public life. They repair the idea of the public good that has been eroded through lack of investment from states and local governments. Minor acts of repair offer a gift, a provocation, and an acknowledgement of our responsibility towards our worlds.

Dilapidation and austerity

Recently, discussions of violence in the humanities have shifted their temporalities away from the spectacular, and towards the slow, the attritional, and the barely registered (Berlant, 2011; Nixon, 2011; Povinelli, 2011). These accounts recognize the wearing, exhaustive forces of neoliberal life, the biopolitical drive of letting die, and the sluggish work of chemical toxicity (Fortun, 2009). A concern with these slow violences has also extended to the world of objects, for example, in Laura Ann Stoler's discussion of post and neo-colonial ruination (Stoler, 2013), and Rachel Pain's work on 'chronic urban trauma' and housing dispossession (Pain, 2019). The term 'urbicide' refers to the deliberate destruction of the urban fabric, for example in regeneration and 'progress' (Berman, 1987) or as a tactic of war, notably in the destruction of cities in Syria, on the West Bank, and during the Balkan conflict (Coward, 2008; Azzouz, 2019; Golańska, 2022). Since 2010 in the UK, the austerity policies of the conservative/liberal democrat coalition government reduced central government grants to local authorities, leaving them with difficult decisions about how to spend what amounts to a 25 per cent reduction in income in the context of ever-increasing adult care and housing bills (Booth, 2017). The subsequent cuts to services have led to a new spatial language of austerity in the UK: one of closed libraries and youth centres, boarded-up shops, poorly maintained housing, cracked pavements, overgrown parks and unswept streets.

As everyday spaces and neighbourhoods become shabbier and shabbier, where shops close, streets go unrepaired and where rubbish is not collected, slow urbicide in the UK operates as a slow, insidious drain on the way in which people experience the world. As public amenities such as libraries, toilets and community centres close, people's worlds contract, turn inwards, and move towards the private and insular, and to isolation and depression (Hitchen and Shaw, 2019). It is within this context of the slow urbicide of austerity economics, a result of both the specific policy initiatives implemented by the UK's coalition government and a more general shift from welfare to neoliberal state, that we visit the public bench as an object of care and reparation.

> Our conversation turns to the subject of dilapidation and decay. Many of these benches have rotten or missing slats. The more recently installed benches, replaced on the concrete slabs where older benches once stood, are made of cast metal. They are cold to sit on, and painted with anti-vandal paint.
>
> Our tour of Bristol's park and street benches highlights the disparities between them. Some of them are carefully welded and wrought or cast, with lions and flourishes adding decoration and finish. Benches whose seats are made of wood, with more slats, and slats that curve to our bodies, are much more comfortable. Many of the wooden benches are rotten, with slats missing. In the absence of public money to look after these amenities, it often falls to community groups such as park 'friends' associations to renovate and maintain them. Several benches had Perspex plaques alerting us to the work of these groups.
>
> Many of the newer benches are fabricated from sheet metal bolted together. They are much colder to sit on in winter. We approach some by the basketball courts that are more like perches than seats. They face away from each other, discouraging interaction. They are impossible to sit on – one has to lean – and water puddles on the flat metal 'seats'. We stop by them for a while.

Linda: It looks like somebody hasn't given it any thought, pure function, but not really even that. If you have a job line that says "bench", it's the cheapest way of getting it done. It could have the name of the park on it, like the others, but it'll knock a few hundred quid off if you don't. … But maybe it's the choice between personalised lettering and keeping a youth club open.

Leila: Well, that's the thing. These are the decisions that had to be made. Youth clubs, adult care, Sure Start.[3] Or a bench.

Linda: We're getting rid of all of it. We're getting rid of all of the things that make us feel held.

Leila: Everything that's public feels like it's at bare minimum, from adult care to the park.

In the UK, we are currently living through an era that struggles to find its way between the promises of the post-war welfare state, as holding environment and safety net, and the neoliberal state that celebrates individual responsibility and the market. We may look nostalgically back at a time when public toilets were open, and when park keepers and teams of gardeners looked after green spaces. And we mourn this time, because we feel that these well-cared-for amenities would make us feel as though we matter. If we can no longer rely on the state to prioritize and look after these objects; the benches, the gates, the water fountains, the toilets; if, as the welfare state, and its concomitant values of care and public life gets increasingly eroded, then who will take on the responsibility to look after them?

It is at this in-between juncture, what Lauren Berlant might call the 'impasse', that the public object, and indeed the very idea of the public, becomes rather shaky. The post-war welfare state, with its investment in public services and amenities, is being dismantled both materially and ideologically. It is in this dismantling – either actively or through neglect and deprioritizing – that we can see the tensions between the residual and emergent cultures constituting this impasse (Williams, 1977). As the relationship between citizen and state weakens, and we encounter a proliferation of voluntary and private bodies that now do the work previously undertaken by public employees, our sense of who is responsible, of who cares for us, of who and what we are responsible for, shifts. Changing patterns of work, too, have reduced the kind of individual responsibility for spaces that comes with intimate knowledge and a sense of custodianship. A depersonalization of relationships between labour and places, tools and objects is also evident across many formerly public industries, including postal work, public transport and parks maintenance, as the economic metric becomes the ur-value (Brown, 2015).

How might we, as fraying publics, might respond to this impasse? While we may both mourn and long for objects that act as monuments to and celebrations of public life, it is increasingly clear that this option is, for most of us and for the time being, unavailable. We will not any time soon see a shift in public spending that values the careful crafting of beautiful public objects, or even their regular care. Taking into account themes of public life, responsibility, care and minor politics, we raise the question of how to take care of public objects in the context of this impasse. Recent community-led initiatives that take on responsibility for formerly public services have been criticized by some as offering a salve, or as bolstering neoliberalization by replacing public provision with voluntary services (Muehlebach, 2011). We, however, take a pragmatic approach to our present predicament, and suggest that taking responsibility and undertaking small acts of care can bring consolation for what is lost, and produce new relations between publics and the objects that help produce them. These relations have the potential to expand our contracted worlds, and foster new forms of life in common. As we care for our everyday infrastructures, we leave a trace on them and substantiate our relationship with them.

Care and repair can foster new alliances with objects that compose into something that is more than ourselves. They are practices of minor politics: 'interstitial' practices that can 'produce alternative subjectivities, spatialities and temporalities' (Katz, 1996: 490, 2017). As a form of love and care for objects, they can generate the different forms of responsibilities that we must (perhaps reluctantly) take in order to keep our worlds in good shape.

In praise of public objects

The political theorist Wendy Brown's *Undoing the Demos* argues that neoliberalism is destroying democracy through the systematic implementation of a normative ordering of life that 'transmogrifies every human domain and endeavour, along with humans themselves, according to a specific image of the economic' (Brown, 2015: 10). She argues that this has resulted in the economic metric being the primary metric by which all spheres of life are judged, and moreover that it is only with reference to the master signifier of the economic that authority is generated. And with this – with the outsourcing and marketization of previously public goods, such as education, infrastructure and public spaces – comes a remaking of the forms and content of these spaces, and of the kinds of subjects that are interpellated through them. Neoliberal life, through its atomizing, individualizing, privatizing and enclosing drives, chips away at our sense of being part of something greater than ourselves. For Brown, this subsumes the idea of publicness, citizenship and politics under the shade of the market. In terms of public goods, she argues that 'when the domain of the political itself is rendered in economic terms, the foundation vanishes for citizenship concerned with public things and the common good'. Neoliberalism 'wages war on public goods and the very idea of a public' (Brown, 2015: 39). Later, she suggests that the saturation of market metrics as the only value of value means

> public goods of any kind are increasingly difficult to speak of or secure. The market metrics contouring every dimension of human conduct and institutions make it daily more difficult to explain why universities, libraries, parks and natural reserves, city services and elementary schools, even roads and sidewalks, are and should be publicly accessible and publicly provisioned. (Brown, 2015: 176)

Brown contends that the form and meaning of public goods and infrastructures are intimately tied to our subjectivity, citizenship and place in the world. They reflect whether we see ourselves as consumers, investors, citizens or publics. Bonnie Honig, in *Public Things: Democracy in Disrepair*, argues that we need to take public things seriously. *Public Things* makes the case for the role that public things play in constituting democratic life and a sense of togetherness: 'public things press us into relations with others. They are sites of attachment and meaning that occasion the inaugurations, conflicts, and contestations that underwrite everyday citizenships and democratic sovereignties' (Honig, 2017: 6). Taking as an example New York's Central Park, she notes how it

> was built in an awful swamp, but on it were lavished incredible skills, craftsmanship, design and materials. This, and, in particular, the use of Alhambra style tiles whose colors do not stop at the surface but run all the way through, stands as a great metaphor for public things run all the way through us. (Honig, 2017: 17)

She reminds us that the lavish care that went into the making of Central Park was a deliberate attempt to produce something beautiful for the people. Her point is that where public things are beautiful, cared for, invested in and good quality, they signify a model of the public itself which is all of these listed: a model which is desirable for everyone, rather than the vestiges of something that no longer matters, for those who cannot afford to opt out or get out, and who will have to make do with the 'bare minimum'.

> In our walks through the city, we pass the boarded up Victorian swimming pool that has been closed for over a year, and whose future hangs in the balance after months of lockdown. The swimming pool whose reception is usually unstaffed, and you change at the pool's edge behind a shower curtain. On the other side of town, can glimpse into the serene, shiny environment of the private spa and gym where, for a significant monthly membership fee, one can feel nurtured by underfloor heating, hairdryers and luxurious showers, and come into contact only with other 'members'.

Yet lamenting the decline in public things is not Honig's main argument. Rather, she sets out to make a powerful case for the role of public things in the maintenance of a healthy democracy. Her argument, that 'public things – objects of both facticity and fantasy – underwrite our collective capacities to imagine, build, and tend to a common world collaboratively' (Honig, 2017: 38) means that investing in and caring for public objects demonstrate a concern for the body politic: they demonstrate that we, as a public, are valued *as a public*. And when these objects are defunded and devalued, this has an effect on the way we see ourselves as a public.

She argues that 'public things stand out as a point worth insisting upon, something that must not be allowed to become part of the morass of disrepair. Their thingness enchants, even as their publicness is under pressure' (Honig, 2017: 32). In her call for a revaluation of public objects, like Brown, she suggests that they "call out to us, interpellating us as a public … that all of us in common get our very sense of commonness from the object" (Honig, 2017: 30–1).

Linda Brothwell's work enacts an ethos of care for objects and for public spaces that speaks to these concerns.

Linda: I think when I read your essay on the commons [Dawney, 2013], and it talks about collective living, or little moments of life lived collectively in small pockets, I didn't realise it at the time, but I'm beginning to understand what that could mean and its relationship to my practice. I'm creating objects and I'm creating the tools to make these objects as well. So I'm creating the tools to replicate them or to take on as your own and to do it somewhere else. I'm proposing a different way: putting something out there as a provocation and saying, you know, this is an important space – this bench is an important space and things happen here. Someone's come and repaired it and created this beautiful, ornate piece – because it has a value. And by walking past it that there's the possibility that you might recognise that value or you take an extra second just to consider why someone's done that. And hopefully, maybe the idea sparks something that makes you think differently about these objects.

We're standing by a bench that was made in the mid-twentieth century. Its frame is solid, cast iron, with lions and flourishes.

Linda: These objects in public space, these benches, and fountains and stuff – as well as them being civic amenities, they have a value, materially, and through their craftsmanship and skill. They ask us to reflect back on who we are. Things that have been made well, things that have been cared for, or things that have been designed beautifully, like these with the flourishes and lions, they reflect back to us our history, and our people. They reflect back the concerns of the time. And so by sitting on them, or having access to them, whether they're water fountains, public toilets, park benches, all of those types of things, we get a sense of the care put into making these objects. When we reduce all those materials to the minimum, and to the budget line, we're reducing our connection with craftsmanship and materials and skills. When you don't go to a museum, or you don't have access to those spaces of high culture, sometimes public space is your only access to something that's been beautifully made or made with care.

If you remove those things, and you only see the cheapest, most basic forms of workmanship, how do you know what is possible? If you want to work with your hands, how do you aspire to act differently? What's the proposal for a different way of living? What's the proposal for taking extra care or paying extra attention?

Leila: It's also a reminder for us that the metrics that we see now weren't always the same and therefore they don't have to be forevermore? That this phase that we're in now, where everything's down to this bottom, might pass.

Linda: That's so true. When it's all you've lived, all your life is where things are now, when we're in times of austerity, then by seeing those lovely park gates and lovely benches … it's different from architecture. Architecture seems other than us, even when it's civic architecture, whereas things you use, and it can interact with and touch and sit on, they're closer to us. Anyone can sit on a bench. We're not in awe, we don't worry, or think we've got it wrong. So when it's beautiful or carefully made or looked after, it gives you a little snapshot of how the metrics could be different.

Public objects that are beautiful, well-made and cared for produce a sense of being valued and held. They highlight the relationship between our subjectivities and our surroundings: how inhabiting ruined or neglected spaces, or sharing our lives with objects that have been neglected, or installed the cheapest, quickest way, makes us feel uncared for. When we interact with a world inhabited by objects that have been looked after, by tended gardens, by tidy and well-maintained streets, we have a sense that we matter.

On holding and being held

The psychoanalyst Donald Winnicott developed the concept of 'holding environments' in relation to the importance of early maternal care – the physical holding, feeding and bathing a baby – to healthy development. In containing the child corporeally, and responding to its needs, the baby is able to experience its own body as a secure world. Holding, for Winnicott, involves consistency, dependability and attunement to one's needs. It is essential for children, and adults, to experience what he describes as 'continuity of being': a sense of being comfortable in one's surroundings and not moving into a 'reactive' (hypervigilant, unsettled) state. Where a holding environment fails, the infant 'disintegrates' as a defence mechanism against the instability of the world. The holding environment is not restricted to the mother: the net is cast wider, and, crucially, includes the environment and mechanisms through which the carer herself is held, or how a therapeutic setting and intervention can recreate a holding environment. The network of care that ensures healthy development involves a much broader framework than the maternal relationship: 'One can discern a series – the mother's body, the mother's arms, the parental relationship, the home, the family including cousins and near relations, the school, the locality with its police-stations, the country with its laws' (Winnicott, 2018: 310). This expanded holding environment sheds light on the non-human providers of care: the objects and spaces that contribute to bodily comfort and ontological stability. Winnicott was concerned with the conditions for a good and healthy life, and mitigating against the alienation experienced from both self and world when these were absent. At the same time, he understood the good life as a life that is interdependent: where relations of care-as-holding are the fabric of ontological security. It is probably for this reason that Winnicott's ideas were so influential during the development of the post war Welfare State in the UK (Alexander, 2012). Winnicott's understanding of the role of authorities in maintaining stability and security aligned with the Keynesian social democratic model that was being developed, and with the ambition of the post-war settlement, which, through its biopolitical management of family, welfare and health, attempted to ensure stability and 'cradle to grave' social security (Gerson, 2005).

If there are clear similarities between Winnicott's networks of care, consistency and stability and an idealized post-war settlement, then perhaps the withering of the welfare state over the past 40 or so years, and the spatial, social and affective impacts of cuts in public service provision, can be understood as a dissolution of a holding environment.

Linda: When a bench is repaired in a really lovely way that is really beautifully and painstakingly done, but also is designed to be incredibly comfortable and welcoming. That is how objects can hold you. And how people who care for objects can hold you via the object.

Leila: Do you think you are trying to create a new relationship between people and their things, that's about taking care of things, and things taking care of you?

Linda: These benches are a way of imagining that things could be done in a different way. I'm giving a living example to people to experience and to sense being held and having this touch of things being done at a level which isn't the bare minimum. So, when the arm holders go down into the lion, and they curve around, and when the back curves round a little bit extra, you have comfort, you are being held, you're being looked after. And when public services have that, like the water fountains and lovely toilets with decorative tiles and hot water and big ceramic basins and all of these other things, it offers this tantalising idea of they don't always need to be what they are now.

When we encounter public things that have had time, money and attention spent on them, when they are beautifully crafted, when they are maintained and cared for, we feel ourselves looked after and cared for; we feel held. When the benches we sit on have been made with comfort in mind, when they are made of wood rather than heat-conducting metal, when they include flourishes, extra struts and slats, lettering that attaches them to a place, we feel held.

Brothwell's work proposes a love for things, and in particular a love for public things. In her painstaking and elaborate repairs, using metal and wooden inlays, engraving and jointed woodwork, she offers an ethics towards the world of public objects that, in part, accepts that we cannot rely on the state to take care of our shared objects, but at the same time takes responsibility, and in doing so, an expansion of her own world, towards these objects. They reconfigure care towards non-human objects, recognizing the role that caring for them plays in the making and healing of worlds.

To begin with care as constitutive of worlds, and to understand how care relies on wider networks, encourages us to consider these broader relations of holding and being held, and to incorporate the worlds of objects into our understanding of care. By this, we mean acknowledging how objects contribute to our comfort, acknowledging what this stands for in terms of the wider networks of support and welfare mechanisms that enabled this, and acknowledging the reciprocal work of caring for objects so that they can care for us. So, we come to a problem, and to a proposal. What if the mundane objects that we encounter in public space were evaluated through different metrics? What if we put comfort and connection before budget and lack of maintenance? What if we cared for our public things as we would like to be cared for ourselves? Maria Puig de la Bellacasa, in her book *Matters of Care*, expands her theorization of care to thinking about more than human worlds (Puig de la Bellacasa, 2017). Steven Jackson, too, in his seminal essay 'Rethinking Repair', understands an ethics of care and responsibility as 'an old but routinely forgotten relationship of humans to things in the world' and asks, 'Is it possible to love, and love deeply, the world of things?' (Jackson, 2014: 231–2).[4]

'I'll go first': caring infrastructures

Leila: Can we talk about the idea of responsibility? Not many people talk about the relationship between care and responsibility, and how taking care is also about making a decision to take responsibility for something. And I wondering whether you have always felt that you should take responsibility for those things.

I read something you wrote about mending the seats on the London Underground? And this struck me as a lovely thing to do. The fact that you had taken it upon yourself to say, this seat's broken. Yeah, I'm going to sort it out. Rather than, 'the seat is broken, why hasn't someone else sorted it out'? At some point, you've taken the decision to think, 'I could mend that – that's on me'.

Linda: Me choosing to do that? Firstly, I'd say it feels completely natural to me. And it doesn't at all seem odd. Even a tiniest bit, it's not an art practice, it's just how I think. There's two sides. One is, I want to care for things. Caring for my own things is one thing, and caring for things that we share gives me a lot of joy. And I feel, why not? I'm skilled, I can make beautiful things. I can share that. Why wouldn't I look after stuff? What's the point in not doing it? But then the other side of it is, that I'm laying a trace, not for the people who see it now really, but I'm sharing I was here, that I existed. I'm interacting with my world and I'm not invisible. I'm doing it for everybody in a way. But there is a part of me that's doing it for me to show that I'm here. And it's my home and I'm connecting and making my home bigger, and expanding what is in my protection, and in my responsibilities. So it's not all for everyone else. I get way more out of it than anyone. You know, it's really fun. It's really enjoyable. I share it with people: people come and then they come and talk to me and they go, Oh, yeah, I'm gonna get into that. How amazing is that?

Leila: Have you ever been on the underground and seen your work again?

Linda: I've never seen it again, but part of that came from this idea that in some places, as a bus driver, you're responsible for your own bus. You look after it. You clean it down at the end of the day. You're not the cleaner but you care for it, you're responsible. And so there is a convention of quick repairs, and you know, you're caring for it all the time. And that idea of responsibility for what you use, even if they're public, in public space, I think is really interesting. And then doing something to enhance it, or to lay a trace of people existing, I think is something that I enjoy doing. But the other thing is I did grow up on a farm. So, the idea of just fixing something, because it's broken is not at all unusual for me.

Leila: There's a confidence in that isn't there? Like at home, I'm thinking how I can either look at something that doesn't work, and complain about it, or I can bodge it. My instinct is always to fix, to bodge some solution. But that takes confidence, especially when, like me, you're incredibly impractical and quite bad at it.

Linda: I think that sometimes the bodged repairs are sometimes the most beautiful. Sometimes, because they show an energy of someone who isn't necessarily the perfect person to do the job, but they've taken it upon themselves to do it anyway. Like we're walking along the waterfront, and if someone said all of that [railings] needed polishing, I would happily just sit and just do it. I guess there's a part of it that's confidence. But then there's a part of it that is, I get to do it when I'm actually working.

So, I can show care and I can be in the world and I can exist and I can have a voice by being a positive force in the landscape by looking after things and repairing things. But I can do it in a way that I'm not making it all about me. I can just do it. The traces of a repair is just like the traces of someone caring. It's quiet but lovely – it's an anonymous act of love.

In this extract, Linda refers to a practice of 'making my home bigger'. Practices of taking care of that beyond the boundaries of our home and property expands our worlds: expands what we feel is ours and what we take responsibility for. If we can no longer rely on the indirect relation of taxation and citizenship to enact this responsibility, then the solution, for Linda, is nurturing a more direct relationship between ourselves and the things we hold in common. Through actively taking responsibility for the places and objects we share, we are reclaiming public things as important to us.

Caring for public objects make people feel part of something, and feel like they have collective stakes. It draws attention to how shared life is produced through objects and spaces: low fences and back alleys were highlighted as ways in which collective life was maintained, in Valerie Walkerdine's research on deindustrializing communities, while my own work draws attention to the role of material infrastructures in a former nuclear town (Walkerdine, 2010; Dawney, 2020). Practices of responsibility and care of our material worlds are a counter strategy to regimes of individualization and neoliberalization, and to the sense of loss brought about by the neglect of public objects.

To some, this might seem like a cop-out, or to excuse states and local administrations from their own responsibilities to take care of public things. To be sure, it fits neatly with a neoliberal push to encourage individuals and community organizations to fulfil roles previously undertaken by the state. It could also fit into the top-down production of neoliberal subjectivities that 'make compassion productive' and offer some kind of salve for the loss of secure jobs and a holding state (Muehlebach, 2011). We do not wish to deny this tension. Nor do we wish to posit DIY repairs and small acts of care as a panacea to neoliberal urban blight. Care for public things consoles, offering a ritual of mourning for a lost public. It also takes the first steps towards recognizing that loss and opening up a space for making worlds collectively. A repair in a public place makes the neglect of public things visible and draws our attention to their value, making us feel cared for, and held. The anonymity of repair itself speaks to the unnamed equality of the idea of the public.

Leila: There's a radical action in taking responsibility and saying that that's mine to look after. There's an act of claiming that needs to take place. That's a really interesting thing to think about in a situation where we don't really think about public things as ours.

Linda: I think the thing about that's mine to look after I would counter that with that's ours to look after, and I'll take it, I'll take the job first. And maybe you guys can take it after. It's a proposal. A provocation to the world. I think the claim is that we can look after things. I'll go first, to take care of it, as opposed to I'm taking ownership of it.

What does it mean to say 'I'll go first'? What does it mean to publicly take responsibility for shared objects? This ethical movement towards the world of objects constitutes a minor politics are not attached to grand narratives, but are instead oriented towards how we live on, in and through times of trouble – how we endure and make lives, how we learn to live differently with what we have. They are about shifting our ethics, sensibilities and practices from within. Brothwell's repairs alter, at the micro level, our relationships with our world, our sense of who we are, and our sense of responsibility and belonging in public space.

Linda: We can care for something and we just give something an extra bit of life. I don't see the harm in caring for something. I don't see why you wouldn't. Feeling an active participant in your landscape gives you a sense of power. It gives you a sense of ownership. It gives you a sense of responsibility. It

> *sounds like it's a negative thing, like you're responsible, but actually it gives you a sense of owning it as well. If I take care of something, I feel comfortable there, I've connected with that part of the city, I've spoken to the people, I've spent time there, I've looked after it.*
> *So, I feel like it's become part of my world, it's come in to my circle. So, if you're caring for something, there is an agency in that, You have a strength, you have a power, you have a way of changing the world to suit how you want it to be in a positive way, or in what I consider to be a positive way. People are pushed to not be allowed to care. But I think people do care, in different ways. They just sometimes need someone to go first.*

'I'll go first' interrupts the current conditions of reliance on structures that no longer hold. It shifts the relationship between individual, state and place. It acts against forces that seek to individualize and privatize; to enclose and to deny a stake. It expands our claim on our worlds through small acts, and alters the need for reciprocity through practising care for objects, and for the world as part of a wider network of holding and mutuality. It opens up spaces and resources, building collaborations and connection and provoking small ways of changing our orientation to the world from one which is privatized and contracting, to one that is expansive. While this may not be perfect, maybe it is all that we have right now.

These interventions, these acts of care for public things, are not just a nostalgic mourning for what is disappearing, but a recognition of what needs repair, and through this, their active remaking. They celebrate what is valued, a desire to hold onto an idea of public life and togetherness: not as a sentimental relic of something gone, but something that retains the potential to ignite a new feeling of public life.

Notes

1. This chapter is the result of a series of 'conversations on benches' between Leila Dawney and Linda Brothwell in Bristol during December 2020. Extracts from the conversations are in italics. The discussion is written by Leila Dawney.
2. Nowhere did this become more apparent than when benches in urban green spaces were taped up during the first Covid lockdown in March 2020.
3. Sure Start is a comprehensive early years welfare and support programme operating in the UK between that was heavily cut during the austerity policy reforms between 2011 and the present day.
4. Donna Haraway too spoke of the task of repairing interdependence, of the 'restoration and care of corridors of connection' as a 'central task' of the communities who 'imagine and practice repair' (Haraway, 2016: 140).

Acknowledgements

The authors are part of a collaboration for the Craftspace 'We Are Commoners' exhibition at Oriel Davies Gallery, Newtown, UK and touring. Funding for this project was through AHRC grant AH/S011986/1.

References

Alexander, S. (2012) 'Primary maternal preoccupation: DW Winnicott and social democracy in mid-twentieth-century Britain', in Alexander, S. and Taylor, B. (eds) *History and Psyche*, New York: Springer.

Azzouz, A. (2019) 'A tale of a Syrian city at war: Destruction, resilience and memory in Homs', *City*, 23: 107–22.

Berlant, L.G. (2011) *Cruel Optimism*, Durham, NC, Duke University Press.

Berman, M. (1987) 'Among the ruins', *new internationalist*, 178: 1–3.

Booth, M. (2017) 'City Council need to close £108m budget gap', *Bristol Post*, 6 November 2017.

Brown, W. (2015) *Undoing the Demos: Neoliberalism's Stealth Revolution*, New York: Zone Books.

Coward, M. (2008) *Urbicide: The Politics of Urban Destruction*, London: Routledge.

Dawney, L. (2013) 'Commoning: The production of common worlds', *Lo Squaderno*, 30: 33–55.

Dawney, L. (2020) 'Decommissioned places: Ruins, endurance and care at the end of the first nuclear age', *Transactions of the Institute of British Geographers*, 45: 33–49.

Fortun, K. (2009) *Advocacy after Bhopal: Environmentalism, Disaster, New Global Orders*, Chicago, IL: University of Chicago Press.

Gerson, G. (2005) 'Individuality, deliberation and welfare in Donald Winnicott', *History of the Human Sciences*, 18: 107–26.

Golańska, D. (2022) 'Slow urbicide: Accounting for the shifting temporalities of political violence in the West Bank', *Geoforum*, 132: 125–34.

Haraway, D.J. (2016) *Staying with the Trouble: Making kin in the Chthulucene*, Durham, NC: Duke University Press.

Hitchen, E. and Shaw, I. (2019) Intervention – 'shrinking worlds: austerity and depression', *AntipodeFoundation. org*, 7. Available from: https://antipodeonline.org/2019/03/07/shrinking-worlds-austerity-and-depression/

Honig, B. (2017) *Public Things: Democracy in Disrepair*, New York: Fordham University Press.

Jackson, S.J. (2014) 'Rethinking repair, media technologies: Essays on communication, materiality, and society', in T. Gillespie, P.J. Boczkowski and K.A. Foot (eds) *Media Technologies: Essays on Communication, Materiality, and Society*, Cambridge, MA: MIT Press, pp 221–40.

Katz, C. (1996) 'Towards minor theory', *Environment and Planning D: Society and Space*, 14: 487–99.

Katz, C. (2017) 'Revisiting minor theory', *Environment and Planning D: Society and Space*, 35: 596–99.
Muehlebach, A. (2011) 'On affective labor in post-Fordist Italy', *Cultural Anthropology*, 26: 59–82.
Nixon, R. (2011) *Slow Violence and the Environmentalism of the Poor*, Cambridge, MA: Harvard University Press.
Pain, R. (2019) 'Chronic urban trauma: The slow violence of housing dispossession', *Urban Studies*, 56: 385–400.
Povinelli, E.A. (2011) *Economies of Abandonment: Social Belonging and Endurance in Late Liberalism*, Durham, NC: Duke University Press.
Puig de la Bellacasa, M. (2017) *Matters of Care: Speculative Ethics in More Than Human Worlds*, Minneapolis, MN: University of Minnesota Press.
Stoler, A.L. (2013) *Imperial Debris: On Ruins and Ruination*, Durham, NC: Duke University Press.
Walkerdine, V. (2010) 'Communal beingness and affect: An exploration of trauma in an ex-industrial community', *Body & Society*, 16: 91–116.
Williams, R. (1977) *Marxism and Literature*, Oxford: Oxford University Press.
Winnicott, D. (2018) 'The Antisocial Tendency 1', in *Through Paediatrics to Psycho-Analysis*, London: Tavistock Publications.

17

Curating Reparation and Recrafting Solidarity in Post-Accord Colombia

Fredy Mora-Gámez

Entanglements of injustice

She was a mother. The past tense is to some extent accurate. Her 19-year-old son left the house one day quite excited about finding a new job and never came back. The next time she saw him, his dead body was wearing strange clothes. 'Guerrillero' the media, the Army and the President called him. The newspapers were wrong, the militaries were wrong, former president Uribe was wrong and she knew it. Twenty years later a controversial Human Rights Watch Report proves her right, as it does to many other Madres de Soacha.[1] Nowadays she marches with other mothers in her situation, even in times of lockdowns. The pandemic has not taken away her pain, memories, and the need for justice.

While marching in the streets of Bogotá, her path comes across other mothers, their partners and children, all of whom embody a different kind of pain. They have walked endless distances to reach this unwelcoming place. They are sometimes mistaken for *Desplazados*,[2] but their Venezuelan accent gives them away. They were displaced from their homes by dispossession, violence, and hunger. They also march, but not to protest as their right to do it is sometimes considered illegitimate. They march to cross the Colombian-Venezuelan border, and to endure subsequent bureaucratic borders. As other people on the move within the country, they arrive in the cities and inhabit the streets of Bogotá where the stories of this chapter become more visible.

State crimes like the ones perpetrated against Madres de Soacha and their sons, also known as the *Falsos Positivos* case, are not new. Falsos Positivos is the name given by the media to at least 6400 young men assassinated by the Colombian Army after being dressed up as guerilla militants (JEP, 2021). It is claimed that these assassinations were motivated by an incentive system promoted during the Presidency of Alvaro Uribe between 2002 and 2008, offering soldiers financial rewards for every guerilla casualty.[3] Similarly, displacements due to violence and precarity are not new either. The increasing cases of Venezuelan nationals arriving in Colombia adds to the existing number of people usually addressed as *Desplazados*, or Internally Displaced People.

State crimes and forced displacement are only part of the entanglements of injustice generalized in the landscape of post-accord Colombia. Post-accord is used here as an intentional term. It acknowledges the contradictory nature of *post-conflict* Colombia and the ongoing violence piercing the everyday life of people, mainly in the rural areas of the country. This ongoing violence is not diminished by lockdowns, quarantines or signed peace accords. Yet, the term post-accord also acknowledges the imaginary around the signature of the peace accord back in 2016 as an inflexion point in the history of Colombia, an imaginary mobilizing hopes for peace and the absence of war. Post-accord also draws attention to the ways in which the landscape after the accord has importantly been shaped by people on the move also inhabiting Colombia after the 2016 peace accord signature.

Within this landscape, governmental reparation and international asylum are deployed as cynical projects that are supposedly designed to reinstate lost rights and conditions, but end up reproducing historical asymmetries of power between the state and those who apply for registration and compensation. Recognition of applicants as *Victims* of the armed conflict is a status granted after long bureaucratic procedures including applicants in the Registro Único de Víctimas[4] (RUV). The material politics of this population management infrastructure unmakes the numerous and divergent stories offered by applicants about the situations of violence they went through, to make and promote the official account of the armed conflict narrated by the state and its representatives (see Brown, 2016). An account that makes invisible the role of the Colombian State as one its main actors and perpetrators. The case of Falsos Positivos is not considered as a part of the armed conflict although it is a consequence of the strategies used by the Army in the war against guerillas. Similarly, other state crimes affecting former militants of guerillas or their relatives, sexual crimes committed by members of the public force and assassination of soldiers who refused to follow orders in the Falsos Positivos case are disregarded from the official records and remain absent from institutional forms of memory of the armed conflict (Mora-Gámez, 2016: 92).

Governmental solidarity, henceforth *asylum solidarity*, towards Venezuelan nationals, as the administrative set of actions to assist asylum seekers entering Colombia, does not escape the cynicism described before. The Registro

Administrativo de Migrantes Venezolanos en Colombia (RAMV)[5] condenses only a portion of the applications processed per day by the National Agency of Migration. The waiting times to be registered in the RAMV reach up to two years, whereas the procedures to obtain a special permanence permit (PEP) are mediated by digital platforms constantly malfunctioning. One of the few successful strategies to access asylum solidarity seems to accept the proposal of civil servants who are capable of producing false birth certificates forging a false Colombian nationality of the applicants' parents.[6] Hence, assemblages like the RUV and the RAMV reproduce and neglect historical and ongoing asymmetric relations of power shaping the identification and registration processes of those who apply. As technologies that make Colombian society durable (Latour, 1990), the RUV and the RAMV materialize a political disagreement (Rancière, 1999), an incapacity of the state to acknowledge the multiplicity of forms of violence affecting the everyday lives of people on the move and victims of state crimes.

Although in other parts of my work I have addressed the paradoxical relations assembled in population (migration) management infrastructures, in this chapter I pursue a different line of inquiry. On this occasion, I wonder about the worlds exceeding governmental reparation and asylum solidarity in Colombia. While engaging with protests in the streets of Bogotá, I have come across material arrangements and experiences which shed light about the ways developed by my interlocutors, mainly applicants and people on the move, to reclaim justice beyond the official channels. The latter channels offer a promise of institutional reparation and solidarity that mainly regulates the temporality of the lives of those who are recognized as victims (Jaramillo, 2012; Ulfe and Sabogal, 2021). This regulation, consisting also of bureaucratic-technological containment, enacts a notion of justice in which the state becomes the unique legitimate repairer and provider of (asylum) solidarity.

Instead of narrating the unsurprising story of the state apparatus of registration, reparation and asylum solidarity, I will share my enthusiasm for particular material transformative practices (Naji, 2009) that I have followed while tracing how non-humans participate in other forms of reparation and solidarity different from the governmental ones. Hence, I will describe those material transformative practices as inter-embodied through the nearness, through the being-with-others, in this case also with humans and non-humans (Ahmed and Stacey, 2001; Puig de la Bellacasa, 2011).

In what follows, I will describe two practices of relevance for the Falsos Positivos case and the struggles of Venezuelan nationals on the move, as two instances in the extensive entanglements of injustice composing post-accord Colombia. First, I will present an ethnographic excerpt about *curating mobile memorials*, which I first came across in Bogotá in 2015 and recently revisited during the protests of November 2019. Second, I will summarize my experience of practices of *recrafting paper* by people on the move finding their ways in the inhospitable streets of the Colombian capital. Drawing on these ethnographic interfaces (De la Cadena et al, 2015), I will reframe them as worlds exceeding post-accord Colombia by reclaiming justice beyond the boundaries of governmental reparation and solidarity. As the ethnographic vignettes will suggest, these bottom- up claims for justice consists on contesting official narratives of the armed conflict, while at the same time opening up spaces for the redistribution of agency among those who participate in such practices of *curating* and *recrafting,* as well as their audiences.

Curating memorials of state crimes

A few months before the signature of the peace accord in 2016, the area close to Plaza de Bolivar was occupied by a unique street arrangement (Figure 17.1). Several pictures of protests, occupations, riots, police
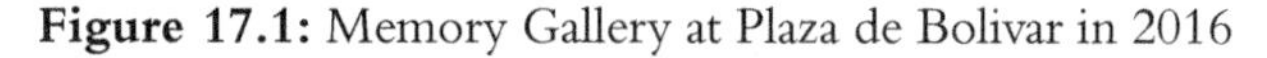
interventions, assassinated politicians and human rights defenders, Falsos

Figure 17.1: Memory Gallery at Plaza de Bolivar in 2016

Positivos, marches and other similar events were exhibited in front of crowds of people. This street exhibition was quite successful in recruiting walking spectators including myself, fascinated by the disposition of the images, the captions accompanying them and the thematic sections. The arrangement recruited different audiences pausing their walks to appreciate the materials, as well as non-humans: lampposts, ropes, multiple pictures, printed quotes protected in plastic hanging from the rope.

After being violently displaced when he was a teenager, the curator of this gallery, Manuel belonged to different peasants' movements and was a member of the M-19, a guerrilla group that surrendered in the 1990s. While living in a rural area, he was again displaced by paramilitary forces in the early 2000s and has lived in Bogotá ever since. For Manuel, instead of helping people, the administrative procedures of registration have become increasingly restricting and discourage applicants from persisting in pursuing compensations. He avoids my questions about his status in the official record of victims, but I manage to understand that he gathered enough money after a few months in Bogotá to buy a camera. 'Nobody was willing to employ me, so I decided to take the camera and show a different lens on reality', which is the purpose of his gallery. The gallery has existed for several years, also on social media, with frequently renovated content. Although most of the pictures are taken by the curator, he has also included pictures donated by people from the social movements that he seldom accompanies during their protests. Manuel also describes the many times that the police have unsuccessfully tried to dismantle and remove his gallery from the public space. and how he has cleverly used those moments to enrich it with more pictures of the dismantling.

I kept in contact with Manuel for some time after that but I lost track of his movements by 2018. Nevertheless, his social media are constantly refreshed with new pictures of protests, marches, headlines of assassinations of social leaders, new developments in the case of False Positives, among other events in the diverse repertoire of ongoing state crimes in the country. I came back to the city centre of Bogotá only until 2019 and I unsuccessfully tried to locate Manuel´s curated memorial. By the time I was actively searching for Manuel and his exhibition, which we will revisit later, a reminder of past state crimes reached the national headlines. A mass grave with at least 50 unrecognized bodies was found in Dadeiba, Antioquia. According to the official reports, these 50 bodies belonged to some of the civilians assassinated by the Colombian Army as Falsos Positivos.[7]

The news of the 50 bodies became one more reason to participate in the protests against the government of President Duque between November 2019 and January 2020. While doing so, I came across another street memorial displayed in the city centre of Bogotá denouncing the assassination of soldiers who refused to engage in the criminal actions of the Falsos Positivos

Figure 17.2: Street memorial in 2020 exhibiting evidence of the murder of a soldier who refused to follow orders in the Falsos Positivos cases

(Figure 17.2). The curator of these images and signs was Raúl, the father of one of those soldiers, and he had been installing this arrangement since 2018.

After following the official bureaucratic routes to report the assassination of his son among other soldiers, Raúl did not receive any official response from the public prosecutor's office. The pandemic has taken Raúl´s life, and yet, his claims and demands have not received any response from the Colombian State. Hence, state crimes are still made invisible in Colombia and arrangements like these challenge the official routes of reparation by making visible some of the accounts against those crimes. Although these memorials address events in the past, the concern of their curators is not about commemorating a violent past, but to cope with the present and protect their stories (Quiceño-Toro, 2016). Despite the ongoing reproduction of asymmetries and injustice by the official routes of reparation, curators of these memorials reinvent, recreate and develop ways to overcome the present injustice by engaging in street arrangements which narrative mainly uses materials as pictures, banners, and short written descriptions, among other objects.

These memorials are never the same and are not finished. Instead, they are continuously transformed by the material addition of new artefacts and only exist due to their constant reconfiguration. Elsewhere, I have addressed these arrangements as *mobile memorials* (Mora-Gámez, 2020) that circulate, share and protect divergent accounts of the armed conflict, those making visible the role of the Colombian State as an actor. Hence, the events commemorated in these memorials are mobilization, protest and the construction of alternative accounts of ongoing struggles with the State, of which the Colombian armed conflict happens to be a part of. While the accounts assembled by official registration insist on the armed conflict as a past event, the images exhibited in this mobile memorial challenge *post-conflict* narratives. Curating mobile memorials produce audiences engaging with alternative accounts of violence diverging from institutional memory.

Curate comes from the Latin *cura* which also means care. In English curate means to select, organize and look after the items in a collection or exhibition, it consists of selecting the performers or performances that will feature in an arts event or programme. Besides having the same meaning in Spanish, curating – *curar* – is the same word for healing. As a material practice engaged by Manuel and Raúl, curating street memorials consists precisely of selecting, organizing and displaying images and captions that will preserve a story made invisible by the official channels. But curating in this case is not only about the images, lampposts, hangers, plastics, and so on. Curating is also about embodying (Courtheyn 2016) and caring for those accounts displaced and neglected by the project of post-accord institutional reparation. Curators of these memorials care about state crimes and claim for justice. By doing so, they perform an attempt to partially *curar* (heal) also the broken past and the ongoing injustice, by promoting fairer stories of the present. As Arthur (2009) claims, the type of connections that memorials make possible and the 'new modes of capturing, storing, presenting and sharing data in people's daily lives' have an effect on 'the way that lives are recalled, reconstructed and represented' (p 46). With their galleries the curators become 'memory choreographers' (Conway, 2010) or 'emplaced witnesses' (Riaño-Alcalá, 2015) who create mundane arrangements in which collective memories can exist. Curating, as a form of reparation, caringly reinstates accounts that were made invisible, mobilizes spaces for usually neglected voices and recruits new audiences interested in those voices and accounts. Curating then challenges, contests and exceeds governmental reparation.

Up to this point, I have described how curating mobile memorials are also practices of care enacting a form of repair and remembrance that exceeds institutional reparation. In the next section, I explore other material practice exceeding the limits of post-accord Colombia by creating alternative forms of solidarity and existence.

Re-crafting paper

During the aforementioned protests in 2019–20 and a few blocks away from the street memorials, my path crossed with those of Natalia, Marco and Mario, a group of three friends who recently arrived in Bogotá with their

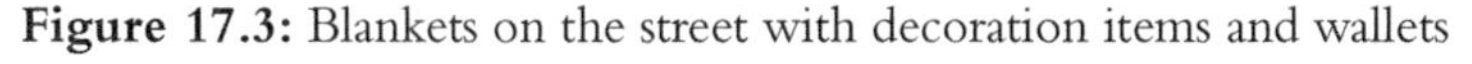

Figure 17.3: Blankets on the street with decoration items and wallets

families months after crossing the Venezuelan-Colombian border. They are not curating an exhibition, they instead use crafts displayed over blankets to claim legitimacy (Zepeda, 2017) on the streets of Bogotá (Figure 17.3). I cannot distinguish the materials of those objects on the blankets first, but I can appreciate the variety of colours and shapes conforming every bag and decoration craft. Natalia uses Venezuelan banknotes as raw materials to craft paper wallets and decoration items.

Despite the pressure from her family back in Caracas to move to Colombia a while ago, Natalia had refused to leave her country until recently. As an industrial engineer, the savings she had gathered from her previous years of work allowed her to still afford basic goods for her family. Nonetheless after the launch of the New Bolivar, the Venezuelan currency, in August 2018, her money started to lose value rapidly. She decided to withdraw all her money from the bank and try to exchange it in Colombia before it lost even more value. By the time she crossed the border in May 2019, the inflation rate dropped dramatically below 1,000,000 per cent and restrictions on the local money supply were put in place. Once in Colombia, her bags were full of worthless Venezuelan notes that nobody would exchange for food, transportation or accommodation. Her requests for asylum and a PEP have not been processed yet and, as it is usually the case of asylum seekers, she was not entitled to engage in paid employment until receiving the right documentation. During her journey, she became friends with Marco, a primary school teacher who went through the same financial situation in Maracaibo.

Marco was not an artisan. However, he remembered some of the activities he engaged with the children at work consisting of folding sheets of paper to tie them with cord into larger, thicker, textured sheets that were painted and reshaped as mats and animal-shaped paper sculptures. As an attempt to do something with the devalued money he was in possession of, Marco started experimenting with different shapes and objects until finding a way to put together a bag and a wallet (Figure 17.4). Since then, his crafts have been well received by the walking audiences of the city centre in Bogotá.

Natalia and Mario, an aficionado painter, decided to put together all the paper they had to experiment with other shapes and objects. When I was introduced to Mario, he was more reluctant to share information about his previous work. He barely engaged in our conversations, so I never came to understand what his previous life was like in Venezuela. He just let me spend time there, next to him, while he was drawing on the sheets of paper made with notes (Figure 17.5). Many tourists were curious about his paintings and constantly asked him for the price of the drawings. Talking to Natalia and Mario, I managed to understand that the money they collected per day was usually enough to pay for food, more cord to tie the banknotes, ink and a room in a neighbourhood – Santafé – close to the city centre.[8]

I witness how the journey of Natalia, Marco and Mario is remade as paper is transformed. The containment imposed by Colombian asylum policies on them is partially overcome by engaging in these recrafting practices. They share their literacy so their recrafting skills are the result of engaging with the very materiality of paper notes, cords and ink. At least to some extent, recrafting worthless paper notes into valuable decoration pieces, bags, wallets and paintings enable people on the move to endure the long waiting times of asylum application. These practices of crafting also reach and produce audiences like pedestrians and tourists learning how bodies on the move recraft what once had enormous financial value into artistic objects. Like

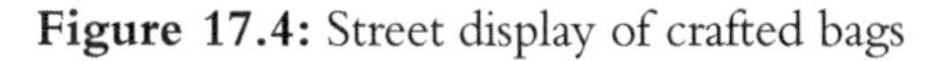

Figure 17.4: Street display of crafted bags

Natalia, Mario and Marco, Venezuelan nationals on the move must engage in inscription practices taking place at border-crossings, also in bureaucratic procedures of registration into systems like the RAMV. At the same time, in a similar way to state reparation (Jaramillo, 2012; Ulfe and Sabogal, 2021) Venezuelan nationals are promised instituted forms of asylum solidarity that follow the guidelines of international humanitarian law. In all these processes, their stories become just codified narrations of their previous lives, they are made visible for the Colombian State as part of larger numbers, and are also prevented from engaging in paid labour which is crucial for sorting out a part of their everyday lives. Simultaneously, Venezuelan nationals on the move are presented in the local media as criminals and imposters.

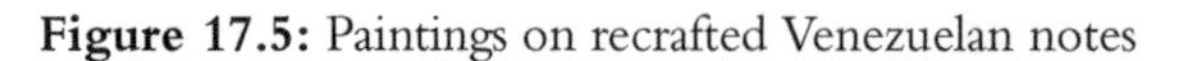

Figure 17.5: Paintings on recrafted Venezuelan notes

Crafting takes an additional meaning in the practices I have described. Recrafting is both remaking money into a new object and at the same time, narrating a story of an almost invisible journey with their own hands as the readiest available tool for someone who is on the move. Instead of neglecting the stories offered by people on the move crossing borders or presenting them as criminals or imposters, the objects recrafted from paper banknotes make visible how Venezuelan nationals on the move overcome their difficult journeys, and how they become part of the entanglements of post-accord Colombia. Although the status of victim, refugee or asylum seeker is granted by legitimized state infrastructures and border keepers, these statuses imply a paradoxical in/exclusion (Mora-Gámez, 2021). But recrafting paper banknotes into other objects like bags and wallets allows people like Natalia, Marco and Mario to survive, establish alternative solidarity networks and challenge both the imaginaries mobilized by the media as well as such in/exclusion in almost imperceptible ways (Wilcke, 2018).

Curating and recrafting: reparation and solidarity

She still marches as a Madre de Soacha. In her pursuit of truth, those who curate street memorials are aware of her painful experiences. Different from the notion of justice offered by the promise of reparation (Jaramillo, 2012; Ulfe and Sabogal, 2021), alternative forms of justice come from the acknowledgement of the suffering produced by State crimes, a contestation of the official accounts of the armed conflict, and the enactment of new spaces where neglected accounts are now witnessed by new audiences. Meanwhile, those who travelled endless distances from Venezuela to reach the cities still inhabit them as unwelcoming places. As 'space invaders' they craft objects to tell their stories by using their hands, but also to survive and establish new life connections with tourists, pedestrians and even with the materials they use. As other recrafting spaces, such mundane arrangements allow people on the move to challenge the official channels of asylum solidarity (Mora-Gamez, 2020) and social imaginaries about themselves.

In this chapter, I wondered what exceeds governmental reparation and asylum solidarity. I inquired about other forms of reparation and solidarity by following material transformations developed by curators and re-crafters in the streets of Bogotá. However, curating mobile memorials and recrafting paper notes are arrangements irreducible to the visible space they display in the streets of Bogotá. Their materiality, trajectories, stories and multiplicity exceed the urban space by condensing also political and historical disagreements between incommensurable worlds. Whereas the material politics of governmental infrastructures like the RUV are technologically shaped by one of the actors of the armed conflict, curating mobile memorials displays a claim for justice by exposing the asymmetry of power assembled in such infrastructure. Similarly, as per many other asylum systems (Esposito et al, 2020), the RAMV imposes containment in those who apply for registration and appeal to the conventional channels of participation. However, recrafting opens up conditions of possibility by offering prospects that allow people on the move to endure extended waiting times.

Curating street memorials promotes alternative forms of participation beyond instituted channels. Curating thus destabilizes the imaginary about the Colombian State as post-conflict nation, as the unique and legitimate repairer fairly acknowledging the consequences of a long-sustained war. Similarly, recrafting paper notes also makes visible the struggles of a journey between countries, facing multiple situations of injustice while interrogating the asylum solidarity of the Colombian State and sometimes destabilizing the imaginaries about themselves as imposters and criminals. Crafting skills hence provide material means of persistence despite the bureaucratic containment unfolded by the RAMV.

Curating and recrafting are mundane practices unfolding intimate affective encounters (Anderson, 2009; Slaby, 2019) embedded in collective efforts, shared literacies (Papadopoulos, 2018) and involving alliances between humans and non-humans. These forms of reparation and solidarity partially challenge injustice expanding beyond the boundaries of state policies. Curating and recrafting thus re-establish some of the loss left by war and border crossing. Despite the precarity in which these mundane arrangements take place, these open-up spaces of everyday affective engagement for their participants and audiences. The forms of reparation and solidarity enacted by curating and recrafting are then relational achievements enacting forms of overcoming, relating to and engaging with the aftermath of war violence (see Cortés-Rico et al, 2020) and border crossing in everyday life. These ontologies produce material, temporary and mobile communities that embody new forms of existence, creating alternative presents in the current entanglements of post-accord Colombia.

Notes

1. Madres de Soacha are a collective of the mothers of some of those who were assassinated in the case better known as Falsos Positivos.
2. Internally Displaced People
3. A recent report by the Jurisdicción Especial para la Paz (JEP, 2021) confronts Uribe's claims about the official numbers of these cases presented by him during his presidential period.
4. Official Record of Victims. See Mora-Gámez (2016) for a detailed reconstruction of this record and its material politics. This record indicates 9,106,309 people as victims of the armed conflict at the time of writing. Source Unidad para la Atención y Reparación Integral a las Víctimas UARIV, 2021.
5. 1,739537 on 31 December 2020. Source: Migración Colombia.
6. Cases reported by El Espectador. 19 January 2021. www.elespectador.com/noticias/nacional/la-red-de-corrupcion-detras-de-la-nacionalizacion-de-extranjeros/
7. El Espectador. 23 February 2020. La verdad oculta en Dadeiba. Available from www.semana.com/nacion/articulo/falsos-positivos-en-dabeiba-jep-identifico-el-primer-cuerpo-exhumado-del-cementerio/651904/ [Accessed on 10 January 2021].
8. My experience of these Recrafting practices in Bogotá is quite connected to others I have accompanied in Athens and Mytilene. My interlocutors in Greece, a collective of people on the move and activists, also re-craft the wasted rubber of their arrival boats in Lesvos

into bags and wallets. Their crafting space in Athens offers their participants the possibility of developing new skills, sharing information about bureaucratic procedures, and coping with the everyday challenges of being on the move in an unwelcoming place after crossing the Southern border of the European Union. (see Mora-Gámez, 2020 and 2021).

References

Ahmed, S. and Stacey, J. (eds) (2001) *Thinking Through the Skin*, London, Psychology Press.

Anderson, B. (2009) 'Affective atmospheres', *Emotion, Space and Society*, 2(2): 77–81.

Arthur, P. (2009) 'Saving lives: Digital biography and life writing', in J. Garde-Hansen, A. Hoskins and A. Reading (eds) *Save as … Digital Memories*, Basingstoke: Palgrave Macmillan, pp 44–59.

Brown, S.D. (2016) 'Violence and creation: The recovery of the body in the work of Elaine Scarry', *Subjectivity*, 9(4): 439–58.

Conway, B. (2010) *Commemoration and Bloody Sunday: Pathways of Memory*, Basingstoke: Palgrave Macmillan.

Courtheyn, C. (2016) '"Memory is the strength of our resistance": An "other politics" through embodied and material commemoration in the San José Peace Community, Colombia', *Social and Cultural Geography*, 17(7): 933–58.

De la Cadena, M., Lien, M.E., Blaser M. Bruun Jensen, C. Lea. T, Morita, A. Swanson H. Ween, G.B., West, P. and Wiener, M. (2015) 'Anthropology and STS: Generative interfaces, multiple locations', *HAU: Journal of Ethnographic Theory*, 5(1): 437–75.

Esposito, F. Murtaza, A. Peano, I. and Vacchiano, F. (2020) 'Fragmented citizenship: Contemporary infrastructures of mobility containment along two migratory routes', *Citizenship Studies*, 24(5): 625–41.

Jaramillo, P. (2012) 'Deuda, desesperación y reparaciones inconclusas en la Guajira, Colombia', *Antípoda. Revista de Antropología y Arqueología*, 14: 41–65.

JEP (2021) Jurisdicción Especial para la Paz, Comunicado 019 de 2001. La JEP hace pública la estrategia de priorización dentro del Caso 03, conocido como el de falsos positivos. Available from: www.jep.gov.co/Sala-de-Prensa/Paginas/La-JEP-hace-p%C3%BAblica-la-estrategia-de-priorizaci%C3%B3n-dentro-del-Caso-03,-conocido-como-el-de-falsos-positivos.aspx [Last accessed 18 February 2021].

Latour, B. (1990) 'Technology is society made durable', *The Sociological Review*, 38(1_suppl): 103–31.

Mora-Gámez, F. (2016) 'Reconocimiento de víctimas en Colombia: sobre tecnologías de representación y condiciones de estado', *Universitas Humanística*, 82(1): 75–101.

Mora-Gámez, F. (2020) 'Beyond citizenship: The material politics of alternative infrastructures', *Citizenship Studies*, 24(5): 696–711.

Mora-Gámez, F. (2021) 'Thinking beyond the "imposter": Gatecrashing un/welcoming borders of containment', in S. Woolgar, E. Vogel, D. Moats, and C.F. Helgesson (eds) *The Imposter as Social Theory: Thinking with Gatecrashers, Cheats, and Charlatans*, London: Bristol University Press.

Naji, M. (2009) 'Gender and materiality in-the-making: The manufacture of sirwan femininities through weaving in Southern Morocco', *Journal of Material Culture*, 14(1): 47–73.

Papadopoulos, D. (2018) *Experimental Practice: Technoscience, Alterontologies, and More- than- Social Movements*, Durham, NC: Duke University Press.

Patarroyo, J., Cortés-Rico, L., Sánchez-Aldana, E., Pérez-Bustos, T. and Rincón, N. (2019) 'Testimonial digital textiles: Material metaphors to think with care about reconciliation with four memory sewing circles in Colombia', in T. Mattelmäki, R. Mazé and S. Miettinen (eds) *Nordes 2019: Who Cares?*, 3–6 June, Aalto University, Espoo, Finland.

Puig de La Bellacasa, M. (2011) 'Matters of care in technoscience: Assembling neglected things', *Social Studies of Science*, 41(1): 85–106.

Quiceño-Toro, N. (2016) *Vivir sabroso: luchas y movimientos afroatrateños en Bojayá, Chocó, Colombia*. Bogotá: Editorial Universidad del Rosario.

Rancière, J. (1999) *Disagreement: Politics and Philosophy*, Minneapolis, MN: University of Minnesota Press.

Riaño-Alcalá, P. (2015) 'Emplaced witnessing: Commemorative practices among the Wayuu in the Upper Guajira', *Memory Studies*, 8(3): 282–97.

Slaby, J. (2019) 'Relational affect: Perspectives from philosophy and cultural studies', in E. van Alphen and T. Jirsa (eds) *How to Do Things with Affects: Affective Triggers in Aesthetic Forms and Cultural Practices*, Leiden: Brill.

Ulfe, M.E. and Sabogal, X.M. (2021) *Reparando mundos: Víctimas y Estado en los Andes peruanos*, Lima: Fondo Editorial de la PUCP.

Wilcke, H. (2018) 'Imperceptible politics: Illegalized migrants and their struggles for work and unionization', *Social Inclusion*, 6(1): 157–65.

Zepeda, H.E. (2017) 'El mercadillo rebelde de Barcelona. Prácticas antidisciplinarias en la ciudad mercancía', *Quaderns-e de l'Institut Català d'Antropologia*, 22(1): 67–87.

PART VI

Representing<>Self-governing

18

Commons-Based Mending Ecologies

Doina Petrescu and Constantin Petcou

Commons and the ecological repair

It appears evident today that the current capitalist mode of production based on the idea of 'eternal economic growth' is incompatible with the ecological limits of the planet. The ecological disaster is too big and complex to be solved by existing capitalist politico-economic structures. This is because, on the one hand, many of these structures have a direct interest in maintaining the current mode of production and, on the other hand, the existing way of making and implementing public policies doesn't have the capacity to follow and react quickly enough to the rapid ecological deterioration of the planet.

In this context, the commons suggest an alternative solution. Commons are defined as a 'social system in which resources are shared by a community of users/producers, who also define the modes of use and production, distribution and circulation of these resources through democratic and horizontal forms of governance' (De Angelis and Harvie, 2014).

Being built on relations of collective care, regeneration, and resilience, the commons can contribute to the planetary ecological repair and provide an alternative to the extractive and exploitative relations of the capitalist economy. Learning how to govern our planet as a commons is part of the imperative of becoming more resilient, but also more democratic. As such, the ecological question should not only be tackled in relation to the environmental crisis, but also, as suggested by Guattari (2000), in relation to the social and politico-economic crisis and the prevailing mental ecology. It also needs to be addressed in post-anthropocentric terms, going beyond the human world towards more-than-human life worlds. An important amount of theoretical work has been done in this sense by anthropologists examining indigenous relations with nature and territory (Viveiros Castro, 1998; Rose,

2013; Tsing, 2015), or post-humanist and vital materialist theorists (Barad, 2003; Bennett, 2010; Papadopoulos, 2010; Puig de la Bellacasa, 2017). Commons can address all these questions in practice at different territorial scales and levels.

Within deprived neighbourhoods for example, commons offer an alternative to the current urban regeneration approaches. This alternative involves not only refurbishing derelict housing estates and renovating physical infrastructure but also repairing broken social relations, altered subjectivities, depleted imaginaries, unjust politics and uneven economics as well as empowering communities to participate in the repairing of their neighbourhoods and at the same time participate in the repairing of the planet. Many civic and protest movements currently ask for the State or the City to adopt quick social and environmental measures against climate change. However, waiting for these measures to be enacted from the top, as a sort of 'reparation' to the damage done to the environment and the society as a whole, takes too long and it is not enough. Citizens, too, can organize themselves to reclaim these rights from below through commoning initiatives. Such processes need agency and architects and designers are well placed in playing a role in enabling it.

R-Urban: a model of commons-based resilience in suburban neighbourhoods

This was one of the stakes of the R-Urban[1] project initiated in 2011 by a collective of architects and urban designers called *Atelier d'Architecture Autogérée/Studio of Self-Managed Architecture* (AAA).[2]

R-Urban is a commons-based network of civic resilience hubs in Parisian suburbs. The 'R' in R-Urban stands partly for 'resilience', understood as a transformative condition, which allows us not only to adapt but also to transform and reinvent our society towards more balanced, equitable ways of living (Brown et al, 2012). The 'R' also stands for 'resourcefulness', situating resilience in a positive light and relating it to the empowerment and agency of citizens and their more than human communities (MacKinnon and Derickson, 2013). Although conceived and initiated by architectural designers and urban researchers, the R-Urban framework is designed to be enacted through co-production with local residents and a wide range of actors, all of whom having a role to play in the process. R-Urban aims to provoke 'commoning', creating spaces where a resilient alternative to the current way of governing resources within a community and beyond can emerge (Ostrom, 1990).

R-Urban addresses a particular territory that is the metropolitan suburban neighbourhood. In the context of a global urbanization rate projected to grow to 70 per cent by 2050 (Department of Economic and Social Affairs,

2011), the metropolitan neighbourhoods become critical sites for urban resilience and climate action.

The project started by proposing ways of ecological repairing and regenerating deprived neighbourhoods in the suburban area of Paris. Most of these neighbourhoods and their mass housing estates are today a derelict remain of the welfare state provision from the 1960s and 1970s. At the time, social programmes and public facilities were directed to these housing estates, which were providing the labour force for the metropolitan centres. These estates are today in decay, and their inhabitants are facing difficult social and economic challenges: unemployment, poverty and exclusion. Many of these problems translate into a social malaise that takes sometime violent forms: protests, youth revolts and urban struggles. As a way of 'reparation', many of these estates have seen heavy top-down regeneration programmes which implied either demolition or forced gentrification.

However, these areas can be understood as a sort of 'Critical Zones' in the sense of Bruno Latour: zones which 'reterritorialise the political questions' in relation with ecology. In fact, many of the social suburbs are already framed as 'zones' by French urban politics: ZUSs (Zones Urbaines Sensibles), ZUPs (Zone Urbaines Prioritaires), and so on, are ways of marking that something special goes on there that needs public attention. Understanding the metropolitan neighbourhoods as 'critical zones' goes beyond criticizing the increasingly diminished State support and point them as places where we can forge 'new forms of citizenship and new types of attention and care for life forms to generate a common ground' (Latour, 2020).

R-Urban recognizes that there is significant social power in these 'zones', which is based on everyone's potential of having an active presence and passing to action. For Foucault, power is never wholly expressed on a global scale, but is diffused within 'micro-powers' in local innumerable points across an 'endless network of power relations' (Waltzer, 1986). Emancipatory politics can use these diffused civic powers and create possibilities of networking with them. This idea is at the core of the micro-political approach of the R-Urban hubs, which offers opportunity for inhabitants to reclaim space and test new ways of living and working together and new forms of citizenship: ecological, social, economic and multicultural.

The R-Urban hubs take advantage of the urbanism of the suburban neighbourhoods based on modernist principles that preserve a lot of empty space between the buildings. This space, which is currently underused and poorly maintained, is an important resource where nature can be nurtured and ecological cycles developed. Also, if these neighbourhoods were built through the urbanization of rural and productive territories, R-Urban is meant to bring the rural back into the city and valorise it as an ecological resource.

Figure 18.1: R-Urban principles diagram

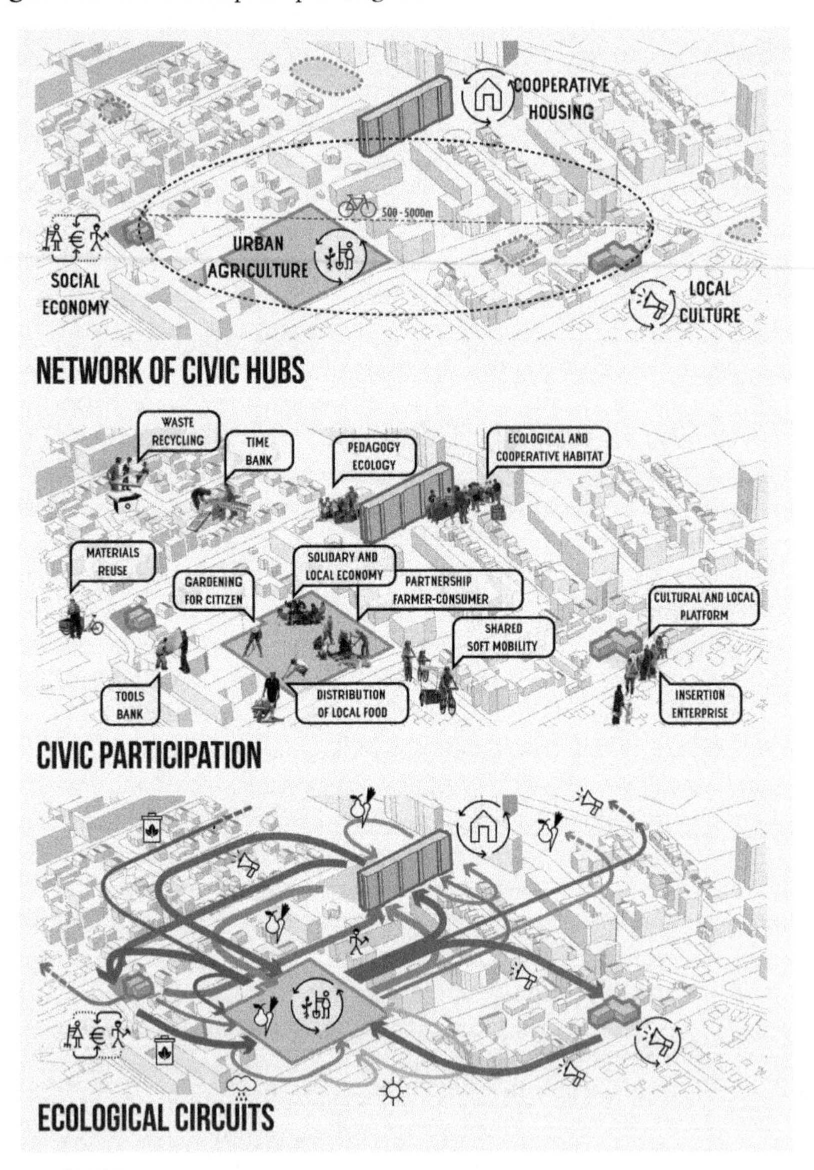

Source: © AAA

The first step in the implementation of the R-Urban strategy is to access urban space and install a physical infrastructure that creates assets for civic resilience hubs where the diffused power of the community is activated and commoning activities can take place (Figure 18.1). The second step involves identifying and enrolling stakeholders, including existing organizations, initiatives and individuals throughout the locality who can use the space and infrastructure and share the resources and the training provided by the hubs. Accessing urban space can be achieved by using available private or public land, including spaces that can be temporarily and reversibly used. The third step is to create networks of stakeholders around these assets and allow them to function through local economic and ecological cycles. As such, energy could be created locally, waste is recycled and transformed locally, food is produced locally, but also nature-based cycles can be developed together with systems of collective governance of community resources.

As famously demonstrated by Elionor Ostrom, the governance aspect is key for the functioning of commons, specifically traditional commons. This concerns exactly the set of collective agreements and shared concerns to not destroy, but to support the resources that a community depends upon (Ostrom, 1990). As such, commons are not only about governance but also about relationships. According to Linebaugh, commons is an activity that, if anything, expresses relationships in society that are inseparable from relations to nature. For Linebaugh, it might be than better to keep the word as a verb, rather than as a noun, a substantive (2008: 279), in order to mark the continuous making and re-making of the commons through shared labours and capacities. Commoning expresses the relation between the community and the resources on which it depends, that is a relation of situated interdependence between humans and non-humans who 'negotiate, imperfectly, the various limits (needs) and possibilities (capacities) that arise in particular contexts' (Bresnihan, 2015).

At a moment in which austerity measures have taken a disastrous toll on public infrastructures (Fritz and Krasny, 2019), the model of the civic resilience hub proposed by R-Urban offers a new type of urban equipment which is self-sustainable and citizen-run. The hubs are managed as commons providing collective resources for the inhabitants of a deprived neighbourhood and enabling them to develop resilience practices.

R-Urban hubs' first implementation

The implementation of the R-Urban framework started in 2011 in Colombes, a suburban town near Paris, and continued in Hackney Wick, London. Three hubs were planned in Colombes: Agrocité, Recyclab and Ecohab.[3] Agrocité was located on a 2000 m^2 plot of vacant land near a large social housing complex, Recyclab was built on one lane of a disused road and the Ecohab housing development was to be built on a vacant plot of land midway between the other two hubs. Hubs were located within easy walking and biking distance from each other to enable the circulation of food, waste, recycled materials, repaired goods, people, knowledge and cultural exchanges.

Figure 18.2: Agrocité – an urban agriculture hub, R-Urban Colombes, 2013

Source: © AAA

From 2011 to 2016 some 6900 citizens participated in the Colombes R-Urban sites from which 400 became active stakeholders. The majority participated in Agrocité's micro farming and community garden plots (Figure 18.2). A building was constructed from recycled wood to house a café, a teaching space, a market, a greenhouse, a kitchen and a workshop space. The site became a hub for ecological education and community learning. Some participants set up small businesses and generated income for themselves and for the R-Urban collective.

The Recyclab was the community recycling and eco-construction hub that was self-constructed with second-hand materials (Figure 18.3). Repurposed shipping containers were used as the ground floor with a first floor built out of wood. The structure housed workshops, materials storage space, a design studio, a café and garden deck. Agrocité and Recyclab established systems to reduce CO_2 emissions, harvest and use rainwater, compost organic waste and collect and recycle other urban waste. Immaterial flows such as knowledge transmission and care practices, social exchanges and green jobs creation were also generated by the network.

Both hubs were organized as civic organizations (that is, associations 1901), with each type of activity (compost, recycling, cuisine, garden, book keeping, repairing etc.) managed by a collective of active members who were passionate about these practices. The thematic collectives organized around each activity were represented in the governance structure of each hub, together with AAA and the City. Decisions were taken during general

Figure 18.3: Recyclab – a recycle and eco-construction hub, R-Urban Colombes, 2014

Source: © AAA

assemblies of all hub members, which took place four times a year. For a designed commons, such as R-Urban, conceiving an appropriate governance format was indeed part of the co-design process.

Unlike the marketized technocentric approaches to resilience, the hubs are run by citizens with the scope of maintaining an ecology of local practices. The democratic governance of the network of R-Urban commons associated with concrete hands-on action becomes a catalyst for urban transformation, innovation, and creativity, facilitating the emergence of new 'commons' within a wider collective urban resilience movement.

Resilience to enclosure

Following the change of municipal representation after the local elections in 2014, and before AAA and the R-Urban team could complete Ecohab (the planned cooperative housing units), the R-Urban hubs were threaten with eviction. This decision generated a wave of protests. In June 2015, the new mayor decided to replace Agrocité with a temporary private car park and also expressed the intention to demolish Recyclab in order to clear the land for future city projects.

This incident confirms that setting up urban commons today is not only a social, economic and ecological project, but also a political project. It is subverting the capitalist order and its dominant modes of production and management. It is reclaiming and capturing resources which are embedded in capitalist transactions and redirecting them to commoning dynamics.

The mayor asked for the demolition of the two R-Urban hubs through a litigation procedure at the Tribunal Administratif. This was a challenging moment for the R-Urban community, and for us as designers and initiators. We had to quickly learn how to organize resistance through press campaigns and civic protests. We realized that creating urban commons does not only mean to support them throughout their development but also to overcome their potential enclosures. We realized also how challenging it is to set up and sustain commons in a capitalist society which is specifically based on principles of commons enclosure, privatization and unrestricted exploitation of the planet's resources.

Many contemporary urban commons around the world are haunted today by new forms of 'enclosure' (Linebaugh, 2008), most of the time driven by capitalist political or economic motivation (Alden Wily, 2012; Bollier and Helfrich, 2012). The enclosure forces could be private developers or property rights holders but also public governments, like in the case of Colombes. Although contested, these new enclosures are most of the time successful because the current capitalist laws support private or state-ownership against the common use. More, all the ecological value produced by projects such as Agrocité is not currently recognized and protected by any legislation.

In general, there is a lack of specific legislation to protect the commons (Capra and Mattei, 2015) and there is no political definition and legislation to protect what could be called a 'right to resilience' through commoning (Petcou and Petrescu, 2015). This right to resilience should be the right of any citizen to have access to the means, conditions and skills to become resilient in the face of climate change and all related crises. The emergence of commons should be encouraged and supported as one of these means.

In the absence of such legislation, the commons depend on the good will of local governments or other external administrations which can refuse to recognize the legitimacy of self-organization (Gutwirth and Stengers, 2016), as it was the case in Colombes. Those contemporary urban commons that succeed to survive is because they manage to find local political arrangements. In the absence of protective legislation, there is a need for commoning communities to collaborate with a 'partner State' (Bawens, 2015), a State which 'assists, enables and supports' (Weston and Bollier, 2013: 199) the institutionalization of commoning for resilience approaches and to sustain human and natures rights as part of this process. In the case of R-Urban in Colombes, after a court appeal and despite a civic protest campaign (Figure 18.4), the case went eventually in favour of the right wing municipality (Tribillion, 2015; Van Eeckhout, 2016).

We learned from this situation that architects need skills not only for designing and building but also for caring for and defending the urban

Figure 18.4: Civic protest against Agrocité's demolition, Colombes, 2015

Source: © Analia Cid

commons and making them resilient in adverse conditions. In the absence of protecting laws, and as an immediate upward, we took inspiration from the self-defence techniques in aikido, which are also indicating ways of being resilient. The act of 'blending', for example, teaches us that in order to stop an attack in aikido, one has to move around the punch. Therefore, we 'moved around the punch' and instead of simply demolishing the building, we have negotiated its dismantling and relocation in Gennevilliers, a neighbouring city with a left-wing municipal council. We used the fact that Agrocité was designed for a reversible installation incorporating resilience principles such as the possibility for dismantling and reassembling. The building was in fact conceived from the very beginning to last beyond the period of land availability and all the construction details were designed to make possible an intelligent demolition and reconstruction. This is a principle AAA has applied in previous projects in order to ensure their sustainability beyond their temporary installation on limitedly available land. The short-term land availability outside the market transactions is indeed an important challenge in metropolitan contexts, which can affect the sustainability of the projects. The AAA projects invented tactics to overcome this challenge, by proposing mobile and flexible building systems that can be easily transported and reinstalled in a different context, allowing the continuation of the project and strengthening the community around it (Petrescu and Petcou, 2013).

This approach is based on a sort of 'rhizomatic' resilience, acknowledging that a project has the capacity to resist in time by transmitting all the

Figure 18.5: Agrocité hub relocated in Gennevilliers, 2016

Source: © AAA

information that is necessary to its survival and dissemination to a new entity which becomes itself a new source of information and transmission. Despite the temporary existence of these projects in different locations, the accumulation of knowledge through experience is transmitted and reproduced in new projects which, although new and singular, are able to take over the same model, using the same protocol and process. The 'rhizome', that we have borrowed from Deleuze and Guattari, became our model of resilience when facing challenging conditions (Querrien et al, 2013).

In February 2017, Agrocité was dismantled and rebuilt in Gennevilliers (Figure 18.5). Also, in the same year, Recyclab was dismantled and rebuilt in Nanterre. Because we have used a reversible approach to building and have imagined the construction details for intelligent demolition, for reconstruction we were able to reuse a 95 per cent proportion of the materials used in the initial construction in a full Cradle to Cradle manner. The R-Urban members were pro-active in maintaining and relocating the functions while the physical infrastructure was not yet set up. The former users became the 'experts' of the project, participating in the reinstallation process. They were motivated by this new opportunity, which offered a new life to the project. Some of them continued to be active in the project in the new location, which was a walking distance away from the initial site, together with new users living in that neighbourhood. Others have joined existing civic initiatives in Colombes or started their own projects (for example new community gardens, collective composting initiatives, organic catering, etc.) keeping in contact with Agrocité, as ongoing R-Urban community members.

The value of commoning

In parallel with these events, we have conducted research to demonstrate the value created by the urban commons which is often neglected by city administrations, politicians and developers, as it was the case with R-Urban in Colombes. We know that the precarity of urban commons is partly produced by the inability to assert their value in the face of the capitalist urban private property market with its inflated financial returns on investment. In order to begin to address this legal and financial precarity, a much more robust understanding of value in social, ecological and financial terms as well as a more honest identification of full costs and benefits are required. With feminist economists Katherine Gibson and Maliha Safri, we have calculated the commoning value created by R-Urban during its functioning in Colombes (Petrescu et al, 2020).

We consider that in a world where we seek to promote ecologically sustainable forms of living, value calculation needs to go beyond financial capital and commodification to include nurturance and eco-maintenance. Our calculation model, using the Community Economy Return on Investment (CEROI) tool, presented one mechanism for doing so.[4] By conceptualizing a diverse (more-than-capitalist) economy, our approach enlarged the scope of what is recognized as valuable, specifically within the context of commons. Both paid (monetized) and unpaid (non-commodified) labour were seen to contribute to living well, together with many other beneficial activities which contribute to expanding commons and building community economies. A community economy is built upon a range of ethical investments – in living well, distributing surplus, responsibly encountering others, consuming sustainably and sharing our planetary commons, all with a view to the future wellbeing of all (human and non-human). To represent the value of urban commoning in R-Urban, our innovation was to use money, admittedly the dominant metric of value in the capitalist economy, to account for the incredible volume of value that can be returned on an investment in commoning. This was a way of comparing and demonstrating in a language that can be understood by the power.

Creating and expanding urban commons requires a lot of input into organization and care of the common space and into the growth of knowledge and skills of the commoners. This is reflected in the huge amount of unpaid volunteer hours put into building, gardening, training, event organizing, managing group activities, administering the hubs and book-keeping. However, the greatest volumes of value generated by R-Urban came via the new capabilities that participant commoners gained and the cost savings to the community, the state and the planet. This latter category included savings in building costs, energy and water use, food transportation, soil remediation, biodiversity decline, waste removal, air pollution, consumption spending, crime prevention, mental and physical health care. As such, we can say that the most significant value creation concerned community wellbeing and mending ecologies, two aspects that are closely connected. We have also showed how this commoning value translates in value for resilience (Petrescu and Petcou, 2020). This value is never considered in current transactions, and projects like R-Urban are usually dismissed because they are seen as not generating direct financial value in the way a developer project does. It was the case with the R-Urban hubs, which were evicted from Colombes. However, this value should be considered an integral part of the diverse economy- ecology of any architectural or urban project and the set-up of urban commons like R-Urban as a valid pathway to the much needed transformational change in times of crisis.

Other municipalities and stakeholders in the Metropolitan Region of Paris started to be interested in implementing resilience hubs and expanding the network. Currently two new hubs were built in Bagneux,[5] while few other municipalities expressed interest. With the construction of the Wiki Village Factory[6] in Paris in 2021, which is a cluster of social and ecological innovation and a sort of headquarters of R-Urban, providing centralized support, the R-Urban network will be able to operate successfully at the regional level in few years' time. In London, R-Urban[7] is collaborating on a new development with Public Works[8] and the Poplar Housing and Regeneration Community Association. Instead of disappearing, the initial hubs have multiplied and the networks have expanded.

The R-Urban experience demonstrates that commons can be a solution for cities mending ecologies and offers insights into how architects can facilitate this process, although this might imply designing, managing but also caring for, sustaining and fighting. This experience shows also that there are still many obstacles and blockages, conflicting interests, gaps in laws and regulations for such practice to become a norm. If opportunities are created for everyone to invest in and collectively reconsider the economic, social and ecological values of their actions, this commons-based mending ecological approach could spread rhizomatically and offer concrete bottom-up solutions to the current ecological crisis.

Notes

1 AAA website: http://urbantactics.org

2 R-Urban website: http://r-urban.net

3 The R-Urban framework's initial implementation was supported by the European programme Life+ through an innovation grant.

[4] For references and detailed calculations see 'Calculating the value of the commons: Generating resilient urban futures' by Doina Petrescu, Constantin Petcou, Maliha Safri and Katherine Gibson published by *Environmental Policy and Governance* available at https://onlinelibrary.wiley.com/doi/epdf/10.1002/eet.1890

[5] R-Urban Bagneux: www.facebook.com/RUrbanBagneux/

[6] Wiki Village Factory: www.urbantactics.org/projets/wiki-village-factory-quartier-saint-blaise/

[7] R-Urban Wick website: https://r-urban-wick.net/; R-Urban Poplar: www.poplarharca.co.uk/new-homes-regeneration/regeneration/project/r-urban/

[8] Public Works website: www.publicworksgroup.net/

References

Alden Wily, L. (2012) 'The global land grab: The new enclosures', in D. Bollier and S. Helfrich (eds), *The Wealth of the Commons: A World Beyond Market and State*, Amherst, MA: Levellers Press, pp 132–40.

Barad, K. (2003) 'Posthumanist performativity: Toward an understanding of how matter comes to matter', *Signs*, 28(3): 801–31.

Bawens, M. (2015) *Sauver le monde: vers une economie postcapitaliste avec le peer-to-peer*, Paris: Les liens qui liberent.

Bennett, J. (2010) *Vibrant Matter: A Political Ecology of Things*, Durham, NC: Duke University Press.

Bollier, D. and Helfrich, S. (2012). *The Wealth of the Commons: A World Beyond Market and State*, Amherst, MA: Levellers Press.

Bresnihan, P (2015) 'The more-than-human commons: From commons to commoning', in S. Kirwan, L. Dawney and J. Brigstocke (eds) *Space, Power and the Commons: The Struggle for Alternative Futures*, New York: Routledge.

Brown, G., Kraftl, P. and Pickerill, J. (2012) 'Holding the future together: Towards a theorisation of the spaces and times of transition', *Environment and Planning A*, 44(7): 1607–23.

Capra, F. and Mattei, U. (2015) *The Ecology of Law: Toward a Legal System in Tune with Nature and Community*, San Francisco: Berrett-Koehler Publishers.

De Angelis, M. and Harvie, D. (2014) 'The commons', in M. Parker, G. Cheney, V. Fournier and C. Land (eds) *The Routledge Companion to Alternative Organization*, Oxon: Routledge, pp 280–91.

Department of Economic and Social Affairs (2011) *World Urbanization Prospects, the 2011 Revision: Highlights*, New York: United Nations.

Friz, A. and Krasny, E. (2019) *Critical Care: Architecture and Urbanism for a Broken Planet*, Cambridge, MA: MIT Press.

Guattari, F. (2000) *The three ecologies*, London: The Athlone Press.

Gutwirth, S. and Stengers, I. (2016) 'Le droit a l'epreuve de la resurgence des commons', *Revue juridique de l'environnement*, 42(2): 306–43.

Latour, B.(2019) *Prospectus of a new exhibition CRITICAL ZONES to be held at ZKM Karlsruhe from May to October 2020*. Available from: www.bruno-latour.fr/node/806.html [Last accessed 6 January 2020].

Linebaugh, P. (2008) *The Magna Carta Manifesto: Liberties and Commons for All*, Berkeley, CA: University of California Press.

MacKinnon, D. and Derickson, K.D. (2013) 'From resilience to resourcefulness: A critique of resilience policy and activism', *Progress in Human Geography*, 37(2): 253–70.

Ostrom, E. (1990) *Governing the Commons: The Evolution of Institutions for Collective Action*, New York: Cambridge University Press.

Papadopoulos, D. (2010) 'Insurgent posthumanism', *Ephemera*, 10: 134–51.

Petcou, C. and Petrescu, D. (2015) 'R-URBAN or how to coproduce a resilient city', *Ephemera*, 15(1): 249–62.

Petrescu, D. and Petcou, C. (2013) 'Tactics for a transgressive practice', *Architectural Design*, 83(6): 58–65.

Petrescu, D. and Petcou, C. (2020) 'Resilience value in the face of climate change', *The Social Value of Architecture, Architectural Design*, 90(4): 31–37.

Petrescu, D., Petcou, C., Gibson, K. and Safri, M. (2020) 'Calculating commoning value for resilient urban futures', *Environmental Policy and Governance Journal*, 1–16.

Puig de la Bellacasa, M. (2017) *Matters of Care: Speculative Ethics in More than Human Worlds*, Minneapolis, MN: University of Minnesota Press.

Querrien, A., Petrescu, D. and Petcou, C. (2013) 'Making a rhizome, or architecture after Deleuze and Guattari', in H. Frichot and S. Loo (eds) *Deleuze and Architecture*, Edinburgh: University of Edinburgh Press, pp 262–75.

Rose, D.B. (2013) 'Val Plumwood's philosophical animism: Attentive interactions in the sentient world', *Environmental Humanities*, 3: 93–109.

Tribillion, J. (2015) 'Why is a Paris suburb scrapping an urban farm to build a car park?' *The Guardian*. Available from: www.theguardian.com/cities/2015/sep/11/paris-un-climateconference-colombes-r-urban-urban-farm-car-park [Last accessed 11 September 2015].

Tronto, J.C. (2019) 'Architecture and care', in A. Fitz and E. Krasny (eds) *Critical Care: Architektur and Urbanism for a Broken Planet*, Boston: MIT Press.

Tsing, A.L. (2015) *The Mushroom at the End of the World: On the Possibility of Life in Capitalist Ruins*, Princeton: Princeton University Press.

Van Eeckhout, L. (2016) 'A Colombes, la lutte d'une ferme urbaine contre un parking'. *Le Monde*. Available from: www. lemonde.fr/planete/article/2016/02/07/a-colombes-la-lutte-d-uneferme-urbaine-contre-un-parking_4860995_3244.html [Last accessed 7 February 2016].

Viveiros de Castro, E. (1998) 'Cosmological deixis and Amerindian perspectivism', *Journal of the Royal Anthropological Institute*, 4(3): 469–88.

Waltzer, M. (1986) 'The politics of Michel Foucault', in D.C. Hoy (ed) *Foucault: A Critical Reader*, Oxford: Basil Blackwell, p 55.

Weston, B.H. and Bollier, D. (2013) *Green Governance: Ecological Survival, Human Rights and the Law of the Commons*, Cambridge: Cambridge University Press.

19

Ri-Maflow: Des-pair, Resistance and Re-pair in an Urban Industrial Ecology

Marco Checchi

12 April 2019. I take the train early in the morning and I arrive in Trezzano sul Naviglio. Right out of the railway station, I find myself immediately in an industrial scenario: warehouses, factories, forklifts, trucks, graffiti. And workers. All warehouses look pretty much the same with their signs reporting the latest name of the company, maybe the name of the latest owner and a vague reference to their trade. I walk through the monotony and the squalor of this half a mile of tired capitalism. The time of thriving entrepreneurs is gone: some businesses struggle to survive, some are tempted by the allure of relocalizing production where labour costs are lower, others have already abandoned their warehouses. Ri-Maflow finally wakes me up from this nightmarish torpor. It looks like just one more warehouse. Its graffiti looks like just the graffiti I have seen on the walls of the previous warehouses. And yet they are different, they speak to me: *'Ri come Ri-Maflow, Ri-uso, Ri-ciclo, Ri-appropriazione, Ri-volta il debito, Rivoluzione!'* (Re as Ri-Maflow, Re-use, Re-cycle, Re-appropriation, Re-verse the debt, Revolution!).

I am here to attend the third Euro-Mediterranean meeting of workers' economy. At the registration desk they immediately find me something I can contribute to: 'You speak English, right? Come here! How do you say "*Dai rifiuti all'arte*?"' (From waste to art). 'Because we have lots of international guests today. If they look at this stuff, they don't understand it was garbage!' Pippo, Mara and Mauro call it *modernariato*, modern antiques: repair and recycling old abandoned objects into authentic pieces of design. 'In Milan, hipsters would pay 300€ for this. Can you believe it?' A few hours and many glasses of wine later, I will walk two miles to the nearest ATM to get cash and buy one of these fabulous lamps (can you believe it?). I meet Abraham: he is probably the youngest worker here at Ri-Maflow, he's from Gambia, he speaks English and he's happy to show me around. I learn from him that I mispronounce the name of the factory: it is not Rima-flow as I thought fantasizing about a romantic association between the factory and poetry (*rima* means rhyme in Italian). It is not a flow of rhymes, despite I still like to imagine this place as a stream of words, signs, machines, human beings and things beautifully and harmonically recombined together. It is Ri-Maflow, as to say it is Maflow once again, a new Maflow.

Maflow was a company producing components for the automotive trade. At the peak of production, the factory in Trezzano counted 300 workers. In 2009, the financial crisis hits hard and Maflow files for bankruptcy. One year later, a Polish corporate group, Boryszew, acquires Maflow and relaunches production in Trezzano. Yet, most of the workers lose their jobs in the process. But this is just an anticipation of what will happen only two years later when the factory shuts down, abandoning all its workers to a state of despair. The factory is quickly emptied: under the eyes of the picketing former workers, the machines are diligently dismantled to be transferred to some other factory of the corporate group. Thieves do the rest: right before the official closure, all the electric system and its kilometres of copper vanish overnight.

Massimo, one of the founders of Ri-Maflow and possibly its most well-known face, remarks the extent of the damage: 'When the factory shuts down, it destroys the territory at multiple levels: environmental, economic and structural' (Lettieri, 2019: 36).[1] On top of the economic and social tragedy, the abandoned factory accelerates those processes of environmental degradation that typically affect industrial areas. The soil was polluted by the presence of toxic agents in the aquifer and the warehouse's roofs had asbestos. It is not hard to imagine what would have happened next if a group of resistant workers did not decide to occupy and reclaim it. As happened to many other abandoned warehouses, it would have not been long before the site would be converted to waste storage and then burned down, provoking an environmental disaster, one of the most lucrative businesses for eco-mafia in the last years, with 690 warehouses burned down in Italy between 2016–19 (Castaldo and Gabanelli, 2019).

In this picture we find this deep interconnection and entanglement between the flows of global capitalism and its strategies of delocalization, the economic, social and psychological effects on workers, their families and the wider community, the damaged and contaminated soil, water and air that endanger the already precarious lives of all beings in the area. The closure of the factory accelerates a process of degradation that affects all these aspects: an ecology of despair. Here I use the idea of ecology as way of thinking transversally throughout all these dimensions. These multiple relations cannot be thought in isolation as separate aspects of life: thinking them separately is not only theoretically reductive, but also socially, politically and environmentally dangerous given the radical urgency that the problem of ecological degradation and extinction of life on our planet imposes upon us.

As Guattari puts it, we need an 'ethico-political articulation', an ecosophy, 'between the three ecological registers (the environment, social relations and human subjectivity)' (Guattari, 2014: 18). This ethico-political take on ecology is the passage from an understanding of the rootedness and symbiotic relations between all beings and their surroundings to a form of resistance that addresses the dangers of degradation and despair by actively engaging with processes of reparation. Re-pair is pairing together once again what has been separated: the workers and their factory, the factory and its territory. Ri-Maflow represents an ecology of repair in an urban context: beyond and against despair and degradation, recuperating a factory means repairing the lives of despaired workers made redundant by the logic of capitalist delocalization, the lives of a community, the environment threatened by an otherwise abandoned factory with its toxic pollutants. But it also shows the other meaning of reparation: demanding a compensation for the damage that the territory and its inhabitants have suffered. In this sense, the occupation of the warehouse is a form of reparation.

The Ri- (the Italian equivalent of re-) in Ri-Maflow stands for an ideal continuity, but also a new beginning, the start of a new adventure: a new enterprise, not only understood as a productive activity, a business or a company, but also in the other, often neglected, meaning of the word that designs an unexpectedly successful achievement against all the odds. Toni Negri says we should reclaim the word *impresa* and take it back from the terminological arsenal of liberalism (Negri, 2017). Enterprise is our word, we are capable of enterprise, we are capable of achieving unexpected results. Enterprise means to put ourselves to work, to put what surrounds us to work, to put an ecology to work, to set it in motion.

Ri-Maflow's achievement is to transform despair into an ecology of repair, turning the residues of the closure (an empty factory and unemployed workers) into resources for creating an experience of autogestion. From the occupation of the factory in 2013, Ri-Maflow has worked towards the ecological reconversion of the factory and the creation of productive activities that could provide work and income free from capitalist exploitation and vertical hierarchies.

Resistant re-pairs

Re-sist, Re-bel, Re-volt, Re-evolution, Re-claim, Re-quest, Re-appropriate, Re-take, Re-cuperate, Re-cycle, Re-use, Re-make, Re-work, Re-compose, Re-connect, Re-store, Re-pair: Re-Maflow, the explosion of this beautiful multitude of Re-s in the *impresa* of a Re-cuperated factory, an urban ecology of Re-pair.

Re- is the hallmark of a fractured temporality. Re- stands for a new opening after a closure, a new beginning after an end, a new present after a deceased past. But it stands also for a form of re-turn, the paradoxical gesture of going back while moving forward. The idea of a re-turn is problematic though: which past do we want to restore? Where do we want to re-turn to? Somehow re- imposes to fracture the past in two: a past to forget and to overcome versus a past to recover and recuperate. The former is the past of the reactionary, the melancholic nostalgia for an order where the few enjoyed and the many suffered. This is the past traversed by power relations, hierarchies and exploitation, and yet liveable. But is this the past we want to re-turn to? Was it not just a milder form of despair? This is why we need to introduce a temporal dimension that precedes this past. Not just as a logical possibility of a radically different time, but a wager on an unknown time that precedes history as we know it. This other past is the past of the revolutionary, the joyous leap to a past that the many imagine despite having no memory of it, a past where cooperation and solidarity prevailed. What are the temporal coordinates of this revolutionary past? Nobody knows, but every revolutionary knows that existed, it must have existed! Erased from history, its memory resists and persists in each struggle, in any tiny act of defiance against authority, in any spark of indignation, in any wild dream of revenge. After all, resistance comes first!

Before all the contractarians from Hobbes to Rousseau started spreading the insidious gospel of law and order through subjection as a remedy to the state of nature, in 1548 Étienne de La Boétie spelled out the primacy of resistance over power (Checchi, 2021). He sees the scattered traces of this (almost) forgotten past in animals. A fish deprived of its freedom when taken out of water prefers to die rather than living a life that is no longer worth living. 'What else can explain the behavior of the elephant who … on the point of being taken, dashes his jaws against the trees and breaks his tusks. … Even the oxen under the weight of the yoke complain, and the birds in their cage lament' (La Boétie, 2008: 51). La Boétie has probably never seen an elephant as he has never met those 'new' species whose stories arrive to Europe for the first time via the colonizers. The tales of these new species and their free way of living trigger the imagination of the young La Boétie. This new way of living, once discovered, will have to be abandoned immediately: bow to the cross, bow to the rifles, bow to the king. But for La Boétie this discovery is the chance to recover that free and cooperative way of living, to recover it as our own (not European, but ours as for all beings) previous way of living. That state of nature, feared by modernity from Hobbes to Rousseau (and then to Rawls and Nozick), for La Boétie corresponds to that revolutionary past to be recovered and hopefully restored in order to end the state of tyranny in which we live: 'we are all naturally free, inasmuch we are all comrades' (La Boétie, 2008: 50). Freedom is a consequence of companionship, a state of solidarity and mutualism, an ecology of natural comradeship, an ecology of re-pair. In the fractured civil state where the tyrant is separated from his subjects (any power is a relation of separation), there cannot be solidarity, there cannot be comradely love. To overcome this denatured society, we need re-pair: ending voluntary servitude with the simple re-solution to serve no more, to pair once again (re-pair!) human beings with their true nature.

Resistance is the beacon that connects this revolutionary past directly to a future of re-pair. At Ri-Maflow, this future is already a present. A true present in the sense that for many, maybe all, Ri-Maflow has been a proper gift.

Ruptures and des-pair

Between the revolutionary past of solidarity and mutualism and the reactionary past of power and separations, there is a rupture, an interruption, a fracture. The rupture has a complex temporality whose limits are only apparently well defined. It begins with an event of transformation that shakes the existent and makes its elements collapse and disperse. Ruptures last: ecologies are often bound to stay with their ruptures for a time that often feels endless. The hope of re-pair is often completely obfuscated by this long duration of the rupture. Likewise, this fog prevents us also from seeing when it started and, more dramatically, it bars access to that revolutionary past we have just recovered through La Boétie. In this sense, the most effective way in which the rupture manages to last for so long is by pretending that it never began: it has always been there, an eternal state of things. It attempts to intervene on the ontology of the possible, imposing closures on alternative becomings, resistant becomings, alternative ways of living, other ways of being together. This is how the rupture becomes tolerable: there was never a rupture, a rupture never happened, this current state of affairs is eternal and static. Here we see the affinity between rupture and power once again: when one becomes accustomed to power (understood with and beyond Foucault as an action that affects the action of others, beings and things alike), it is easier to tolerate its effects and even its abuses. Masked by its continuous and eternal presence, power/rupture becomes an unchallenged destiny. The rupture creates a hopeless time, a time of des-pair.

Nevertheless, not all the temporality of rupture is marked by des-pair (how could there be re-pair otherwise?!). If we rewind to the start of the time of rupture, we should be able to find that event that fractures the past in two and initiates the time of power and separation. Can we find it though? And what would be the point of individuating this event? The story of Ri-Maflow can help us to reflect on this point. If Ri-Maflow is an enterprise of reparation, when did the rupture begin? At first glance, we might think that the event of rupture is when Maflow goes bankrupt, as the occupation is the immediate response to the closure of the factory. And yet, the closure is just the peak of a longer process of des-pair and degradation. Contemporary capitalism is a game of separations: a BMW car is the result of multiple separated processes of production dispersed all over the globe according to a logic that makes it more convenient to have an engine, a tyre and even a screw travel thousands of miles. As the costs are externalized to our lungs, oceans and ecosystems, so the production of a specific part is externalized to a factory in Trezzano. Once BMW chooses another supplier though, Maflow has to shut down. Is this the event of rupture? It is easy to get confused here. That moment is extremely dramatic and des-pair reaches an intensity that makes it even more manifest: 'We were kicked out and disposed as human waste' as Stefano puts it in Azzellini and Ressler's film *Occupy, Resist, Produce: Ri-Maflow* (2014). Deprived of their jobs, of their source of income, of their material infrastructure that sustained their sociality, their families and also the wider reproduction of a whole community, we are confronted with the tragedy of a broken ecology, an ecology of des-pair.

But before the bankruptcy of Maflow what was that ecology like? The grey concrete jungle of Trezzano is a brutal reminder that even when the flows of capital are not interrupted, we are still confronted with des-pair. Any urban industrial ecology is always one small step away from its catastrophic collapse: its workers are always one small step away from being disposed of as human waste. Is this not already a broken ecology? Despite the factory poisoning the soil with its harmful waste and workers being exposed to asbestos and to the toxicity of the rhythm of work, des-pair is somehow kept at bay through a fragile equilibrium that makes this ecology tolerable. Once the factory shuts down though, des-pair becomes even more manifest. Resistance then becomes urgent and necessary. The bankruptcy of Maflow is not *the* event of rupture: it is just a further step in that time of degradation and despair that initiated when La Boétie's revolutionary past of solidarity and mutualism was abandoned. If there ever was such an original rupture, it is lost in time and it is no use to look for it. Yet, we need it somehow even if it is only to imagine that revolutionary past that gives us the spark to resist and re-pair the present!

From des-pair to re-pair

Resistance is the leap from des-pair to re-pair. It is the moment when we fully accept and engage with our broken ecology, our predicament of being (human) waste. Fragility and precariousness are the ontological conditions of any ecology. The risk of extinction is always present. But this is the hallmark of a dynamic ontology, always on the verge of transforming itself, opening to unexpected changes, experimenting with improbable possibilities.

We should not have needed that rupture to discover that despair is not an inescapable predicament, that re-pair can at times rescue endangered ecologies and prevent their ultimate destruction. The rupture was unnecessary, an accident, a tragic misfortune (Clastres, 1994). And yet that misfortunate rupture is there: it is time to work it out, let's re-pair it! Those who enjoy the rupture, who take advantage of a broken ecology do not realize what a re-paired ecology can do. They content themselves with an illusion of wealth and bar themselves (and, most importantly, all others) from enjoying the full potential of an ecology no longer separated. Their illusion of wealth is static, barred from becoming, sterile: 'Dai diamanti non nasce niente. Dal letame nascono i fiori' (Nothing grows out of precious diamonds, out of dung, the flowers do grow, De André, 1967). From (human) waste to Ri-Ma-*flowers*!

How have these Ri-Ma-flowers blossomed? A variety of material practices, dynamic intra-actions, unexpected encounters, petals that fall, stems bent by the cold wind, new re-pairs: a veritable ecology that dynamically reworks its connections, disconnections and re-connections between humans, spaces, plants, things, machines, technologies, flows of money, paperwork, waste, concrete, love, solidarity. 'We thought that those materials that others considered waste, for us could have been resources. We were determined to reconvert the factory ecologically by recuperating waste materials, recreating 300 jobs, as many as there were before the crisis. We started walking on a path of possible utopia' (Lettieri, 2019: 35).

Re-cycling becomes the main source of income that re-produces the life of this new fragile ecology. It is not only a productive choice, but also an ecological stance, an ethico-political gesture. All activities at Ri-Maflow are oriented towards recuperation, repair and recycling: from house removals they gather the materials that will be either repaired and sold or reconverted into artworks by one of the many artisan laboratories that Ri-Maflow hosts. Another source of income comes from the ecological dismantling of washing machines. Even one of the warehouses is the object of repair and recuperation: 'We ventured on the roof and cleaned up the obstructed gutters that had caused the flooding of the warehouse. We then cleaned the floor and convert the warehouse into a covered marketplace open to the public' (Lettieri, 2019: 59). Through these steps, Ri-Maflow opens up to its territory, developing a symbiotic relation with the wider community through an expansive process that is also regenerative as it reawakens a desolate industrial area from its torpor: 'Ri-Maflow is the active core of the town, despite Trezzano sul Naviglio is the typical dormitory town of the Milanese hinterland' (Lettieri, 2019: 60).

This complex ecology continuously renegotiates its boundaries with an environment that can also be hostile. The occupation of the warehouse is the first construction of boundaries. It is an occupation only if we make the effort to understand and accept the perverse mechanism of private property, debt and legal enforcement. This is not an occupation, but a re-occupation that follows an illegitimate and unfair eviction. We cannot tolerate and accept the right of a bank (where was the bank while workers were sweating hour after hour in that factory?) to keep a space shut and empty! Contrary to what the bank and the tribunals say, the re-occupation is not just legitimate: it represents a form of reparation for the damage these workers and this territory had to suffer for the faults of others. This is the other way in which Ri-Maflow embodies an ecology of re-pair.

This meaning of the term reparation first appears in the late fourteenth century as 'reconciliation' and has then been applied after World War I to define the economic compensation owed by the aggressor, a meaning lately associated with the decolonial demands of Indigenous and Black communities (Figueroa, 2015). Here the aggressor has no intention to pay for the damage and the harm done though. The occupation is the way to forcefully exact the reparation: 'The project Ri-Maflow was born … [with the objective] to reclaim the use and the management of the area as a social compensation [risarcimento sociale] for the entire community' (Lettieri, 2019: 43). This is also present in one of the many '*Re*' that constitute Ri-Maflow: 're-vert the debt' (Lettieri, 2019: 91): as the closure of the factory is the effect of the debts generated through capitalist accumulation, these debts should not be paid by the people. The occupation of the factory is a way to revert the debt and let the owner of the factory pay, asserting the moral right to reparation. The claim to reparation is affirmed also in the lengthy negotiations with the bank that in the meantime had acquired the property rights of the factory. Ri-Maflow affirms their right to own the warehouse on the basis of all the work done over the years to repair the warehouses and prevent them from abandonment and degradation. Here both aspects of reparation reinforce and sustain each other.

The re-occupation prepares the ground to welcome the first new seeds. Solidarity waters the ground through flows of donations that allow the implementation of all the safety measures necessary to remove the toxicity that the previous owner had left for lack of care. Multiple material practices of re-cycling: a recuperated warehouse to host new activities, recuperating workers who learn new skills. The old capitalist division of labour is gone and workers finally re-pair their intellectual and manual potential: who would have thought that workers do not need managers?

Ri-Ma-flowers continue to blossom. Ri-Moncello is the first liqueur they produce. And then the Vodka Kollontai: a feminist, anti-sexist spirit. Amaro Partigiano is possibly the drink that mostly embodies the multiple re-connections of this expansive ecology. It is a herbal digestive liqueur, good for re-pairing your stomach after a heavy lunch (and lunches are particularly long and heavy at Ri-Maflow: Abraham tells me that lunch is usually a collective moment of joy that resembles more a collective theatrical piece rather than a mere break between working hours). The herbs have been carefully selected in the 'resistant woods of Lunigiana', a mountainous region between Tuscany and Liguria: the perfect territory for the partisans of Italian Resistance, a nightmare for the Nazi invaders and their coward fascist allies. The label on the bottle beautifully remarks this re-connection between nature and its political spirit: *Naturalmente di parte*, naturally partisan, naturally taking sides. The sales of Amaro Partigiano do not support only Ri-Mallow, but also the Archivi della Resistenza (the Archives of Resistance) in Fosdinovo, a collective that re-covers and conserves the memory of these years of fierce struggle against fascism, a struggle to re-pair the country from the violence and the trauma of fascism.

In a way, Ri-Ma-flowers are today's partisans: they (re-)pair themselves with that resistance and with the whole tradition that traces a red line between the revolutionary past we found in La Boétie and an impatient present that wants to liberate its potential of solidarity, mutualism and comradely love. It's a red line of resistant memory that needs to be protected with care and often restored against the assaults of reaction. Ri-Maflow intervenes in one of the most contested memories of Italian history. On 12 December 1969, Piazza Fontana, Milan: a bomb kills 17 people and injures another 88. After more than 50 years, the truth about that bombing and many other terrorist attacks of those years is officially still veiled by mystifications that cover the political responsibilities of the attacks. Three days after the Milan bombing, Giuseppe Pinelli, an anarchist railway worker, dies after falling from the fourth-floor window of the police station where he was held as a suspect for the bombing. The investigations for his death concluded that his fall had been caused by fainting and losing balance. This is the limit where re-pair cannot arrive: 'Who will give life back to the railway worker Pinelli?' (Banda Bassotti, 1995). Nobody. No ecology can re-pair death. But his memory can be rescued. A plaque in his memory is vandalized and Ri-Maflow materially contributes to its restoration to remember that Pinelli, innocent, was killed in that police station.

After that plaque, Ri-Maflow participates in a project for restoring and preserving the wider memory of antifascism in Milan: the red line of memory needs continuous re-pairs. To materially sustain the project, this time the idea is to produce a beer. The hops come from a project of social agriculture inaugurated by Ri-Maflow a couple of years earlier: re-claiming and re-cycling land to recuperate a memory of resistance. Even the name of the beer is emblematic: 'La staffetta', the courier, a figure often forgotten by the history of Resistance to Nazi-fascism. The partisan couriers were mostly women, carrying weapons or information to the other partisan clandestine.

Each element of this ecology, from the hop to the smile of the partisan courier on the bottle, re-pairs the red thread of resistance.

This red thread is also weaved through the re-connection with several other laboratories of resistance both at the local and international levels. The conference where I first encountered Ri-Maflow featured other recuperated factories from Greece (Vio.Me), France (SCOP-TI), Bosnia and, in particular, from Argentina, where the movement of *empresas recuperadas por sus trabajadores* (worker-recuperated enterprises) has represented an inspiring and effective response to the crisis in the mid- to late 1990s and into the early 2000s (Azzellini, 2018, Ruggeri, 2016).

And it also welcomed other laboratories of resistant production, such as those that compose the network 'Fuori Mercato' (Outside the Market): re-claiming a new agriculture that promotes ecological practices and abolishes labour exploitation and environmental destruction. What was the purpose of this conference? It is partly a means to generate multiple encounters that reinforce that red thread: when many threads find each other and commit to interweave together, they form a rope. Solid, robust, hard to break, resistant.

But the conference is also an end in itself. For Ri-Maflow it is the occasion to experiment with new practices of organizing: how to host hundreds of people in Trezzano, how to provide simultaneous translations to facilitate communication between speakers of different languages, how to transform the space of the warehouse and re-cycle it for yet another purpose. Another *impresa*, an achievement that displays one more potential so far untapped, not yet explored. It adds up to the ordinary life of this ecology by experimenting with new possibilities, new practices.

Ri-Ma-flowers and their ecology

The expansiveness of Ri-Maflow ecology lies in the strength of its openness, its continuous movement that reworks its boundaries, that rejects the static repetition of an everlasting present and experiments with the extra-ordinary. Re-pairing is the continuous coupling with the outside: porous boundaries open to experimentation and yet capable to set a protective membrane to deal with the ever-present threat of rupture.

Because in the world of capital ruptures are always there, always on the verge of cracking the borders of ecology. Processes of degradation and destruction at environmental, social and political level are everywhere. It is easy to succumb to des-pair, not just as a subjective psychological state, but as a collective resignation, as if we were unable or unwilling to come together and intervene to stop those processes. As Guattari warns us: 'It is not only species that are becoming extinct but also the words, phrases and gestures of human solidarity' (Guattari, 2014: 29).

Ri-Maflow instead recuperates those words, phrases, and gestures, it puts them into practice in a radical mutualism that resists ecological degradation. The closure of the factory formed an ecology of des-pair: all registers (environmental, social and mental) were damaged. The occupation of the factory addresses simultaneously all these levels. It manifests an ethico-political stance that pairs once again (human) beings and their surroundings, affirming their interdependency and the urgency to resist degradation. Resistance marks the passage from des-pair to re-pair, the ethico-political take on ecology that affirms a way of living based on solidarity and mutualism. Resistance is the moment when what the capitalist system considers waste (redundant workers, a factory, toxic pollutants) can be turned into a resource to create new pairings, new ways of living, new forms of life: a new ecology of re-pair. Reparation is a practice that engages with something broken, experimenting with its possibilities in order to find a new form: re-pairing a warehouse to turn it into a laboratory of autogestion, re-pairing workers and turning them into entrepreneurs, managers of their own enterprise, re-pairing a territory by preventing the factory to contaminate the soil, the aquifer and air. But reparation is also a way for compensating a community and a territory for the damage it suffered, all the wrongdoings perpetrated in the name of capitalist accumulation.

Today Ri-Maflow is a proud and resistant ecology of re-pair. Eight long and dense years from that re-occupation, that first reparation. Eight long and dense years that show how to resist des-pair. Ri-Ma-flowers keep spreading their seeds, fostering a fertility that brings oxygen and regenerates life in the midst of the concrete jungle of that urban and industrial space that happens to be Trezzano, but we can also find in the periphery of every city. We still do not know what else Ri-Ma-flowers can do. We still do not know how many other ecologies of re-pair can flourish.

Note

1 All subsequent quotes from this source are my translation.

References

Azzellini, D. (2018) 'Labour as a commons: The example of worker-recuperated companies', *Critical Sociology*, 44(4–5): 763–76.

Banda Bassotti (1995) *Luna Rossa*. Propaganda Pel Fet.

Castaldo, A. and Gabanelli, M. (2019) 'Il traffico dei rifuti è meglio della droga: 690 roghi in 3 anni', Corriere della Sera. 6 October 2019. Available from: www.ambiente.it/informazione/focus-on/il-traffico-di-rifiuti-e-meglio-della-droga-690-roghi-in-3-anni.html

Checchi, M. (2021) *The Primacy of Resistance: Power, Opposition and Becoming*. London: Bloomsbury.

Clastres, P. (1994) 'Freedom, misfortune, the unnameable', *Archaeology of Violence*, translated by Jeanine Herman, New York: Semiotext (e), pp 171–88.

De André, F. (1967) *Via del Campo*, Nuvole Ed. Musicali Sas, Universal Music Publishing Ricordi.

Figueroa, Y.C. (2015) 'Reparation as transformation: Radical literary (re) imaginings of futurities through decolonial love', *Decolonization: Indigeneity, Education & Society*, 4(1).

Guattari, F. (2014) *The Three Ecologies*. London: Bloomsbury Academic.

La Boétie, É. d. (2008) *The Politics of Obedience: The Discourse of Voluntary Servitude*, Auburn: Ludwig von Mises Institute.

Lettieri, M. (2019) *RiMaflow. Storia di una fabbrica recuperata*, Palermo: Aut Aut.

Negri, A. (2017) 'Chi sono i comunisti?' *Conferenza di Roma sul comunismo*. 18–22 January 2017. Available from: www.euronomade.info/?p=8701

Occupy, Resist, Produce: Ri-Maflow (2014) Directed by Dario Azzellini and Oliver Ressler. Available from: www.youtube.com/watch?v=sX0Qga5BC68

Ruggeri, A. (2016) 'Workers' economy in Argentina: Self-management, cooperatives and recovered enterprises in a time of global crisis', in C. DuRand (ed) *Moving Beyond Capitalism*, London: Routledge, pp 49–57.

20

Chilean Streets: An Archive against the Grain of History

Cristobal Bonelli and Marisol de la Cadena

Chilean streets, October 2019

Estallido social: these are the Spanish words that people use to refer to what in this article we call 'Chilean streets'.[1] *Estallido social*, a social outburst: on 18 October 2019 crowds of protesters took to the streets in the defence of life and against everyday abuse and death in Chile. They wrecked sites that they considered denied their lives and also transformed street walls with words, images and art that proposed hope for a different life. They expressed what we think of as 'historical wounds' that the state – colonial and modern – inflicted on the Chilean territorial body as it made a nation on it. In response, the government declared both a 'state of emergency' and 'a war', bringing the military onto the streets for the first time since the Chilean military dictatorship ended in 1990.

We translate '*estallido social*' as 'outburst': people loosely organized burst out, away from a suffocating grip. In Chile heterogeneous desires and frustrations 'outbursted' away from a recent history of neoliberal suffocation; against it 'the streets' propelled a demand to rewrite the neo-liberal constitution and obliged the government to accept the project. But 'the outburst' also expressed wounds grounded in history, current and past; it emerged against the official historical oblivion of those wounds. As event, the outburst expressed the long durée that birthed it and actively cracked the dam that contained it. In so doing, the outburst may be an event beyond, and perhaps because of, its ephemerality. This eventfulness may also have an oxymoronic and bold future long durée that asserts a possible that while still unknown, may now pulsate as imaginable and against the history that in suffocating them, also allowed their becoming. 'Chilean streets' may project a proposal away from the grip of neoliberal politics and the long History that made it. Surpassing the limit that denied them (Guha, 2003), 'the streets' became an avalanche. Thus, we propose Chilean streets – and perhaps streets in other parts of Latin America – become in an outside in which they ferment and self-transform in constant and irreducible excess to politics as usual. The streets may be subject to brutal state repression and/or conversely their demands may be accomplished by the government or its opposition, and their force placated. However, that force cannot be assimilated to either left or right: ephemeral, impermanent, leaderless and disorganized, the *estallido social*, the outburst slips from any institutional political grasp, and vanishes from their view, while the potential presence of 'the streets' strongly remains – even if suppressed.

We think 'the streets' possess what Walter Benjamin would call 'retroactive force' (Benjamin, 1968: 255): their becoming event may 'flash up as a memory in a moment of danger' the German philosopher would say. And perhaps Chilean streets would rephrase the aphorism as 'a moment of danger as flash back that impelled us – the streets – to outburst'. In October 2019, the Chilean streets were sustained by the force of a vitally present past. *They have taken everything away from us, even fear* – reads an inscription that repeats itself on many walls that participated in the *estallido social*. Fearlessly, the streets persisted for a month, then it ended; however, in the inscriptions and images with which participants in the outburst dressed the street walls, their retroactive force persists and projects a future. In what follows we think with those inscriptions to imagine the possibilities they express. Inspired by Haraway's (2016) string figures, we read the writings, drawings, paintings on the walls as proposals by heterogeneous street authors who take relays to think a collective imaginary in a process of contiguous ongoing-ness: they are once more the expression of the Latin American historical mantra *la lucha continúa* (the fight goes on): the inscriptions on the walls express a permanent emergence of vital connections that continues the fight for always better worlds. In what follows we first offer an incomplete (how could it be complete?) description of what we think happened; then we walk the streets and talk with the inscriptions as we also show them to the reader. What we say does not follow a script and does not represent either; ours is a streets-inspired storytelling that does not offer explanations, final or transitory. Our purpose is to share with readers the force of the Chilean streets to affect, us in this case, but also others. We end with a brief commentary about the historicity of that force, and the alter-archive that the inscriptions on the walls may propose.

El derecho de vivir en paz

> Sigan ustedes sabiendo que, mucho más temprano que tarde, de nuevo se abrirán las *grandes alamedas por donde pase el hombre libre*, para construir una sociedad mejor.
>
> Salvador Allende, September 1973

Figure 20.1: Las mil guitarras de Victor Jara

Source: Luis Poirot

> Keep on knowing that, much sooner than later, *the great avenues (sp. grandes alamedas) through which the free man passes will open again*, to build a better society.
>
> Salvador Allende, September 1973

On 25 October 2019, responding to a call from the collective *Mil guitarras para Víctor Jara*, thousands of people met at the entrance of the national library of Santiago, Chile. The desire to sing 'The right to live in peace', a *canción protesta* (protest song), moved them. It was composed by Victor Jara, a troubadour that dictatorship of Augusto Pinochet assassinated on 16 September 1973; his torturers cut his hands before killing him. The death of Jara and thousands of others is among the wounds 'the streets' were moved by and expressed on their walls in October 2019.

Singing the protest together made possible this photo, authored by the Chilean photographer Luis Poirot; it records an occasion that was unimaginable until it happened. Víctor Jara was the force that orchestrated this chorus. The musical genre to which this song belongs was embodied by the streets; chanting was one of the myriad forms that the protests of 'Chilean streets' adopted. The demands were also heterogeneous; this, and that they were leaderless moved us to call the protests 'Chilean streets'.

Demanding free education, students protested the debts incurred to access universities; seniors wanted the end of the private pension system that had destroyed social security and imposed post-retirement poverty; small farmers demanded free irrigation water, or even access to it; the sick demanded prompt and decent health care; inhabitants of environmentally damaged and polluted regions – infamously known as 'sacrifice zones' – demanded the right to breathe without getting intoxicated. 'Until dignity becomes habit' (sp. *hasta que la dignidad se haga costumbre*) was the chant unifying protests; it met Víctor Jara's claim for the 'right to live in peace'. A government declared increment in the price of the subway fare catalysed these demands:

> a rebellion that added all the accumulated discontent, when young high school students said enough! to the rise of the passages of the Metropolitan subway train (the Metro) of Santiago. It became clear that the massive evasion of paying the tickets, jumping the turnstiles to enter the Metro stations, was no longer a claim of the students by and for them, but was a general malaise in the population, against abuses from the State and the business community and, ultimately, a struggle to achieve their deserved dignity. (In 'Chile: Genealogy and Challenges of the Constituent Process | NoticiasSER', n.d.)

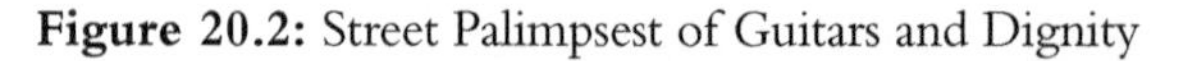

Figure 20.2: Street Palimpsest of Guitars and Dignity

The 25 October saw more than a million and a half persons marching together in the largest peaceful demonstration in the history of Chile. Salvador Allende, the Chilean president assassinated by the dictatorship in 1973, would have been joyful: almost 50 years ago and before dying he predicted that people would march across the *grandes alamedas* creating possibilities for new ways of living together. In October 2019, these possibilities were declared feminist and were called revolution: 'the revolution will be feminist, or it will not be!' the streets asserted. Moreover, sustaining those words was the name and face of Camilo Catrillanca, a Mapuche person the Chilean State killed with a shot on his back in 2018. Revolution-feminist-Mapuche overflowing the avenues and occupying the nervous centre of Santiago, the Plaza Italia, in the middle of which is a monument honouring General Manuel Baquedano, a military who won battles that established the supremacy of Chile vis-à-vis Peru and Bolivia in the modern history of the Andean region. Thus, represented on this Plaza is the official History of Chile.

Figure 20.3: Baquedano displaced by Dignity

Remarkably, the force of October 2019 renamed it the *Plaza of Dignity*. Massively occupying this plaza, the streets with their bodies and its walls with words, drawings and posters, people protested current dire conditions of life and profoundly rejected the capitalist-colonial political tools the State used to write them into national history. In October 2019, the streets exposed the wounds the state as maker of History excised across the territories it occupied to make 'a nation' starting with the early separation of Mapuche collectives from their land. The word Mapuche can translate as 'people of/with the land'; their uprooting provoked a wound that the State both deepened throughout history *and* (in a bid to anaesthetize its citizens) also ignored as history. As an event, the streets of Chile in October 2019 may reclaim the condition of 'people of the land' for a 'complex we' – a *nos-otros* in Spanish – that draws from Mapuche and non-Mapuche histories and practices alike, to care together (and in difference) for the wounded territories they heterogeneously inhabit and are. *Nos-otros* of/with the land, singing to the wounded earth and against the commodification of life the dictatorship inaugurated with Chilean neoliberalism and boasted about an 'economic miracle' for the few that was a disaster for the many.

The wound unites us

Figure 20.4: La herida es lo que nos une (The wound unites us)

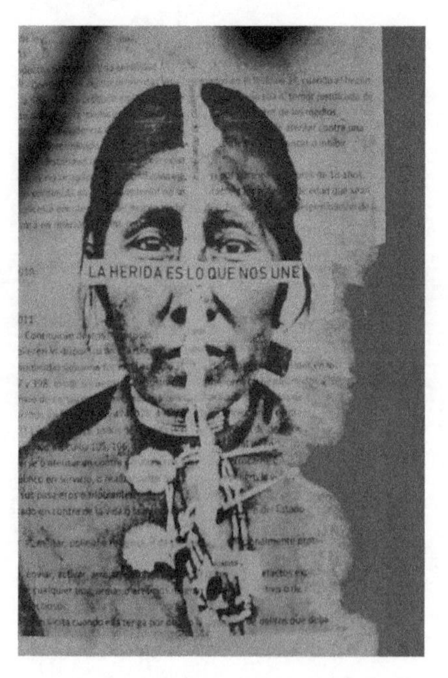

Enough. They – the police, the state and capitalism – chased us to death to turn us into the productive bodies they yearned for – but they failed. They were never where we ever started. Ignoring us, they took great pains to move us to where they wanted us to be, as if history were theirs. As if history were an armoured artifact blind to the force of our inheritances. Their history was stubborn and simple: a traction that propels one forward and prevents any other will, like the waves of a powerful sea that prevent a drowning swimmer reach a shore. That history wanted to direct us, all of us, to move forward they contended, to improve ourselves and even *the species*. To be, in fact, the best: Chile, the best species-country in the continent. So they said. Enforced, that history made the Nation; it also uprooted us from land and exiled us. Enforced, that History made a wounded us, Mapuche and non-Mapuche, bodies that are also territories, together as a wound.

★

Nos-otros somos, we are a deep wound that does not want to be called this Nation – it does not do good. This nation makes wounds and then it wants to close them with burning alcohol and stitch them. It imposes an allegedly form of healing that hurts. A wound has to be cared for, not forced to heal. But it is the will of the state to defend itself from everything; even from the wound on which the nation was founded and its categories distributed: ethnicity, race, gender, sexuality, age. All those categories can be used as tools to distribute tools – speech, for example – to forcefully suture the wound, preventing it to scream its pain.

The wound needs care (*cuidado* in Spanish). The word *cuidado* comes from the Latin *cogitatus*; usually translated as 'to think', it also means to have 'an interest in something that makes us think'. Our interest in the wound obliges us to *cuidarla* (to care for it): it moves us to think, feel and act in its presence. Each of us differently. The *cuidado* of a wound is specific to how it feels to those who bare it, to those care for it; that *cuidado* relates specifically. It does not work with a clear and uniform suturing prescription. The wound cannot be captured by the State: sooner or later, the wound will reopen and express its pain. Like in October 2019 when the wound occupied the streets of Chile singing demands to live in peace.

Wounded s-tree-t

This is the sacred tree of Chile, its painter says. Its lifeblood is the names of those 'detained-disappeared' (*detenidos desaparecidos*) that the Chilean dictatorship killed. Absent in official records, their more than 1200 still missing bodies may have also disappeared into the soil that now flourishes with

Figure 20.5: Enduring Presences in the Sacred Tree of Chile

them. Their names remain present; perhaps living with that soil, their bodies also remain present, defying the will of the state that dictated their end and would dictate their oblivion. But it cannot: the state has limits. It can only dictate over what it can accept. Thus, as it persists (regardless how), the life of those the state killed escapes its will; their names make their way to history, violating state sovereignty. The adage that 'history is written by the victors' may be only for the State to believe, or … the State may not be the victor.

Figure 20.6: The Sacred Tree of Chile outside the Centro Cultural Gabriela Mistral (GAM)

But this tree is also a wound – it is neither a monument nor a symbol! Made of names of blood, this tree is a people assassinated. Stripped of their right to live in peace, they do not rest in peace; they left in violence and in such they do not rest. The tree that is a wound (*los cuida*) takes care of them. Painted on the street – it inspires our calling it s-tree-t – the tree is an invitation to be with those who give it life. It compels our interest in their care.

In the 1970s, Chile was simultaneously the territory where many were executed and the delivery room of neoliberalism; the same practices and materials made both. The lands that feed Chile are wounded; yet having inherited the missing they may also root a democracy that reinvents politics in the presence of the disappeared, those whose names flowing through the tree as its sap, make it sacred and heal the soil. Inspired by Vinciane Despret's (2017) musings on the dead's capacity to do things, we write:

Honoring the lives of the dead and the living
Nos/otros face the duty to offer the dead
-those other that we are-
esos otros que nosotros somos
Another way of existing

Helia Witker, a Chilean artist created *the sacred tree of Chile*. Helia exhibited her sacred tree in the 'Alameda Art Center' days before the Chilean outburst. On 29 October, days after the outburst, Helia took the tree to the street and made a tree-street fusion *-the s-tree-t-*. She offered it as an altar 'where

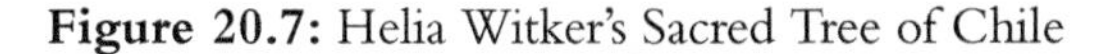
Figure 20.7: Helia Witker's Sacred Tree of Chile

nos/otros can honor all the people who died, not only during the military coup; also the Mapuche who the Chilean state killed during an allegedly democracy (since 1990)', and to all those who have fallen in this *estallido*.[2]

At the base of this tree, where the roots – with a musty smell – look for deep nutrients, the name of Camilo Catrillanca inhabits. In October 2019, attacked by the Chilean police (*los pacos*) during the protests, Fabiola Campillay and Gustavo Gatica lost their eyes on the street. Their names approach the tree from its sides. The names that run through the veins of the wounded *s-tree-t-* transgress the history that the state dictates. They are there to make nos/otros respond.

Kiltro politics

Figure 20.8: Pelea como Kiltro (Fight as Kiltro)

I am a dog (*un kiltro*). In Chile a *kiltro* means a mutt, more than a mixture, and more (or less) than a breed. I live on the streets making and being made by relations, freely breathing and walking around with my four paws on the ground, touching the soil. On the street, people (*me cuidan*) care for me; I eat what they give me, for free, in exchange for nothing, or as little as my looking at them, my walking with them for a little while. Their care makes me – *soy con el cuidado de la gente*. I live happily outside capitalist transactions, outside of capitalist logics of productivity through relations that have life-value but not price. I am loyal to life and life is priceless; I fight for life. And in fighting like the *kiltro* I am, I never give up – that is what the words on my picture want all to know: *pelea como kiltro*. Never give up on my fight for life, mine and of those who make me live.

Like other dogs, we are a companion species. But being *kiltro* on the streets we are companion to many and to none, and that, perhaps, makes us strange as 'companion species' for we are not owned. Those of us living on the streets are many, and many of us die young, even before we turn one. We may be run over by a car, or get sick. And then, there is the state: many of *nos-otros* also die assassinated by the state; we are a hindrance, because we do not abide by their categories. In some cities (like Valparaiso) the health service kills us in mass when they need to clean the streets for military parades that celebrate the heroes of their history. The history the state makes is for humans; it kills us.

I do not simply represent protesters because being made of mixtures I cannot be a unitary referent. Maybe that is the reason why protesters have put me at the centre of the *Plaza de la Dignidad*, at the very place occupied by the statue of one of the heroes of the state. Instead of him, a helicopter installed a sculpture with my shape!! I displaced General Manuel Baquedano. A full squad of *pacos* deployed to protect the emblematic place did not react to this intervention. It was a digital intervention to be seen online, virtually.

Figure 20.9: Snapshot from digital intervention CIMA Gallery

Source: CIMA Gallery; to watch the whole digital intervention, please visit Galeria CIMA YouTube Channel www.youtube.com/watch?v=hXmr9JeBlw4

A group of activists from CIMA, an art gallery located in a building facing the *Plaza de la Dignidad*, broadcast my virtual statue as well as other events since the beginning of the Chilean outburst. I think those activists are also *kiltro*: they mix art, archives and activism. I call them *art-chivists* and I think they have a force that I have, a *kiltro* force.

Kiltro – the force

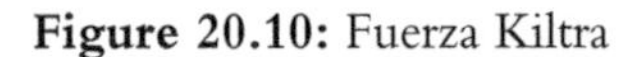

Figure 20.10: Fuerza Kiltra

I am *kiltro*: a force that displaces usual politics by improvising, accompanying, protecting, caring, altering. I am the permanent presence of the ephemeral: connections that appear and disappear on the streets. I am the companion presence that moves street protesters from a politics that hinders life to a *kiltro* politics that makes possible what seems impossible. I am made with the absence of fear, with the strength of mixing, a companion species of commitment and affection. And yes: made of care as relation. So being *kiltro* force, I am unlike power. I do not seek the state; my unruliness shakes its need of taxonomies. I am not 'resistance' if it demands exclusive allegiance. For you see, on the streets, my feed is the freedoms (many!) of relating, connecting and spreading the care that makes my life which is not mine alone. I am *kiltro* because I am with others, a street ecology of care. I am recalcitrant to evil because I have seen it kill.

Daughters of Witches

> As for witches, I think not that their witchcraft is any real power; but yet that they are justly punished, for the false belief they have that they can do such mischief, joined with their purpose to do it if they can.
>
> Hobbes, 1963 [1651]: 67, cited in Federici, 2004: 143

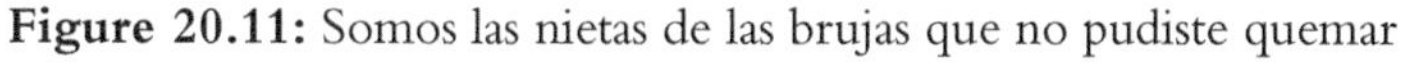

Figure 20.11: Somos las nietas de las brujas que no pudiste quemar

Somos las nietas de las brujas que no pudiste quemar ... Witches are the nightmare of the state – they have been so for centuries, and they still are. The state, since its conception as Leviathan, wrote their history. Hobbes was scared of them we think; he dismissed their force and wanted them punished – perhaps 'just in case?' Upon arriving in what they gradually invented as the Americas, fresh from witch hunts in Europe, colonial friars and the Inquisition persecuted what they translated as witches, obscuring to these days their practices. We do not know who they were, we call them healers: pillars of rebellions, they staunchly guarded the mountains of the Andes, the rivers of the rainforests, the sands of the oceans – they have not ceased at it. Their actions certainly instilled fear in their wilful eradicators – just read what follows:

> there was one kind of sorcerer that the Inca kings permitted, and these are like witches, and they take any appearance they want, and they fly through the air ... and they see what will happen; they speak with the devil, who responds to them by means of certain stones or other things they highly venerate ... and they say that it is usually old women who perform this act. (De Acosta, 1954 [1590]: 172)

Acosta was a colonial friar, and this is what he wrote. But far from vanishing, in October 2019, they wrote themselves on street walls: 'we are the grand daughters of the witches that you [the State, past and present] could not burn'. They are even of those that the State did burn but continued living in ways not known. That is not unlikely: witches have their own ways of living. It was said they communed with the dead, remember? But ... what if it also worked the other way around: as the dead communicating their breath to the living? Imagine (and feel!!) thousands of voices chanting Víctor Jara's song. Yes, it could be just people singing an old song, and making it new; but this very same thing could also be the breath of the dead – of Víctor Jara – sustaining the life that dressed up the great avenues of Allende for the occasion with chants and writings. Sustained by the dead – witches and not – chanting with all their grand-daughters (who were also grandsons!), the life occupying the streets was extraordinary: *kiltra* life emerging to touch (and shock) with feminism the patriarchal nerves of history and all those nerves sustain.

Authored by a female collective self-named *Las tesis*, a chant resonated for months:

> *El patriarcado es un juez que nos juzga por nacer y nuestro castigo es la violencia que no ves.*
>
> *Patriarchy is a judge who judges us for being born and our punishment is the violence you don't see.*

Son los pacos. Los jueces. El Estado. El presidente. El Estado opresor es un macho violador.

They are the cops. The judges. The state. The President. The oppressive state is a macho rapist.

The song travelled the world, shocking patriarchal nerves and histories in many languages. Casting circles of protection where it was chanted, it must have made the witches happy – those dead and alive.

Join the dance of the left over, *Únanse al baile de los que sobran*[3]

Figure 20.12: Somos los nietos de los obreros que no pudiste matar

Witches and workers have more in common than the 'w' that starts the words. A narrative of their togetherness may go back to the bursting period of global capitalism and aggressive expansion of the world that saw itself as modern. Arbitrarily – but not incorrectly – starting with the enclosure of the commons, the narrative would explain that, dispossessed of the collectives that sustained them, families were translated into units of 'workers' and their wives. Accordingly, the former received a salary, and the latter worked to make the salary meet the ends of the family in more than 'many' ways. Granted the power to expand along with modern capitalism, such narrative became universal; it blurred the alternative stories it may have encountered *and* effected patriarchal subordinations of women to men, indeed, but also men invented as lower owed to men invented as higher: peons to landowners, for example.

This narrative made suspects of women without patriarchs, or those who challenged those they should obey … they could be witches. Without a man to control or satisfy their lust they were said to copulate with the devil and, perhaps in exchange, received powers from him. Moreover, those satanic unions resulted in offspring – it was all too dangerous; they had to die. Such deaths would cancel superstitions the Leviathan read, and then 'men would be much more fitted than they are for civil obedience' (Hobbes, 1963 [1651]: 67, cited in Federici, 2004: 143).

Working families? Other than as labour they 'counted as not counting'; they were *los que sobran*, the 'left over', as in the lyrics of the best known protest song by the Chilean rock band *Los Prisioneros* (The Prisoners) Rancière would, perhaps, sing along sadly. Workers could be replaced when they died of hunger or disease; destitute, their families follow the same fate would then also die abiding by patriarchal rules which always changed to meet the ends of capitalism. Transgression of those rules was heterogeneously punished, including the death of the transgressors. Society had to be defended, we learned with Foucault: it was all for the common good. Everything worked like a clock except for upheavals: moments that may last into events, when persons of all colours, genders, sexualities, ages, 'aware that they were about to make the continuum of history explode', may arrest the clocks in all Main Squares 'to stop the day' – the words in quotes are Walter Benjamin's. He was writing about workers that fascism – in Chile embodied in Augusto Pinochet and his military crew – killed because they were dangerous to the origination of neoliberalism. Yet like witches, *obreros* (workers) survived their death even if they were killed: their *nietos*, offspring of their offspring, gather force looking at that past moment as it 'flashes back' to the present, 'calling into question the victories of the rulers'. Again, readers will recognize Walter Benjamin in those words. In Chile flashbacks are permanent and those flashbacks are not a mere re-production of a lost past. These flashbacks flash forward as they may propose to reconstitute what people in Chile call memory, thus opening up the possibility of what kept existing as unruly presences that took the streets.

Kiltra Revolution

The revolution will be feminist, or it will not be.

The revolution will dismantle Unity and Hierarchies; it will be a revolution without a central command, without the virility of the mandate: there will

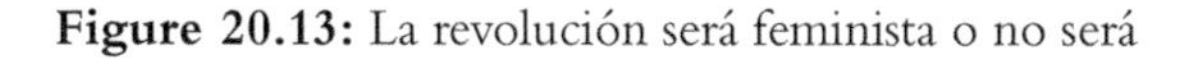

Figure 20.13: La revolución será feminista o no será

be no minorities because there will be no majorities, and men and women will free themselves from the patriarchal order and its capitalist ultimatums and illusion made out of infernal alternatives.

The revolution will be feminist: it will liberate all intimacies from the rule of the rapist authority. This revolution which will not occupy the state; it will affect it dissolving its patriarchal (in)morality. Morals will be made by people free from objectified bodies emerging in relations of joyful care and caring joy. Morals will be made by people's capacity for joy, for what Beatriz Preciado (2013) has called *potentia gaudendi*: the force [potentia], of total excitation of bodies, that *kiltro* force the state failed to expropriate.

The feminist revolution will be made of fragments, incomplete connections and inconclusive conclusions, going viral without command throughout the Plazas de la Dignidad in all countries: its extraordinary circulation will infiltrate all corners of the planet.

This will be a *kiltra* revolution, or it will not be a revolution! A revolution where there are no Others without *nos-otros*: where *we* will be mixed, and at the same time, *we* will be autonomous, but not individually autonomous. This *kiltra* revolution will rise as self-defence, but also as an affirmative proposal to transform life. We name this a *kiltra* revolution. Names are important; and with the force of witches and their spells, they make things appear. We need new languages to protect us from the words produced by the masculine mythology that haunts us.

The revolution will be feminist
or it will not be

The phrase is historically transformative. Like the image on the wall where it appears along with the spectre of Camilo Catrillanca, a 24-year-old Mapuche murdered by a Chilean antiterrorism police squad in 2018. Yes, the *kiltro* is strange: it mixes everything, and strange things come out of that mixture. Because Camilo could also be any of *nosotras*, in our dividual roots; his name is capable of mutating into other names and struggles, as part of these others that *nosotros* are. When the others are *nos-otros*, no one needs to think for others.

But *cuidado*! (in Spanish you can also call someone to 'pay attention' when we say *cuidado*!)

This *nosotros*, and the others that *nosotros* are, will always be haunted by what it rejects, and can make *not be*: the patriarchal revolution, made of *uni*ty, and derivatives: *uni*form, *uni*vocal statements. And it is from this patriarchal stalking that we have to defend and *cuidar* this revolution, because by undoing the Nation and the Historian State with its transgressive force, the *kiltra* revolution is exposed to constant danger; it remains vulnerable like the *kiltro* that is not afraid of being so. That is why we will have to defend it everywhere and ways Rita Segato (2018: 206) calls amphibious: 'The way forward is amphibious, involving work inside and outside … in both intra and extra-state politics, reconstructing communities that are under attack and that have been dismembered by the colonial and state interventions called modernization'.

#noseborra: an archive against the grain of History

Figure 20.14: Speaking Walls

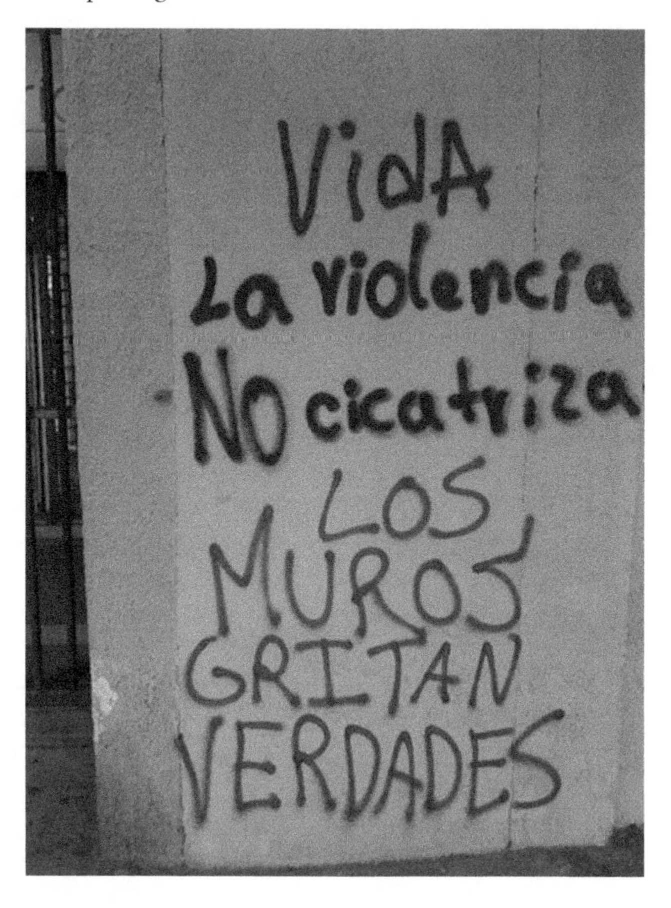

> The city walls are the blackboard of frustrated aspirations, hatreds and loves of the common citizen. This was the case during the Bolshevik revolution of 1917, the Mexican agrarian revolution of 1910 or the uprising of Paris in 1968. … [N]ow, the walls of our Alameda de la Delicias would be depositories of the waterfall of those longings people hang onto, after dozens of years governed by an exclusionist economic-social model.
>
> Miguel Lawner Steiman, cited in Ureta, 2020: 52

This picture reads 'the walls yell truths' and 'nothing sutures violence' – the word *Vida* (Life) acts as title of these phrases written on a piece of paper pasted on a street wall. This, along with the others that we have talked to and about in this piece, is a document that along with myriad others that we have not included here, and still many others that we have not even seen, compose what we imagine as an archive against the state. Following traditional state-fascism, the Chilean dictatorship burned large numbers of books about Marx and Marxism, even about cubism: they could be about Cuba! Hence Walter Benjamin's apposite warning: 'There is no document of civilization which is not at the same time a document of barbarism.' Therefore, he proposed the task of historical materialism – in which Benjamin includes spiritual things, humour, creativity and strength – would be 'to brush history against the grain' (1968: 256). This was one of the tasks of the Chilean streets. The writings that dressed them displaced the public archive and turned its practice inside out. Achille Mbembe, a philosopher of colonialism, describes the state archive as both a building and the documents it houses (Mbembe, 2002). The walls of Chilean streets offered both: building and documents, they could be produced by anyone and belonged to no one, they could be accessed by all. Transgressing the city ornate, the outburst wrote a differently public archive on the outside of the state, and of the politics it sanctions: the walls *exscribed* what the state inscribed and offered an alternative historical source, one that the state cannot assimilate because it exceeds its possibility of recognition. The Chilean estallido social *exscribed* wrote an outside in defiance to the limits of the inside – demands that aspired and inspired a different life, a life lived with habitual dignity. Throughout the country the outside read: *hasta que la dignidad se haga costumbre.*

With the quote that opens this section, we suggest the Chilean streets offered a site against the dictatorial silence cloaking Chile for decades. The voices erupting with the *estallido* wrote their rejection on the walls of the street, and this time, because the revolution would be feminist or it would not be, the exscriptions read: 'Down with patriarchy (and its state-violence, taxonomic control and binary mandates)!' 'Down with capitalism (and its extraction of surplus value and natural resources)!' 'Down with human-centered history (and its speciesism)!'

Countering the ephemerality of the outburst, the documents of this archive have the will to be durable. As soon as state agents moved to delete the walls, counterefforts to maintain them were displayed. Mediated by *#noseborra* people communicated to protect the street documents with their bodies and also digitally. Currently the wall-documents proliferate, virtually creating an archive that is 'open source' and 'open access'. We got this quote from one of these efforts; it is a *footnote* on the virtual walls of a digital book created to overlap images on the physical walls that flank more than two kilometres along the central Alameda of Santiago.[4] Unlike usual footnotes

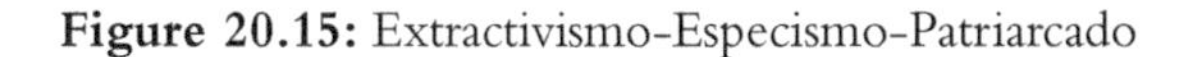

Figure 20.15: Extractivismo-Especismo-Patriarcado

that add information to a master narrative (for example, the history of the state embedded in the official archive), these street footnotes do not have a central text. The eventfulness of the Chilean streets does not have a master narrative. Unlike the public archive, the street archive lacks classificatory codes and chronology; its documents are ephemeral figures that can be read and interpreted in the manner of cartomancy, as the same image blends with diverse images in different streets: there is no rule sequencing their reading. Moreover, these images neither 'prove' nor offer 'evidence': the stories they tell are immanent to the publics that authorize them. As an alternative archive, the walls of the street express uncountable interrelated truthful wounds that have spun out new relations between them, displacing the history of the authoritarian state – that the displacement is fleeting may be irrelevant.

Relentlessly feminist, the eventfulness of the streets challenges the terms of officially public history: it disarranges time and geographies, confronts grammars of difference. Sometimes it brings the dead to life. The eventfulness of the streets is without teleology; its history exists only as it is experimented. A *kiltro* history, emerging from *kiltro* politics perhaps. Like the Zapatistas, this politics asks its surroundings before moving along: *preguntando caminamos*. Asking how to move, what to do, where to go, how to speak, what to say, this politics opens room for the reality of dreams: oneiric images whose force remains when we extract ourselves from of categories of state appropriation. Streets offer dreams that allow us to imagine *even* a state that is not yet: 'a state that knows how to give space while our worlds and our imaginations regenerate' (Stengers, 2020). That state is not yet – might it be? Ushering our walk, Chilean streets have asked the question.

Kiltro politics/ecological reparation

Figure 20.16: Minute of Silence, Inauguration of Chilean Constitutional Convention

Source: Cristina Dorador. Available from: https://commons.m.wikimedia.org/wiki/File:Convenci%C3%B3n_Constituyente_Chile_4-julio-2021.jpg

This is a picture of the inaugural session of the Chilean Constitutional Convention. Cristina Dorador, a member of the Convention, a scientist and our friend, took it on 4 July 2021. She wanted to capture the moment when Elisa Loncón, its President and a Mapuche women, called on all participants to observe a minute of silence to honour the presence of the dead. This is what Elisa Loncón, said:

> 'This is a historic moment for us. We know that we come from a history of denial, especially denial of the native nations, denial of many rights of the people of Chile, denial of many rights of women, of children, and in this context we also request a minute of silence for the memory of our dead, we must honour our dead, of the 500 years, the dead of the native nations after with the state of Chile, the dead, the women who have been victims of femicide, we should ask for a minute of silence for the comrades who also died during the dictatorship, we should ask for a minute of silence for the dead comrades of the revolt, we should ask for a minute of silence for the dead of the military occupation of the *wallmapu*, for all the dead of the native nations, the *yagan, selkman, quawascar, aymara, quechua*, for all of them, we as a convention shall honor the history of great fighters, we are going to take in this minute, a minute of silence.'

Requesting a historically powerful minute of silence, the President of the Convention brought an unprecedented will within the Chilean state as it opened possibility for the final constitutional document to be feminist and indigenous and with the dead. When we wrote the body of this piece, we only knew that the Constitutional Convention would happen – now, we feel the Chilean streets are resplendent, and its archive prescient: the centuries-long denial of life – the killings – their walls denounced, was invited to the Convention. Following Elisa Loncón's words, the minute of silence may have transpired as inhabited by a wound, that while heterogeneous to all those invited to the convention, was nevertheless collectively felt: a wound that divergently connected, a wound ecologically experienced.

Thus, Chilean Streets walk into the core theme of this volume, 'ecological reparations', a phrase that invites us to think about territories wounded by centuries of old extraction of minerals, water and labour. Thus, thought through the previous pages, the notion of 'ecological reparations', and the current Constitutional Convention, these territories may appear as having been ecologically damaged – or wounded – and will be repaired by a historically different, because left-leaning, Constitutional action. That this happens is seemingly a current possibility. However, we want to combine ecological reparations and our previous pages to conceptualize Chilean territories as an ecologically experienced wound heterogeneously embodied – even housed – in a complex *nos-otros*, bodies that partake also heterogeneously its cure. Being such, this wound houses its own cure and enacts a displacement of the separation between a subject (that cures) and an object (to be cured). Thus thought, this wound can also dislodge the epistemic mandate – effected by the colonial state and its requirement of history – that denies the possibility of humans and other than humans to be in a relation of mutual inseparability. Dislodged by ecological reparation, the notion of curing bodies *and* damaged territories would be replaced by organic practices relating wounded human bodies and territories that emerge together through its cure, and conceivably with the vocation to house the dead. Thus, the silence Elisa Loncón requested might have exceeded its one-minute moment and acquired a weird temporal strength able to enact a 'now' that includes in the future it makes, the words, actions, emotions of the dead – their music too.

The streets of Chile proposed – we think – an *ecological practice of reparation* articulated by a feminist disposition to narratives that while assertive do not want to be univocal; instead, they are inclined to constant negotiations and political improvisation. In improvising they are like a street dog, a *kiltro* skilled at crossing heavily trafficked streets because they know how to wait; willing to be afraid of what may poison them and to doubt what seems extremely appealing; accepting that their singular lives are with others. Albeit their skills, *kiltro* dogs do not know ahead of time: they always wonder where to go, where not to go, how to go, when and how to walk in a group, when and how to be alone. For a politics of improvisation engage in a practice of ecological reparation, the *kiltro* dog of Chilean streets is not only a metaphor, it is also a presence to learn from.

Notes

1. All photographs in this text, expect for Luis Poirot's one, were taken by Cristobal Bonelli and Marina Weinberg, and are part of their collective archive, which does not trace, nor claim, individual authorship.
2. www.laizquierdadiario.com/El-arbol-de-la-vida-una-obra-monumental-en-los-muros-de-la-resistencia-chilena
3. From 'El baile de los que sobran' by Los Prisioneros, www.youtube.com/watch?v=XYAnmsbnKM
4. https://archive.org/details/laciudadcomotexto/0_LCCT_LibroDigital_Sept/page/n23/mode/2up

References

Benjamin, W. (1968) *Illuminations*, New York: Shocken Books.

De Acosta, J. (1954 [1590]) *Historia Natural y Moral de las Indias*. In F. Mateos (ed) *Obras del P. José de Acosta, de la Compañía de* Jesús, Madrid: Ediciones Atlas.

Despret, V. (2017) *Au bonheur des morts. Récits de ceux qui restent*, Paris: La decouverte.

Federici, S. (2004) *Caliban and the Witch: Women, the Body and Primitive Accumulation*, New York: Autonomedia.

Guha, R. (2003) *History at the Limit of World History*, New York: Columbia University Press.

Haraway, D.J. (2016) *Staying with the Trouble: Making Kin in the Chthulucene*, Durham, NC: Duke University Press.

Hobbes, T. (1963) *Leviathan*, New York: World Publishing Company.

Mbembe, A. (2002) 'The power of the archive and its limits', in C. Hamilton, V. Harris, J. Taylor, M. Pickover, G. Reid and R. Saleh (eds) *Refiguring the Archive*, Dordrecht: Springer.

Preciado, B. (2013) *Testo Junkie: Sex, Drugs, and Biopolitics in the Pharmacopornographic Era*, New York: The Feminist Press at the City University of New York.

Segato, R. (2018) 'A manifesto in four themes', *Critical Times*, 1(1): 198–211.

Ureta, C. (2020) *LCCT la ciudad como texto*. Available from: www.laciudadcomotexto.cl.

PART VII

Isolating<>Embodying

21

(Un)crafting Ecologies: Actions Involving Special Skills at (Un)making ~~Things~~ Humans with Your hands

Eliana Sánchez-Aldana

Instructions to read

Dear reader:

When reading this text, I invite you to remember two things. First, if you are not familiar with weaving, I would like to show you two components of woven textiles: the warp and the weft. That is the first thing you should bear in mind.

The second is also an invitation to think of this content as a textile in four dimensions:

1. In a concrete material dimension, it is textile because it produces a textile piece.
2. It is also textile because it follows textile logics during the process of material making. The textile making process is a social practice weaving. Imagine the textile making process as the warp that gathers together and assembles, diverse species – humans and more-than-humans – as part of a textile ecology and other ecologies.
3. The text itself is an attempt to weave different voices. The narration is interweaving diverse voices as threads. You will find these voices in different styles. In these brackets [·] you will find my thoughts, my emotions, connections I make as a voice off; you can skip these parts if they get too noisy for your taste. *In these brackets < · > and in italic you will find the participants' voices*, and **in these brackets {·} and in bold**

Figure 21.1: Warp: the threads on a loom – a machine used to make cloth – that other threads are passed over and under to make cloth. //Weft: the threads that are twisted under and over the threads that are held on a loom

Source: Oxford English Dictionary

Figure 21.2: A detail of the textile woven

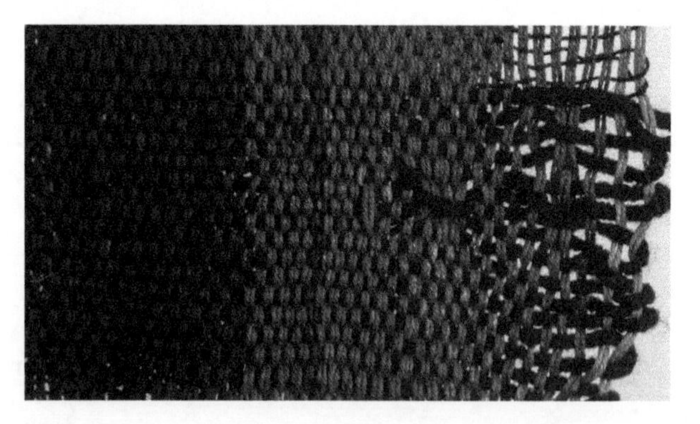

you will find fictional voices as part of academic worlds. Along with the text, you will also find ~~crossed-out~~ words. I would rather keep them than erase them; if I took them out, you would not notice their absence, but their crossed-out presence will add material meaning.

4. Finally, all that is here is part of a giant fabric whose [I would say]. Its materiality is semiotically textile. As noted by John Law when explaining material semiotics, 'practices in the social world are woven out of threads to form weaves that are simultaneously semiotic … and material' (Law, 2007: 7).

Figure 21.3: The textile logics for the textile making process

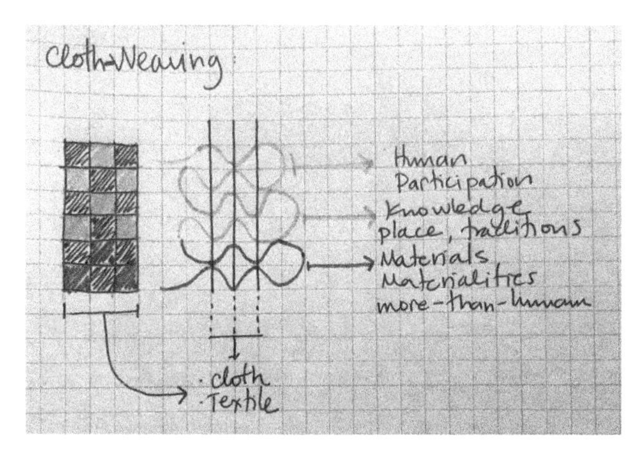

Figure 21.4: The textile logic for the text

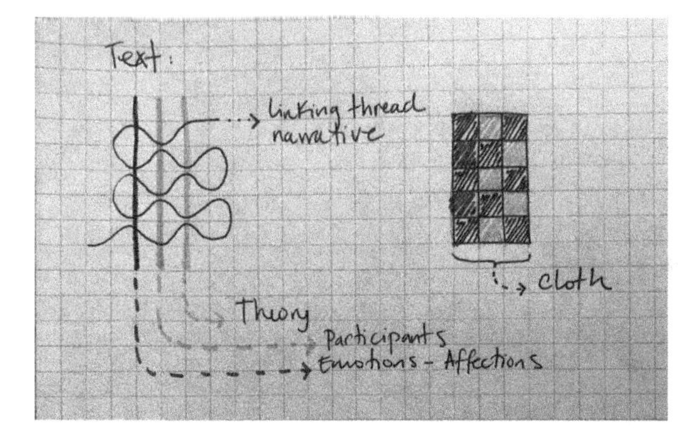

<*Look, she is Eliana*> Salvador told his workmate. He had his hands on my shoulders, holding me carefully from my back, while he gazed at his friend with a look that told us all what we needed to know. A story of a year was said with a look in just a second. Salvador is a metal crafter. Even though he is not what is traditionally known as a crafter. I call him a crafter because of his high metalwork skills, dedication to his work, profound material knowledge, and his connection with metal that allows him to speak on its behalf. He works with metal lathes; although they are semi-industrial machines, they require to develop manual expertise, especially considering that the quantity of what he makes is not large, he works with small series. Salvador is a technician with years of experience. Many other metalworkers will refer you to him if you need detailed, precise and on-time work. The workshop is placed in a popular neighbourhood in Bogotá called Samper

Figure 21.5: Material-semiotic in textile logics explained by Law (2007) and drawn/woven by me

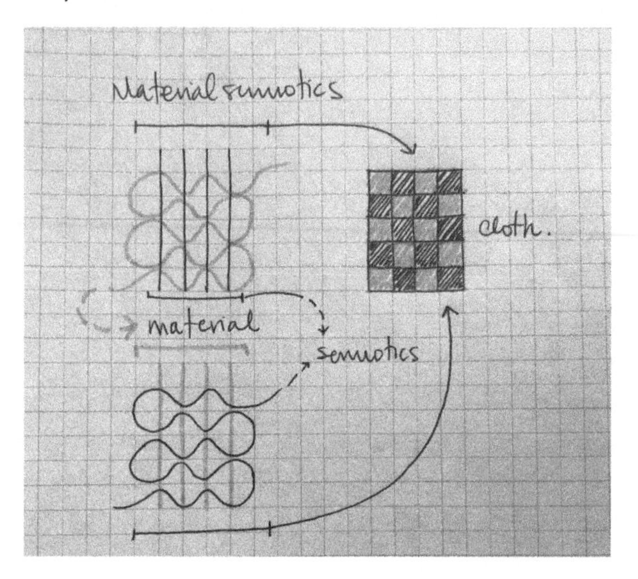

Mendoza. Located close to once was Bogota's railway company until the 1960s, this neighbourhood inherited the knowledge and the materials related to metalwork that the locomotives, wagons and tracks required. The district is not safe, at least not for me: a woman, an outsider. Visitors have to know how to walk the streets; they have to know people to comprehend what to do there. It is a very masculine area; it was the place where Marco had made his lamps, furniture and home appliances. Marco worked with Salvador. He trusted him with his most exclusive bronze and copper pieces. And now, I needed very detailed, precise and on-time work, and Marco, my husband, was not here with me anymore. I had to do it by myself. I was there, by the door of Salvador's metal workshop, bringing with me my spinner and some metal ideas to hold my textile pieces. To hold what I weave, as once Marco held me.

I started to tell this story from the end, but let's go back to the beginning. This text(ile) was born one year ago.

In May of 2019, I set my loom with 456 black and grey cotton threads as part of my textile explorations. I had decided to start a new textile to observe what happens and intervene while weaving. It was a small component of my PhD studies [by that moment, my research focused on something different]. At that time, I was working with a team of two textile crafters, Margarita and Robert. A manual loom ~~labelled~~ called THL105 [Although it could be complicated to name every human or more-than-human present in this story, I instead do so to acknowledge their presence] together we

Figure 21.6: Street view of Salvador metal workshop (left) and inside the workshop (right), 19 June 2019

designed and set the threads that form this text(ile). We used to meet every two weeks in the textile studio at Universidad de los Andes [That place is unique because it keeps memories of the Textile Arts programme closed at the end of the 1990s]. During our meetings, we talked about the people we knew that were involved with textiles, the wool and cotton producers and suppliers – they ended up being the same one – and the stories and travels of the former university's looms. Each one told a different part of the same story, and in that room, we assembled it together. {**Levi R. Bryant: That reminds me of what Deleuze said, it could be a sort of multiple spaces locally integrated (2011).**}

We also talked about what happened during the week, our families, our work and our issues. While making the textile, we shared details of our lives, the stories of how we learned to weave, we discussed the best way to knot, cut, comb. Making together, we created another environment, a subjective one, where our affections, emotions, textile knowledge/language could inhabit safely. As part of an ecological way of thinking, we were situated in a giant textile net of people, textile materials, tools, looms, knowledge, and emotions that resist the fast production flow in which handcrafts are endangered. {**Felix Guattari: The three components of the ecosophy: environmental ecology, social ecology, and mental/subjective ecology (1996).**} The closeness wove us, forming the trust, reviving the memories of the space; it was a way to keep alive the textile knowledge. By weaving, we enacted and kept alive a textile ecology.

The warp was set. And I left it ready. I planned to weave after a couple of weeks. But, by then, I had to wait, other parts of my work needed all my attention.

28 June 2019

[Marco, my husband, visited me at work. He went to my office and saw a textile piece I had hanged on one of the walls. He was surprised; it was

Figure 21.7: Notes taken the day we calculated and designed the textiles, a textile in numbers and text (top-left); Margarita, Robert, THL105 and I (top-right); cleaning THL105 (bottom-left); the warp (bottom-right)

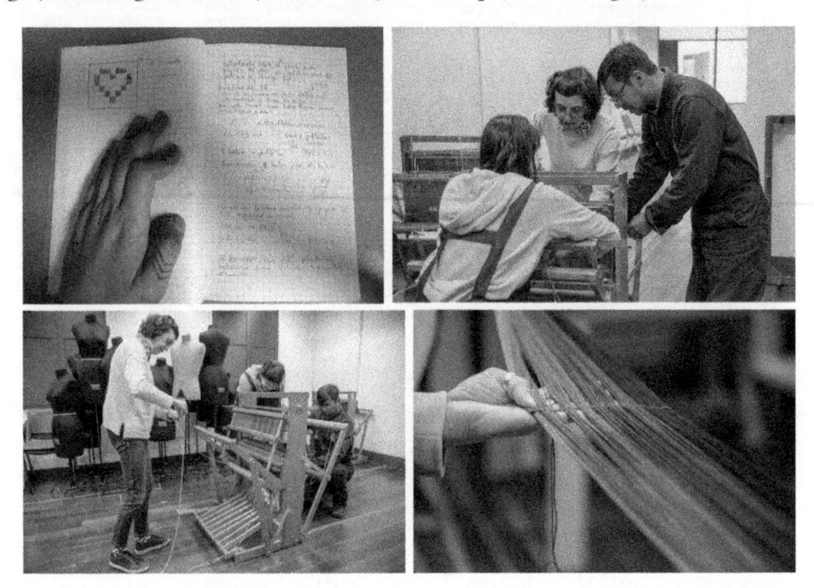

Source: Alejandro Barragán (photographer), Faculty of Architecture and Design University of los Andes

the first time he saw what I wove, and he was amazed. Although I always talked about what I did and wanted to do, I did not feel my textile pieces were worth being shown.]

21 July 2019

I had a clear idea of what I wanted my future to be alike. But now, it was forced to change. Two weeks before, my partner had died. His absence altered what I had in mind for me. [To go back in time is still painful. As soon as I start thinking of those days, I start crying. But the presence of the pain reminds me of all I had been through, how I have changed. It helps me to find meaning in things I did not see before. It is not easy to put my story into words; it is not easy to make it public, but why would it be better to make others' stories public instead of mine? This autoethnography sets a tension between the inside-outside observation, between who researches and who is at the aim of the research. The location/position of THE researcher is visible, and my own experience is analysed in relation to others' experiences. Nevertheless, although this autoethnography is a reflexive and individual practice in which meanings are produced collectively, in relation to others in particular socio-cultural contexts (Chang et al, 2016).]

I decided to move out of my house, too many material memories there. I stopped working for some weeks. My friends took care of me, they established shifts to make sure I was always accompanied and cared for by someone. His death tore me apart, unmade me {**Sara Ahmed: it was your body responding to your emotions, your emotion became an attribute of your body (2007)**}.

On 21 July I decided to go back to work [I actually do not know why did I go back, I guess it seemed the right thing to do] although I tried to focus on the simplest things I had to pay attention, I could not get out of my mind the fact that Marco was with me no more. I could see every space that now was empty, the evidence of his absence. My thoughts were a tangle. I remember I felt my head filled with a dark cloud of entwined threads. I could feel the weight of the pain in my body, it made me tired. It was the physical injury that the sorrow left {**Elaine Scarry: Those physical signs were the certainty of your pain (1985)**}. I was broken. It was the consequence of having lost him, my tears were tangible in my sleepless nights, my dry skin, my swollen eyes. {**Sara Ahmed: it was your body responding to your emotions, your emotion became an attribute of your body (2007).**} I could not concentrate. So, I paused most of the projects I was working at. And that very day, I asked two friends to help me to take THL105 – my loom – to my office. [I do not remember why I thought it could be a good idea to have THL105 close. I just remember it was the first thing I did when I went back to work.]

As a researcher, I have worked with women who have told the Colombian Conflict by creating handmade textiles (Cortés Rico, Patarroyo, Pérez-Bustos and Sánchez-Aldana, 2020, Patarroyo, Cortés-Rico, Sánchez-Aldana, Pérez-Bustos and Rincón, 2019, Sánchez-Aldana et al, 2020). Textiles have been a more-than-human companion during their mourning, a material voice to make memory and denounce what they have had to suffer in the Colombian war (Gonzalez Arango, 2015). I have worked with them on two projects. I have listened to their stories. We have embroidered and knitted together. I have felt ~~empathy~~ __________ for their pain. But I could have never ~~imagined~~ felt the pain they had to go through, until now. [It is important to highlight that my reality is different, Marco's death was not a consequence of the war. I am privileged, I can count on an economic and social network of support to face the death of a beloved one.]

Within this story, many other stories, times and places come to mind and become one. The time I was working with the textile collective of women Costurero de Tejedoras de la memoria de Sonsón[1] in 2016 [First time-space interwoven: 2016/Sonsón]. I remember the story Blanca – a woman from Sonsón, Antioquia – told us once. The story when she, after the murder of her husband 15 years ago, she started knitting a blanket. She told us she almost couldn't stop knitting during two weeks, only until she finished the textile piece, and decided she then could go back to her daily labours [Second time-space interweaved: 2000/Blanca's house]. The time when, as Blanca but without trying to replicate her decision [I didn't remember her story at that very moment], I had the need to start weaving. As an impulse, I looked for my loom and put it close to me [Third time-space interweaved: 2019/ my office]. And this very time [Fourth time-space interweaved: 2020/My computer-my house] I am writing to weave those times together and make sense through my own experience [I understand the story of Blanca from/ with/through my body (Scarry, 1985), this multiple is now singular (Mol, 2002)]. Weaving is a movement of the body, and it is while weaving, both the threads and the memories, that it becomes easier to move among times and places {**Jussi Parikka: It happens because the body *becomes a hinge that itself enables more than meets the eye at first* (2019, p 125), and when it moves, thoughts follow**}.

With THL105 in my office, I started weaving. Along the day there were time gaps I could focus in work duties, but when my thoughts turned into a dark heavy tangle, I jumped to my loom and started moving my feet to take up the frames and repeatedly pass a thread from side to side. The set warp allowed me to weave as I felt I should. We turn out to be a temporary assemble – THL105, me, the threads, my mourning – that became my place to be. The feeling of the materiality of the yarns, the sound of the frames going up and down, brought me back from the past. And it prevented me from going to the future. I had to anchor in the present to be able to weave. My sadness did not go away, but the mess was combed by the rhythms of the textiles making. Sometimes, when you are combing the threads, you do not know if you are disassembling, unmaking, something existing or making something new {**Iris Van der Tuin: The making and the unmaking stop being an antagonist to happen simultaneously (2019)**}. With time, I perceived my physical and emotional experience of weaving as a process of making and unmaking of my subjectivity. It became the unquestionable starting point of what before I was only observing in others {**Sara Ahmed: The stories you told were forced to stop othering the pain and emotions, this time it was you (2007)**}.

The quarantine began on 25 March in Colombia. Moments before leaving my office I looked at THL105 and thought: you are leaving with me. But I did not have what it needed to take it with me. So, I left and decided I would come back to get it.

It was already two months since the lockdown. I embroidered for a while, knitted a couple of raws. But for me, there is nothing like weaving. I could not wait any more, I wove through drawings. When I drew, the absence of my loom became an enhanced presence. {**Jean-Luc Nancy: It is a RE-presentation, not as a repetition but as a stressed presentation a more intense presence (2006).**} When drawing, I could listen to the woody

Figure 21.8: Textile sample woven with the same warp set. Details of what I wove in the closest time to Marco's death

Source: Photographer: Alejandro Barragán, Faculty of Architecture and Design University of los Andes

sound of the shafts moving. My hands felt the tension of the warp's threads, and I wondered about the groups I would form. I visualized the edges and the spaces I would weave. THL105's absence allowed me to understand it better. I could fill the empty space left for my loom with vivid memories that made me remember/feel weaving and our relation better. I did not realize how deep I knew THL105, its mechanisms, its possibilities before. We assembled back from a distance. We reconnected from afar.

THL105's absence helped me to understand Marco's absence. Now I knew why I remembered moments that once were insignificant but became answers to everyday issues. My body remembered Marco's materiality, I could hear his voice, smelled his essence and felt his body. Marco's absence was also an absence that was now enhanced. Often presences are taken for granted, and memories are used up to the last piece, regardless the significance or size, a small memory could become a long presence.

2 June 2020

Finally, the conditions are right to go and pick THL105 up. I decided to change the warp. It is time to start a new one. But first, I will have THL105 serviced before beginning a new project. I have the opportunity to send it straight away to Señor Arana [Jorge Arana is a loom's artisan who was in charge of the looms of the closed Universidad de los Andes' Textile programme]. I have no time to take the warp off the loom, so I send it as it is.

13 June 2020

THL105 is back. Excited, I bring it home. I check it and start unwinding the warp I did not have time to throw it away. Those threads had been inactive

Figure 21.9: Textile woven with pencil lines

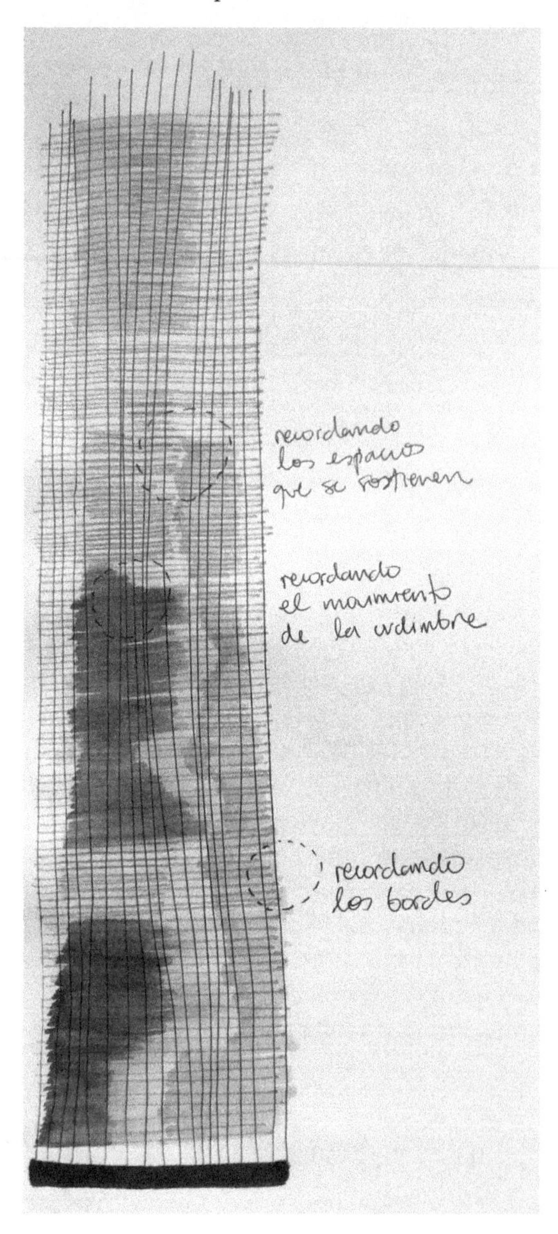

and unchanged since I stopped going to my office. But when I unwrap it, some pink stains start to appear. The cotton threads that I thought were immobile in my absence had changed without me. Something had changed their colour:

What looks still is not. {**Bruno Latour: Did you think the threads of the warp had to be honoured by your vision to live? No, they**

Figure 21.10: THL105 as I received it; the warp rolled; once I started unrolling it, I started seeing changes; thread under a stereoscope

got by very well with-out you (1988, p 193).} I thought my loom's world would stop without me, but no, it does not need me. So, I decided I would send the warp back. I know it is a difficult task, but the beauty of the unexpected cannot be ignored. I thought I had the control on the design decisions, but the ecology assembled in my absence by THL105, the environment, the time, the threads, the cotton, the dying acted to choose their own way. How can I call this encounter as inanimate?

With my loom, the place I am living starts to get small. I do not fit in there anymore. It is time to expand. I need a bigger home. [When I moved out of the apartment – our apartment, Marco and I apartment – I felt I shrank. I stored most of my belongings. I decided to live in a shared flat. But, at this moment I thought I could start living on my own. My loom was asking for space, it helped me to make my mind. I need more space.]

9 July 2020

New home, old threads. It is the second day at my new place. I set the threads back. I have to comb many times to put the threads back in order, their order. Unravelling, combing, unravelling, combing. Not an easy process, to have a new warp would be simpler. Moving away from the logic of one (Van der Tuin, 2019), assembled again, we – THL105, the warp, the weft, and me – weave a place to dwell. We weave our new home; we together re-make the warp and myself. {**Iris Van der Tuin: You all were a whole that did not unify the singularity of each, you were a becoming, rather than a being (2019).**} Sometimes, we tend to see the effects of the unexpected as damage, as trouble. It is easier to get rid of what is different to what we expected. But we can stay with the it, we can stay with trouble (Haraway, 2016). Without attempting to erase or ignore it, from the possibility of embracing it, it is not necessary to hide what had happened. It would be impossible to take the tear or stamp off. The warp should not be impeccable, I should not either.

[A long process of re-making myself was made into a fabric.]

The fabric is ready, I have taken it off the loom. But I feel it unfinished. The day Marco saw what I wove, we talk about making something together. I design three metal hangers; one will be for this textile. I remember I have the phone number of Salvador, the metal crafter Marco used to work with.

Figure 21.11: Combing, unravelling the warp (left); re-threading (centre); weaving at my new place (right)

Figure 21.12: The textile piece out of the loom; detail of the hanger (top); detail of the unexpected woven (bottom)

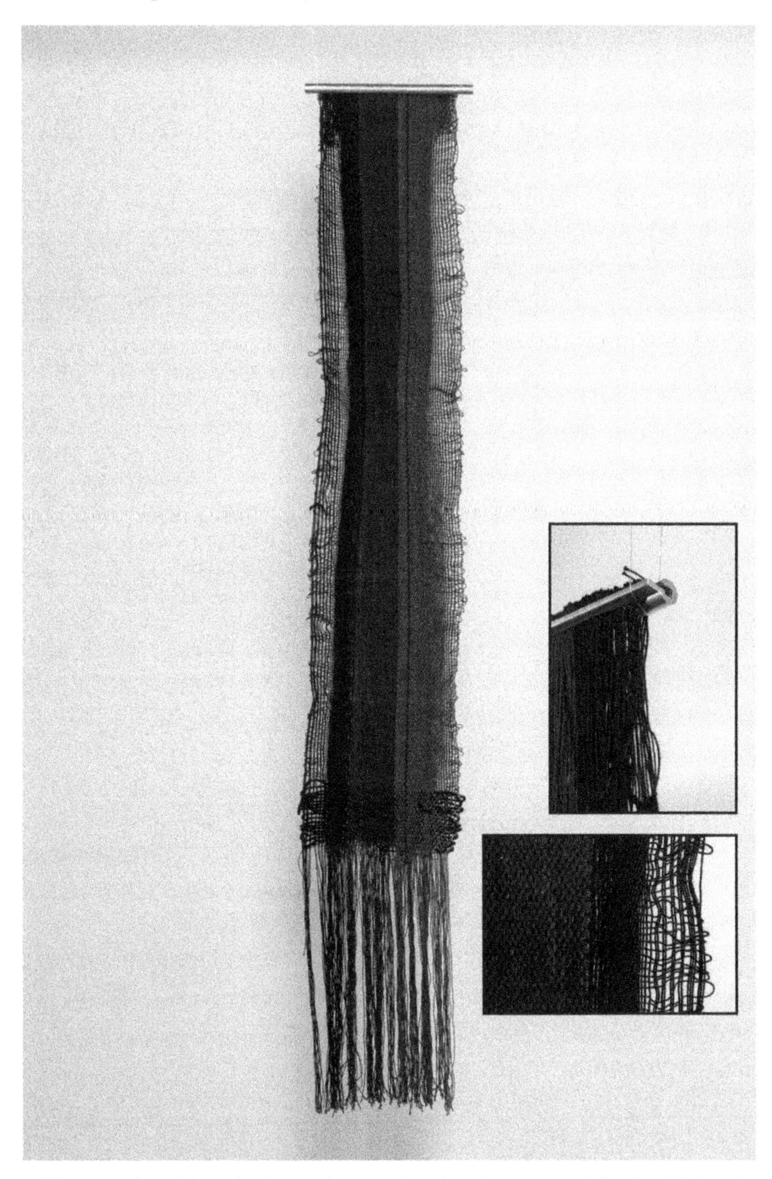

Source: Photographer: Alejandro Barragán, Faculty of Architecture and Design University of los Andes

I call him and ask if he could make a copper holder. I am nervous about meeting Salvador, it is going to be the first time I see him. But It was not only the right place to talk about well-done metalwork, but it was also the place to feel one more time the absence of Marco. Weaving has been the place to bear with my troubles, with his absence. His absences have been filled with my memories, but I know there will be a moment I will let them go too.

An assembling, disassembling and reassembling weaving ecology has given me the strength to stand by myself. I have started to hold what Marco hold once.

Salvador and I schedule a meeting. We will meet on 28 September.

[Gracias Laura Morales por revisar este texto y ayudarme a que las emociones que puedo narrar fácilmente en español tengan sentido cuando las escribo en inglés.]

Note

[1] El Costurero is a self-managed and continuous process born the Asociación de Víctimas por la Paz y la Esperanza de Sonsón (Association of Victims for Peace and Hope of Sonsón) in 2009. Every week its members meet to knit and embroider, giving way to the voice of the victims who, among threads and needles, like a ritual of healing and healing (Gonzalez Arango, Palomino and Junco Gómez, 2020). For more information about this textile/memory collective please visit: www.textilestestimoniales.org/creadores/2

References

Ahmed, S. (2015) *The Cultural Politics of Emotion* (2nd edn), Edinburgh: Edinburgh University Press.

Bryant, L.R. (2011) *The Democracy of Objects*, Michigan: Open Humanities Press.

Chang, H., Ngunjiri, F.W. and Hernandez, K.-A.C. (2016) What is collaborative auto-ethnography?', *Collaborative Ethnography*, 17–36.

Cortés Rico, L., Patarroyo, J., Pérez-Bustos, T. and Sánchez-Aldana, E. (2020) 'How can digital textiles embody testimonies of reconciliation?', *ACM International Conference Proceeding Series*, 2: 109–13.

Gonzalez Arango, I.C. (2015) 'Un derecho elaborado puntada a puntada. La experiencia del Costurero Tejedoras por la Memoria de Sonsón', *Revista de Trabajo Social*, 18: 77–100.

Gonzalez Arango, I.C., Palomino, S. and Junco Gómez, L.C. (2020) 'Archivo Digital de Textiles Testimoniales'. Available from: www.textilestestimoniales.org/ [Last accessed 12 August 2021].

Guattari, F. (1996) *Las tres ecologías* (2nd edn), Valencia: Pre-textos.

Haraway, D. (2016) *Staying with the Trouble,* Durham, NC: Duke University Press.

Latour, B. (1988) *The Pasteurization of France*, Cambridge, MA: Harvard University Press.

Law, J. (2007) 'Material Semiotics', *Heterogeneities*. Available from: www.heterogeneities.net/publications/Law2019MaterialSemiotics.pdf [Last accessed 12 August 2021].

Mol, A. (2002) *The Body Multiple: Ontology in Medical Practice*, Durham, NC: Duke University Press.

Nancy, J.-L. (2006) *La representación prohibida: seguido de La Shoah, un soplo*, Madrid: Amorrortu.

Parikka, J. (2019) 'Cartographies of Environmental Arts', in R. Braidotti and S. Bignall (eds) *Posthuman Ecologies: Complexity and Process after Deleuze*, London: Rowman & Littlefield, p 294.

Patarroyo, J., Cortés-Rico, L., Sánchez-Aldana, E., Pérez-Bustos, T. and Rincón, N. (2019) 'Testimonial digital textiles: Material metaphors to think with care about reconciliation with four memory sewing circles in Colombia', *Proceedings of Nordes 2019: Who Cares?*, 8(8): 1–7.

Sánchez-Aldana, E., Patarroyo, J., Villamizar, A., Padilla Casas, C., Rincón, N., Junco Gómez, L.C. and Saldarriaga, M. (2020) 'Time (s) to listen: A collection of asynchronous experiences', *Participatory Design Conference 2020*, 3: 207–9.

Scarry, E. (1985) *The Body in Pain: The Making and Unmaking of the World*, Oxford: Oxford University Press.

Van der Tuin, I. (2019) 'Deleuze and diffraction', in R. Braidotti and S. Bignall (eds) *Posthuman Ecologies: Complexity and Process after Deleuze*, London: Rowman & Littlefield, p 294.

22

Cultivating Attention to Fragility: The Sensible Encounters of Maintenance

Jérôme Denis and David Pontille

Maintenance practices have been an object of growing interest in recent years, and numerous researchers working in a variety of disciplines have made stimulating forays into a vast continent of jobs and ordinary activities which still appears largely unexplored (Jackson, 2016; *continent*, 2017; Strebel et al, 2019; Denis and Pontille, 2020b). This body of research works has notably played an active role in acknowledging the importance of material 'reproductive work' in the constitution of social life, bringing to light the omnipresence of mundane practices dedicated to making things last. Among these 'maintenance and repair studies', some have also initiated a fruitful dialogue with feminist theories of care, especially the work of Tronto (1993), whose definition of care, articulated with Fisher, famously refers to 'everything that we do to maintain, continue and repair our "world" so that we can live in it as well as possible' (Fisher and Tronto, 1990: 40). Extending the empirical and analytical gesture aimed at 'surfacing invisible work' (Star, 1999), this conversation between maintenance and care has made two important contributions to the ways in which social sciences approach the relationship between humans and artefacts. On one hand, it has participated in decentring the traditional focus on stability and persistence towards material fragility and its various manifestation (Denis and Pontille, 2015; Domínguez Rubio, 2020; Henke and Sims, 2020). On the other hand, it has emphasized, and documented, the ethical and affective dimensions of maintenance (Houston and Jackson, 2017; Puig de la Bellacasa, 2017).

In this chapter, we propose to further these reflections by exploring a less known, though complementary, aspect of maintenance: its sensorial dimension. If maintenance is a matter of affect, it is also a matter of attention. Material fragility is indeed anything but a transparent feature. In fact, when it appears obvious to anyone, this generally means that maintenance has failed. Fragility of things is thus something that one needs to become sensible to. Making things last entails cultivating a special relationship with them, keeping careful eyes and hands on their condition, and scrutinizing, even sensing, their transformations. Even though some scholars have highlighted the sensory skills involved in specific maintenance activities (Dant, 2008, 2010; Cállen and Sánchez Criado, 2015), little is known yet on how maintainers concretely become attentive to fragility. What is this attention made of? What does it take? How is fragility experienced? How is it dealt with?

Answering these questions implies describing maintenance activities as closely as possible, taking maintainers' preoccupations seriously, as mundane as they may seem. This is what we propose to do throughout the next sections, drawing on two recent investigations, the first one on graffiti removal in Paris (Denis and Pontille, 2020a), the second one on water systems management in France (Florentin and Denis, 2019). By looking at contrasting maintenance situations (urban maintenance articulated to city cleanliness on the one hand, infrastructure maintenance in utilities on the other), our aim is not to produce a systematic comparison, obviously. Rather, we take the opportunity of these two ethnographic inquiries to put to the fore various ways of cultivating attention to things, and of enacting fragility during maintenance interventions. As we will show, in both cases, maintenance operations do indeed take the form of situated explorations through which maintenance workers become sensible to the state of things, deal with their tendency to deteriorate and disintegrate, and foresee their capacity to progressively mutate. Such 'material politics' (Gregson, 2011) paves the way for a particular 'ecological thinking' (Puig de la Bellacasa, 2016), which in the case of urban settings and infrastructures broadens two well-known approaches: the human ecology of the Chicago School (Park, 1936) and the ecology of infrastructures (Star and Ruhleder, 1996). Accompanying maintainers at work, and describing the encounters they cultivate within emerging material entanglements, enables us to consider further how humans interact with things and their environment. Such an ecological approach accounts for the daily presence and vitality of heterogeneous biological, animal, mineral entities in mundane urban practices, and the active 'capacity of relation-creation, to how different beings affect each other, to what they do to each other, the internal "poiesis" of a particular configuration' (Puig de la Bellacasa, 2016: 52).

Sensorial explorations

In order to understand how the maintenance workers' explorations are performed, and to familiarize the reader with the domains in which they take place, let's start by following the course of two interventions.

> [Intervention 1]. Étienne starts the visit of a site on a mountainside where a reservoir, a valve room and a technical facility are located. For the past two years, he has been travelling the roads in the area to inspect the drinking water system works that are managed by the public utility hiring him. His visits also have an inventory function: the agency has recently incorporated many small municipalities and faces a considerable lack of knowledge about the nature of the facilities and their condition. His intervention starts outside. He goes around the elements that are directly accessible, inspecting the walls, ventilation ducts and roofs when they are visible. He takes pictures with his phone and uses a voice recognition program that allows him to dictate his remarks: a moved tile, a crack in a wall, a spot of moisture. He enters the valve room and roughly draws the installation on his phone. He then goes down inside the reservoir, which has been emptied in preparation for cleaning by two agents who greeted us on arrival. Once downstairs, he explains: 'The first thing to do is to stay put and look down, all the way across. Sometimes there are slopes or trenches where you can fall and get hurt. Then I look at the walls.' He sweeps the beam of his flashlight across the floor, then over the walls. He approaches and passes his hand in several areas (Figure 22.1). He then concentrates on the ceiling and finishes by sweeping the raft, which he walks over, still holding the flashlight. Once back up, he adds a few comments to his list and then takes some time to talk to the workers who are former employees of the municipality and who are still operating the facilities on site. He asks them some questions about the history of the site and the interventions that have been carried out in the past. Back at his office, Étienne transfers his photos to his computer and prints out his notes and sketches. He slips everything into a folder on which the name of the site is written, and which already includes a description form that he will fill in later. The file joins the stack of 'to be processed' files that Étienne will eventually record in a database that he himself configured and which gathers all the items of each visit. This data informs the other people of the agency of the condition of the structures and makes it possible to prioritize interventions throughout the year.

> [Intervention 2]. This morning, Tom starts his round at Place de la Nation. After a quick glance at the printed map of the neighbourhood

Figure 22.1: Inspecting a water reservoir: touching

Source: © J. Denis

that he has taken out of his pocket, he chooses a street, then he goes into it, phone in hand, staring at the building facades. As a full-time 'detector', he has been roaming the streets of Paris daily since 2005, in search of all forms of graffiti. Each of his detections initiates another round, that of an operator in charge of removing the unwanted inscriptions he has noticed. That's why Tom scrutinizes the facades so meticulously: the more accurate the information he provides, the easier it will be for his colleagues to proceed. As soon as he spots graffiti, he opens the dedicated application on his phone: 'I create a file for each intervention by filling in the different categories," he explains. 'The address and its geolocation, the time of intervention (including opening and closing times for stores), the type of vehicles to be used (for parking), the surface properties, the size of the graffiti calculated by square meters, the recommended intervention technique ... Then I take a photograph, and I validate my entry (Figure 22.2). Everything is directly transmitted and connected to the database.' Despite the experience accumulated over the rounds, detection is not always so straightforward. Depending on the spread of the graffiti, Tom sometimes takes some time to reflect further before indicating the total surface to be processed. He may also come closer to the facades and frequently touches the graffiti, passing his hand over

Figure 22.2: Reporting graffiti presence

Source: © J. Denis and D. Pontille

> several times, before filling in his file. Some stone surfaces may be covered with a thin layer of almost invisible paint and require a specific removal technique. Others, made of porous material, may have been degraded over removal interventions and require careful reflection to avoid further deterioration.

Apart from their specificities, these two scenes share several important features. They show inspection sessions, common to most maintenance occupations, which consist for the maintainers in making sure that 'all is well' and, if not, in identifying what is called in the vocabulary of professional maintenance 'disorders'. These inspections are linked to a variety of documents (laws, standards, contracts, guides, etc.) that define some of their aspects, and guide part of their accomplishment.

Regarding water networks in France, for instance, the so-called Grenelle I (Article 27) and Grenelle II (Article 161) acts, which were followed by local regulations, have designated water network maintenance as a major national issue, notably by setting objectives in terms of reducing leaks and developing a sound knowledge of the network's components. Together with a profound reorganization of local governance that has redistributed the responsibilities among authorities at different scales, this legislative framework has contributed to transforming maintenance practices by placing them at the forefront of the management of water infrastructures. Etienne's activity is an integral part of this political and managerial agenda, as it draws on numerous forms and texts that identify a series of elements he has to pay attention to, and the main signs of failure he must take into account.

Graffiti removal in Paris is part of a municipal policy that is closely connected to the 'broken windows' thesis initially formulated in the United States by Wilson and Kelling (1982). This doctrine instantiates a very particular maintenance epistemology (Denis and Pontille, 2020a) that makes any visual sign of disorder (a broken window, litter on the street, graffiti and so on) a disturbing element which, albeit minimal, must be taken care of as quickly and systematically as possible, at the risk of their proliferation. This is translated in legal documents and contracts in which it is stipulated that removal must be carried out at a steady pace: every day Tom's inspections are mandatory, and each graffiti he reports has to be erased within ten days.

These political, legal and technical documents, together with the situated inspection activities, emphasize how maintainers' attention combines a slow and routine mode of perception during rounds and a much more pressing and tense organizational rhythm. These temporalities of maintenance point to distinct, and potentially conflicting, ways of paying attention to fragility. They are an important aspect of the heterogeneous assemblage through which specific features of things come to matter more than others. The expert exploration that maintainers engage in to appraise the condition of things is crucial in this process. Ethnographic observation is a good way to appreciate this expertise. Indeed, the maintainers' ability to perceive certain marks or minute traces contrasts sharply with that of the observer, who is absolutely incapable of distinguishing certain imperfections on the surface of things they struggle to examine scrupulously. During their rounds, maintainers demonstrate 'connoisseurship', a form of knowledge, both intimate and skilled, which tends 'to an appreciation of details rather than of the [thing] as a whole' (Ginzburg, 1979: 274). They spot sometimes infinitesimal clues, 'details usually considered unimportant or even trivial' (p 280), thanks to which they manage to assess the gaps between what things are and what they should be. This ability to explore the condition of the world is practiced through professional experience, sharpened into a 'skilled vision', much like the cattle farmers observed by Grasseni (2004), who learn to estimate herds and cows by breaking away from the habits of ordinary perception.

As the very idea of 'rounds' suggests, inspection is also a matter of both proximity and displacement. To take care of things, maintainers get close to them, which tells us that maintenance goes beyond the loose vocabulary of observation. The sensibility to fragility is activated though the encounter between the maintainers' body and that of the things. A matter of co-presence, of 'staying with' (Jackson, 2016: 183), such proximity is all the more valuable as it goes with constantly reiterated displacements. The attention cultivated during the rounds is indeed enacted in motion: Etienne walks around the site he is visiting before moving into the tank, and once inside, he never stops walking around; Tom roams streets, his gaze meticulously focused on one building facade after another. These displacements are an attention operator in their own right. Repeated regularly, they fuel the emergence of salient features among things and the identification of material variations, and eventually fragility to manifest itself. In our cases (and in numerous other situations), maintainers' attention relies thus heavily on what Gibson (1986) termed the ambulatory dimension of perception.

Importantly, as researchers who have focused more specifically on diagnostic practices have shown (Orr, 1996; Dant, 2010; Sanne, 2010), such expert exploration is largely multisensorial. The use of hands, of course, is crucial for most maintenance workers (one of the origins of the verb to maintain, *maintenir* in French, is the Latin verb *manūtenēre*, which means 'to hold with the hand'). In our cases, touch allows to appraise substances through manipulation, specify the identified problems, and anticipate further interventions. Though hearing and smelling are more rarely involved, they sometimes allow the detection of signs to which one must be able to pay attention to (an unusual smell that indicates infiltration in a structure for instance), or to supplement the first inspection steps (for example by striking against a wall whose resonance helps to identify its composition). Apparently insignificant and characterized by the monotony of their repetition, these gestures belong to what Dant (2010) calls 'sensual knowledge', which both aims to produce knowledge about what happened and to guide the flow of the intervention as the close examination goes on.

The expert exploration at play during inspection sessions is actually twofold. On one hand, it is oriented towards the series of features and signs that are more or less precisely identified in technical documents. Never fully exhaustive, of course, these lists nonetheless constitute important resources for developing a careful scrutinization. On the other hand, maintainers cultivate a more 'floating' kind of attention, which is genuinely open to discoveries.

> Étienne has just finished his round. He is talking to the two cleaning agents who washed the tank after he made his inspection. One of them is leaning against the outer wall of the valve room. As he lifts his back from the surface of the wall, Stéphane stops in the middle of a sentence and looks at him, puzzled. He exclaims, 'Wait, what's that noise there? Do it again …' The agent puts himself back in place, amused, and then pulls himself off the wall again. A loud 'ploc' resonates. Étienne steps forward and feels the surface of the wall in several places with his hand, reproducing the suspicious noise (Figure 22.3). 'I can't believe it. What the hell is that thing? The paint's completely peeling off.' He turns around and explains to the observer: 'See, the colours are a bit different, here, it goes up very high. I would never have guessed.

Figure 22.3: Ploc, ploc, what's that sound?

Source: © J. Denis

> Maybe it means that there is an infiltration problem, or simply that the paint has been badly applied. It's outside, so it's not a big deal anyway. I still need to discuss this with my colleagues at the headquarters.' He then grabs his phone to take pictures. After several shots, he hesitates, then: 'Actually, I'm going to film, it's the only way to make them understand.' He records a video of his hand pressing on the wall and reproducing the sound.

What is at stake in this form of attention is to let unexpected events happen. Taking into account how material objects 'surprise' the people who deal with them, Yaneva (2008) shows, allows to reconsider material agency, beyond the idea of inert artefacts that humans can turn into powerful and stable resources to do 'politics by other means' (Winner, 1980). Surprise emerges from 'what is being added, above and beyond what is done and expected' (Hennion, 2015: 50). By leaving room for surprises, maintainers let walls, paint, doors or pipes act and 'speak' (Sanne, 2010). They acknowledge the capacity of things to demonstrate 'a heightened presence' (Hennion, 2015: 39), and disturb the course of human action in manifesting their own signs of existence and transformation. Each maintenance inspection is therefore both a test of conformity and an opportunity for the thing to 'makes itself knowable and lets itself being known' (Yaneva, 2008: 17).

Ecologies of fragility

Once inspections are completed, maintenance interventions go through even closer relationships to things, in the course of which maintainers encounter a great variety of materials: plastic, wood, marble, concrete, plaster, bricks, water, sand, cement, stones … as well as soil, plants, various pollutants, in the case of water networks management; and inks and paints (from graffiti itself or already on surfaces) in the case of graffiti removal. Among these heterogeneous materials, some are clearly part of the thing that has to be maintained, while others compose its environment. Some others, in contrast, cannot be easily distributed into this binary categorization beforehand. Much more than identifiable objects with stable properties, maintainers deal with what Bennett (2010) and Ingold (2012) call an 'ecology of materials', in which they try to sense the behaviour of material entities.

For instance, before removing a graffiti, the workers we observed spent several minutes touching the graffitied surface so as to estimate the degree of penetration of the graffiti paint into the stone, examine the invasiveness of an acid-based ink on a shop window, or anticipate how well the covering paint will hold in contact with the surface's components depending on the weather conditions (Figure 22.4).

Materials are apprehended not so much for what they are, their physical-chemical properties at rest, but for what they do and, even more so, for what they generate together and might become, as part of what Edensor (2011) terms 'entangled agencies'. Conversely to a sign of passivity or a lack of

Figure 22.4: Touching the graffitied surface

Source: © J. Denis and D. Pontille

agency, material fragility is thus apprehended during these sensible encounters as the result of the active, uncertain, and even surprising relationships between materials. From this point of view, maintainers come close to the figure of the 'craftsman' that Deleuze and Guattari (1987: 409) have defined as 'one who is determined in such a way as to follow a flow of matter', rather than interacting with previously stabilized entities whose characteristics are clearly circumscribed. They cultivate a careful attention towards material variations, modulations and movements, feeling these singularities in the making, engaging with the flow of matter so they can temporarily reorient its always transitory course.

The metaphor of flow 'monitoring' should, however, be used with precaution. It could indeed give the impression of an asymmetrical confrontation between a human being in a situation of detached 'perception' on the one hand, and an inert world that would be 'apprehended' and transformed by the former on the other. In fact, each intervention involves not only the eyes and sensory organs of the maintainers, but also their flesh, muscles and bones, which are entirely part of the material ecology of the situation. These material relations between workers' bodies and things become particularly salient when tensions arise. For example, removing graffiti involves handling tools (high-pressure water spraying guns, extension arms for paint rollers, spatulas for manually mixing sometimes very thick paints, etc.) which, as for many 'physical' jobs, can lead to pain or even musculoskeletal disorders, notably in the back and hands. To these potential sources of pain is added the harmfulness of the solvents the removal workers sometimes use without wearing their protection: without a mask, their prolonged inhalation causes lung problems, and without a cap or gloves, the burn is instantaneous on the skin and hair. Water networks management comes with its own physical stresses, starting with those generated by the very access to the facilities and infrastructures: going down ladders, moving in confined spaces, walking on soils with uncertain surfaces, and so on (Figure 22.5).

Just as the bodies of the maintainers, the tools they handle and the products they use obviously participate in the material ecology of maintenance. In the case of graffiti removal, taking into consideration the particular sources of transformation they may generate is a constant concern. For instance, a table included in the particular technical specifications of the call for tenders of the City of Paris (Annex 1, for the 2018 version) lists compatible materials and helps to select the appropriated removal technique by identifying forbidden combinations. Beyond these main principles, the practical implementation of techniques and the use of instruments require adjustments to be made in situation. From one intervention to another, some maintainers pay special attention to the reactions that the tools and products they use are likely to trigger among the heterogeneous elements at play.

Figure 22.5: Inspecting a water reservoir: the descent

Source: © J. Denis

> Jules walks up to a door full of graffiti, and touches it: 'That's baked paint, it's strong, it's factory-made! Nothing has been added to it, I can go straight with bubblegum [one of the four chemical solvents at his disposal].' He pours some product on his scraper and carefully rubs the door, alternating quickly with a cloth pass, without pressing, while specifying: 'It is necessary to wipe quickly so as not to let it act too long. Even on this paint it would make a much lighter stain.' While he continues (Figure 22.6), he insists on the quality of the rags: 'The choice of rags is important. They must not scratch the paint. The recent purchase order they made doesn't work, the new cloths damage the surfaces. It's the same with the scrapers, by the way. I keep some old, tired ones that hardly scratch anymore, for cases like this one. Because on a dark paint, black like here, it scratches very easily. You have to be really careful.'

Solvents, paints, rollers, rags: maintainers' 'intuition in action' (Deleuze and Guattari, 1987: 409) unfolds from the fine-tuning of the gestures that accompany the action of tools and products, to the accuracy of the rhythm of their passage and coating, to the adjustment of their own material qualities. This is what makes Jules prefer here the use of certain worn scrapers whose softened action makes it possible to preserve the paint on the door. The

Figure 22.6: Chemical solvents and the dance of rags

Source: © J. Denis and D. Pontille

choice of removal instruments and their careful handling are intended to 'leave as few traces as possible of my passage', as Jules later explained to us. This meets a professional requirement: the technical specifications of the contract prohibit the removers from producing 'cleanness stains' (p 14, for the 2018 version), under penalty of sanction by the municipality or litigation on the part of the facades' owners. But their level of precision is also a matter of personal meticulousness. On several occasions, Jules showed us traces of intervention by colleagues that he described as less 'serious'. Whatever the fine-tuning of these gestures and the level of operators' personal commitment, it is in the course of this complex choreography that the 'material politics' (Gregson, 2011) of graffiti removal is actualized, in the intertwinement of facades, graffiti, workers' bodies, the instruments they handle and the products they use.

The attention to fragility that unfolds during maintenance interventions is thus a matter of pace, gestural coordination, balanced rotation between the action of tools and products, and the reactions of the maintained thing itself within a material ecology whose boundaries are never fixed once and for all. During these reciprocal transactions where sensibilities are distributed, a lot of materials may interact with each other, those that are maintained as well as those that maintain, those that must be preserved as well as those that must be eliminated. Far from being a stable and intrinsic quality against which maintainers should struggle, material fragility is ecologically enacted through these close encounters during which the decay of some entities (here, Jules' rags) and the destruction of others (the graffiti) are used as means to preserve a thing (a surface in the city), while the degradation of yet others (the very maintainers' body) appears as the unevenly distributed price to pay for this thing to last. In this sense, these ecological encounters are always simultaneously political, affective, sensual and, ultimately ontological.

Thinking with maintenance

By sticking as close as possible to the gestures and words of the maintenance workers, we seek less to produce knowledge *about* maintenance, though, than to develop a way of thinking *through* and *with* it. What does it mean concretely? If 'thinking-with' is a way to making worlds happen (Puig de la Bellacasa, 2012), what kind of world emerges when thinking with maintenance? Mierle Laderman Ukeles' work offers a fantastic inspiration to answer this question. Author of the pungent *Manifesto for Maintenance Art, 1969!* Ukeles, a feminist conceptual artist, dedicated her whole career to bring maintenance in the realm of arts, questioning in return what has so far prevented modernity from taking maintenance practices into consideration. In the catalogue of the major retrospective that the Queens Museum dedicated to her work in 2016, Phillips (2016) describes Ukeles' performances as a gesture which incites spectators to pay attention to the maintainers' own attention. Placing maintenance in its very mundaneness operates thus what she calls a 'transfusion of attention' (p 99).

Ukeles' very first performance is particularly telling in this regard. Entitled *Transfer: The Maintenance of the Art Object: Mummy Maintenance: With the Maintenance Man: The Maintenance Artist, and the Museum Conservator*, it consisted in inviting two museum employees in front of the glass vitrine protecting a 5000-year-old Egyptian mummy at the Wadworth Atheneum in Hartford, Connecticut. The performance was organized in three successive moments. It began with the cleaning of the vitrine surface by the maintenance worker, who did it as he had done every day since the piece was present in the museum. In a second step, the worker's gestures were repeated identically by Ukeles, who had carefully observed them. Ukeles then stamped the vitrine with the words 'Maintenance Art Work' written in ink. Such a gesture, directly inspired by Marcel Duchamp's work, turned the pristine vitrine into evidence that this protective furniture of art display had become itself a work of art. In a third step, the conservator was then compelled to authenticate it and write a report. Above all, he became the only person responsible and authorized to carry out future cleaning operations. Through this performance, attention has thus been concretely and even legally transferred: the conservator had now to perform the careful gestures of maintenance the maintenance worker used to do.

Such a posture is highly political, since attention engages specific power relationships. Just as the responsibilities between those who 'bear the burdens (and joys) of care' and those who 'escape them' are distributed according to social and institutional contexts (Tronto, 2013: 32–33), there is a big difference between those who can afford not to pay attention to the fragility of things, and those who have no choice but to be sensible to it. Ukeles' work invites to reduce these asymmetries and redistribute both attentional abilities and responsibilities.

This is also what we try to do, in humble terms, in our own research work. By making ourselves attentive to the attention of maintainers, by conducting inquiries alongside their own inquiries, and then write about it, we strive to produce the conditions for what Citton (2016) calls a 'meta-attentional engagement', in which the reader 'is found to be plugged into the attentional experience of another more or less strongly subjectivized perception of the world, through which a certain reality is revisited' (p 139).

In the particular context of this collective volume, we believe that exploring maintainers' attention helps to grasp two important aspects. First, it outlines ways to assume a specific ecological posture. As a situated and political process that implies materials, bodies, instruments and texts, maintenance also draws on a delicate art of provoking and cultivating encounters within emerging material ecologies wherein what acts, and what interacts, is never completely known in advance. This consideration for material ecologies implies to become aware of the constant mutations things are made of, from micro-variations to dramatic transformations, and the flow of materials that animate them. Maintainers, we argue, can help us to cultivate 'deep attention to materiality and embodiment in ways that rethink relationality, in ways that suggest a desire for tangible engagements with mundane transformations' (Puig de la Bellacasa, 2017: 112). They are among the best positioned to shed light on the incessant material metamorphosis that characterize every object, which 'are continually becoming in the course of their lives' (Gregson et al, 2009: 250).

But cultivating such attention to fragility is also ecological in the sense that it is a matter of relationships and connections. Understanding how maintainers engage with the objects they take care of leads to reconsider how humans interact with things, even the most insignificant ones, way beyond 'cultural' or 'symbolic' relations. Maintenance both acknowledges and enacts the vital interdependencies and 'attachments' which hold that or who is cared for and the ones who care for them together (Hennion, 2017).

Since it invites to be simultaneously attentive to the maintainers' careful physical explorations and to the degradation of their bodies, this ecological approach resonates with Jackson and Houston's call to 'think [the] poetics and political economy [of maintenance] together' (Jackson and Houston, 2021: 246). Thinking ecologically helps indeed to identify the intricacies of material fragilities and thus to tackle some of the ambivalences of care (Puig de la Bellacasa, 2017). In our case, this consists in exploring how urban maintenance articulates capitalist and disciplinary considerations with the aesthetic and affective dimensions of maintainers practices. Such posture has theoretical, but also methodological implications. As Domínguez Rubio (2020) has stated in his careful exploration of the worlds of art conservation, because it evolves in the 'ecological nexus' (p 8) through which things are made to last, maintenance (and care) requires imagining ways of conducting ecological inquiries. And what better guides to initiate this kind of inquiries than those who engage in it on a daily basis, those who have become experts in material mutations and entanglements by constantly taking care of things?

Finally, the second aspect maintainers attention invites to reflect on concerns the idea of reparation itself. Round after round, intervention after intervention, maintenance performs a kind of reparation that has no end, and cannot be 'settled' once and for all. Lying at the core of maintenance, such a form of repair is always in the making (Denis and Pontille, 2022). Fragility requires a continuous attention: routine and uneventful work, which gains its value from repetition and constancy and acknowledges its never-ending modest, though vital, contribution. What maintenance can teach us is that, just like the artwork Souriau (2009; Latour, 2011) wrote about, things from mundane artefacts to spectacular pieces of urban infrastructure always remain 'to-be-done'.

Acknowledgements

We warmly thank the people who agreed to answer to our questions and to be shadowed during their working days. We are particularly grateful to Annemarie Mol, Mandy de Wilde, Antoine Hennion, and Alexandre Monnin for their insightful comments and suggestions on an earlier version of the text. We also thank Sara Aguiton, Marc Barbier, and Soraya Boudia for their invitation to the 2019 edition of the Institut Francilien Recherche Innovation Société summer school during which an early draft has been discussed. Finally, we are extremely thankful to Maddalena Tacchetti, Dimitris Papadopoulos and Maria Puig de la Bellacasa for their careful reading and stimulating comments.

References

Bennett, J. (2010) *Vibrant Matter: A Political Ecology of Things*, Durham, NC: Duke University Press.

Callén, B., Sánchez Criado, T. (2015) 'Vulnerability tests. Matters of "care for matter" in E-waste Practices', *Tecnoscienza*, 6(2): 17–40.

Citton, Y. (2016) *The Ecology of Attention*, New York: Polity Press.

continent. (2017) *R3pair Volume*, n°6.1. Available from: www.continentcontinent.cc/index.php/continent/issue/view/27

Dant, T. (2008) 'The "pragmatic" of material interaction', *Journal of Consumer Culture*, 8(1): 11–33.

Dant, T. (2010) 'The work of repair: Gesture, emotion and sensual knowledge', *Sociological Research Online*, 15(3): 97–118.

Deleuze, G. and Guattari, F. (1987) *A Thousand Plateaus*, Minneapolis, MN: University of Minnesota Press.

Denis, J. and Pontille, D. (2015) 'Material ordering and the care of things', *Science, Technology, & Human Values*, 40(3): 338–67.

Denis, J. and Pontille, D. (2020a) 'Maintenance epistemology and public order: Removing graffiti in Paris', *Social Studies of Science*, 51(2): 233–58.

Denis, J. and Pontille, D. (2020b) 'Why do maintenance and repair matter?', in A. Blok, I. Farías and C. Roberts (eds) *The Routledge Companion to Actor-Network Theory*, London: Routledge, pp 283–93.

Denis, J. and Pontille, D. (2022) 'Before breakdown, after repair: The art of maintenance', in A. Mica, M. Pawlak, A. Horolets and P. Kubicki (eds) *Routledge International Handbook on Failure: Critical Perspectives from Sociology and Other Social Sciences*, London: Routledge, pp 209–22.

Domìnguez Rubio, F. (2020) *Still Life: Art and the Ecologies of the Modern Imagination*, Chicago: University of Chicago Press.

Edensor, T. (2011) 'Entangled agencies, material networks and repair in a building assemblage: The mutable stone of St Ann's Church, Manchester', *Transactions of the Institute of British Geographers*, 36(2): 238–52.

Fisher, B. and Tronto, J.C. (1990) 'Toward a feminist theory of care', in E. Abel and M. Nelson (eds) *Circles of Care: Work and Identity in Women's Lives*, Albany, NY: State University of New York Press, pp 36–54.

Florentin, D. and Denis, J. (2019) *Gestion Patrimoniale des Réseaux d'Eau et d'Assainissement en France*, Research Report, Caisse des dépôts – Institut pour la recherche et Banque des territoires. Available from: https://hal-mines-paristech.archives-ouvertes.fr/hal-02391959

Gibson, J.J. (1986) *The Ecological Approach to Visual Perception*, London: Routledge.

Ginzburg, C. (1979) 'Clues: Roots of a scientific paradigm', *Theory and Society*, 7(3): 273–88.

Grasseni, C. (2004), 'Skilled vision. An apprenticeship in breeding aesthetics', *Social Anthropology*, 12(1): 41–55.

Gregson, N. (2011) 'Performativity, corporeality and the politics of ship disposal', *Journal of Cultural Economy*, 4(2): 137–56.

Gregson, N., Metcalfe, A. and Crewe, L. (2009) 'Practices of object maintenance and repair: How consumers attend to consumer objects within the home', *Journal of Consumer Culture*, 9(2): 248–72.

Henke, C.R. and Sims, B. (2020) *Repairing Infrastructures: The Maintenance of Materiality and Power*, Cambridge: MIT Press.

Hennion, A. (2015) 'Paying attention: What is tasting wine about?', in A. Barthoin Antal, M. Hutter and D. Stard (eds) *Moments of Valuation: Exploring Sites of Dissonance*, Oxford: Oxford University Press, pp 37–56.

Hennion, A. (2017) 'Attachments, you say? ... How a concept collectively emerges in one research group', *Journal of Cultural Economy*, 10(1): 112–21.

Houston, L. and Jackson, S.J. (2017) 'Caring for the "next billion" mobile handsets: Opening proprietary closures through the work of repair', *Information Technologies and International Development*, 13: 200–14.

Ingold, T. (2012) 'Toward an ecology of materials', *Annual Review of Anthropology*, 41(1): 427–42.

Jackson, S.J. (2016) 'Speed, time, infrastructure: Temporalities of breakdown, maintenance, and repair', in J. Wajcman and N. Dodd (eds) *The Sociology of Speed: Digital, Organizational, and Social Temporalities*, Oxford: Oxford University Press, pp 169–85.

Jackson, S.J. and Houston, L. (2021), 'The Poetics and Political Economy of Repair', in J. Swartz and J. Wasko (eds) *MEDIA: A Transdisciplinary Inquiry*, Chicago: University of Chicago Press, pp 244–64.

Latour, B. (2011) 'Reflections on Etienne Souriau's Les différents modes d'existence', in H. Graham, L. Bryant and N. Srnicek (eds) *The Speculative Turn: Continental Materialism and Realism*, Victoria, Australia: re.press, pp 304–33.

Orr, J.E. (1996) *Talking About Machines: An Ethnography of a Modern Job*, New York: Cornell University Press.

Park, R.E. (1936) 'Human ecology', *American Journal of Sociology*, 42(1): 1–15.

Phillips, P.C. (ed) (2016) *Mierle Laderman Ukeles: Maintenance Art*, New York: Prestel.

Puig de la Bellacasa, M. (2012) '"Nothing comes without its world": Thinking with care', *Sociological Review*, 60(2): 197–216.

Puig de la Bellacasa, M. (2016) 'Ecological thinking, materialist spirituality and the poetics of infrastructure', in G. Bowker, S. Timmermans, A. Clarke and E. Balka (eds) *Boundary Objects and Beyond. Working with Leigh Star*, Cambridge: MIT Press, pp 47–68.

Puig de la Bellacasa, M. (2017) *Matters of Care: Speculative Ethics in More Than Human Worlds*, Minneapolis, MN: University of Minnesota Press.

Sanne, J.M. (2010) 'Making matters speak in railway maintenance', in M. Büscher, D. Goodwin and J. Mesman (eds) *Ethnographies of Diagnostic Work: Dimensions of Transformative Practice*, Houndmills, Palgrave Macmillan, pp 54–72.

Souriau, É. (2009) *Les différents modes d'existence*. Suivi de *Du mode d'existence de l'œuvre à faire*, Paris: PUF.

Star, S.L. (1999) 'The ethnography of infrastructure', *American Behavioral Scientist*, 43(3): 377–91.

Star, S.L. and Ruhleder, K. (1996) 'Steps toward an ecology of infrastructure: Design and access for large information spaces', *Information Systems Research*, 7(1): 111–34.

Strebel, I., Bovet, A. and Sormani, P. (eds) (2019) *Repair Work Ethnographies: Revisiting Breakdown, Relocating Materiality*, Singapore: Palgrave Macmillan.

Tronto, J.C. (1993) *Moral Boundaries: A Political Argument for an Ethic of Care*, New York: Routledge.

Tronto, J.C. (2013) *Caring Democracy. Markets, Equality, and Justice*, New York: New York University Press.

Wilson, J.Q. and Kelling, G.L. (1982) 'The police and neighborhood safety: Broken windows', *Atlantic monthly*, 249(3): 29–38.

Winner, L. (1980) 'Do artifacts have politics?', *Daedalus*, 109(1): 121–36.

Yaneva, A. (2008) 'How buildings "surprise": The renovation of the *Alte Aula* in Vienna', *Science Studies*, 21(1): 8–28.

23

Technological Black Boxing versus Ecological Reparation: From Encased-Industrial to Open-Renewable Wind Energy

Aristotle Tympas

Retrieving the historical contrast between open and closed wind energy structures

Our present perception of wind energy technology is dominated by the version of wind turbines that we find in 'wind farms' ('wind parks'), which is associated with landscape degradation and local resistance. Locals don't want these turbines installed in their back yard, we don't ever see them posing happily in front of them. Even their most ardent supporters, who promote them as an unavoidable necessity in the face of the global environmental crisis, agree that there is nothing aesthetically appealing about these wind turbines. They just charge the local opponents of these wind turbines with suffering from the 'not in my back yard' syndrome. The reference to such a syndrome is by itself an acknowledgement of the negative aesthetic impact of these wind turbines. The 'energy landscapes' produced by the installation of wind farm turbines are certainly not attractive. In fact, if the criterion for the evaluation of the merits of wind energy turbines is their impact on the landscape, wind farm turbines score no better than fossil fuel energy generation plants (Pasqualetti and Stremke, 2018).

There were, however, in the past versions of wind energy technology that people were looking forward to be in a picture with. Going through the album of pictures that T. Lindsay Baker collected in *American Windmills: An Album of Historic Photographs* leaves no doubt about it (Baker, 2012). The owners of the kind of wind energy technology that we find in this album, together with a crowd formed by their relatives and friends, posed happily in front of it on all important occasions: from a baptism ceremony that was taking place at a water tank filled through the use of a wind pump to a wedding ceremony that was bringing together a comparable crowd, which also posed in front of the wind pump and the farm house that it was right next to (or just behind of). As we see in the picture that Baker chose for the cover of his book, people could not simply pose *in front of* this wind energy technology; they could actually pose *in* it (and *on* it). We will here refer to this past farm wind energy technology as 'open', while we will argue that the wind farm energy technology of present day is 'closed'.

It is estimated that about five million open wind energy structures were installed in the western part of the United States in the decades between the late nineteenth century and World War II. While most of them were wind pumps, by the late interwar period exemplary durable wind electricity generators also became available, especially after the replacement of the traditional wheel by turbine blades. They could be of a size comparable to wind pumps, like the ones used to supply with wind electricity all the lamps and the appliances of a household, thereby making them independent from the grid (transmission network). They could also be very small, like the ones used to run a radio set. Next, we see typical interwar advertisements of small wind structures for radio sets (for an introduction to the history of these structures, see Righter, 1996).

Image 23.1

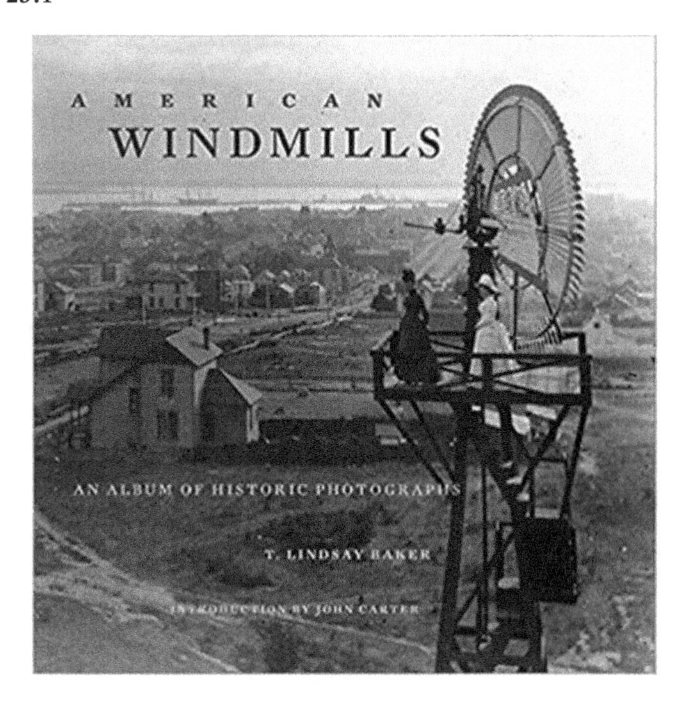

Image 23.2

Several scenes in movies of the immediate postwar period that were made by a generation of directors who had first-hand experience of the usefulness and appeal of open wind structures, are very suggestive. This is the case with a scene in the 1956 *Giant* (directed by George Stevens) – a movie that, for an historian of technology, may be read as a history of the transition from open water pumps to closed oil pumps (the screenplay was adapted by Fred Guiol and Ivan Moffat from an Edna Ferber 1952 novel). I am referring to the scene where the protagonist, played by James Dean, climbs on and stands in a wind pump. There is also the duel scene with Charles Bronson in the Sergio Leone 1969 spaghetti western *Once Upon a Time in the West*, which takes place in front of a wind pump like the ones installed by railway

Image 23.3

companies at stations along the line. Like the turning of the wheel of the traditional water mill, the turning of the wheel of this wind pump is symbolic of the unfolding of the movie narrative.

I expect that studying how Hollywood depicted these open wind structures will tell us a lot about the independence afforded by the ownership of one of these wind structures, from pumps to turbines, which will help us understand the associated attachment to them. And reversely, about how devastating the loss of such ownership could be. A wonderful scene on this central point can be found in the Arthur Penn 1967 *Bonnie and Clyde*, where the series of robberies by the infamous couple comes as a response to the 'robbery' of a farm household by a bank in the context of the interwar economic crisis. While narrating the devastating loss of the farm family house to the bank, the director zooms in on the wind structure right behind the house. This zooming suggests that the farm wind technology was a prime symbol of social independence.

Research is needed in order to learn the precise connection between the social fabric lost parallel to the loss of the ownership of wind energy technology by interwar American farmers. And, further, to learn about the involvement of the state in the unrooting of this wind energy tradition by force by the 1940s, in order to support the connection of farms to the grid (the transmission network). What we know seems to suggest that the capitalist state has been an indispensable institution in pushing towards closed wind structures. Notably, by the 1970s, when the first crisis of the fossil fuel energy economy sparked a new interest in wind electricity, the US and some major European states completely ignored the successful interwar experience with open wind structures and moved on to fund projects for the design of closed wind energy structures of the largest scale possible. The scale was so large – a historian of wind energy called it a 'hubris' – that these projects did not succeed (Heymann, 1998). Although the experimental pursuit of wind technology of the largest scale possible appeared to be a failure, it was in line with the approach that led to closed wind turbines of the large size that we find in contemporary wind farms.

The exact opposite historical example to the state support to an extreme scaling up of closed wind turbines was community tinkering with open wind turbines of the smallest scale. I have had the privilege to work with Kostas Latoufis on the history of one such case, that of the small wind turbines constructed at the UK Scoraig community by relying on scrapped/recycled material (Latoufis and Tympas, 2020). Latoufis is an engineer who is involved in worldwide non-for-profit initiatives/networks that seek to advance a similar wind turbine design philosophy. Noticeably, the members of such networks call their turbines 'open source' because they can be improved and adjusted to concrete use through the sharing of knowledge about them.

While the interwar open wind turbines of the American West were commercially manufactured and therefore relatively standardized, the Scoraig ones, also open, were idiosyncratic and constructed by a community. For an historical example of a very small idiosyncratic wind energy structure, of the kind used to run a radio but constructed by a community (as opposed to being commercially manufactured), I would refer to an artifact constructed by a group of Greek soldiers who were imprisoned in a remote north Africa desert camp by the British Army during World War II. While the Greeks and the British were allies during the war, these soldiers were imprisoned for having joined the left-leaning Greek resistance movement in the demand for the postwar democratization of Greece. Against their imprisonment, they managed to be updated on the latest news by constructing a radio set that was run by a very small wind energy arrangement. The radio was made by a wireless radio that they took from an airplane that had crushed in the desert. The wind arrangement was of their own construction. The Greek prisoners were presenting this wind arrangement to their guards as a toy by hiding its connection to the wireless radio that it run. When the guards were moving the prisoners around in order to check their stuff, the prisoners were disassembling the wireless radio into many small parts that they were placing inside their metallic water flasks. The guards assumed that the metal detector that they were using was activated by the flasks. Upon their release by the end of World War II, when asked by their guards to reveal how they were updated on the latest news, the prisoners enigmatically replied: 'we received the news from the wind' (Μακριδάκης, 2006).

I will conclude this attempt at a critical introduction to the history of wind energy technology with the case of the (approximately) 15,000 wind pumps that were constructed and used at the Lasithi Plateau of Crete, Greece, during the interwar and the early postwar decades, which is when they were dismantled. The water pumped through their use was stored in an open tank built on the ground under the wind pump so as to irrigate the land at the appropriate time. They were built and maintained by local artisans, with materials within their reach, like wood and iron, as in the case of the interwar US wind structures as just mentioned (Hoogervorst and Land, 1983). The design of this locally and easily reparable Cretan wind pump was in the 1970s and 1980s used in many locations at poor regions worldwide (Van de yen, 1977; Mann, 1979; Morgan and Icerman, 1981; Fraenkel, 1986). Interestingly, upon copying the traditional Greek windmills, their sails were made by cloth (see picture that follows, from Evangelodimou, 2018 and sketch from Van de yen, 1977; the Lasithi Plateau with the windmills operating is shown in a scene of *Signs of Life*, a 1968 movie that was directed and produced by Werner Herzog).

It was the use of cloth, a time-tested and fully repairable and recyclable material, that caught the attention of Gordon G. Brittan, Jr., a philosophy professor who has experimented with the construction of a cloth-based sails for the wind turbine of his American West (Montana) house (Brittan, 2002). This practical experimentation was inseparable from his philosophical elaboration on the difference between open and closed wind structures. Based on this, Brittan moved on to propose wind structures that contain no concealed parts. In the context of doing so, he insightfully compares wind structures with open and concealed parts from the perspective of their diverging emphasis on repair. This can be linked directly to the issue of ecological reparation, to which we shall return. In the following section, I will argue that his elaboration offers an appropriate starting point for what interests us here.

Image 23.4

On technological black boxes: between industrial wind turbines and digital computers

In the context of arguing that 'different views of technology are embodied in different types of wind turbines', and, further, 'certain kinds of design might more naturally, and hence, aesthetically, be introduced into the landscape' (Brittan, 2001: 169), Brittan follows what philosopher Albert Borgmann has written about 'devices'. As he sees it, devices offer an exemplar of how contemporary technology conceals and, in the process, 'disengages'. 'The more advanced the device', writes Brittan, 'the more hidden from view it is, sheathed in plastic, stainless steel, or titanium'. 'Moreover', he continues, 'concealment and disburdening go hand in hand. The concealment of the machinery, the fact that it is distanced from us, ensures that it makes no demands on our faculties. The device is socially disburdening as well in its isolation and impersonality' (Brittan, 2001: 174).

I would argue that we can keep the Borgmann and Brittan point about the concealment of a version of the wind turbines without having to call them devices because the concept 'device' refers to technologies of comparative small size, which we usually find inside laboratories and homes. I find more appropriate the shifting of the emphasis to the 'black box' concept – a shift that Brittan has endorsed:

> We could summarize Borgmann's position by referring to the familiar theoretical notion of a 'black box'. In a 'black box' commodity-producing machinery is concealed insofar as it is both hidden from view

Image 23.5

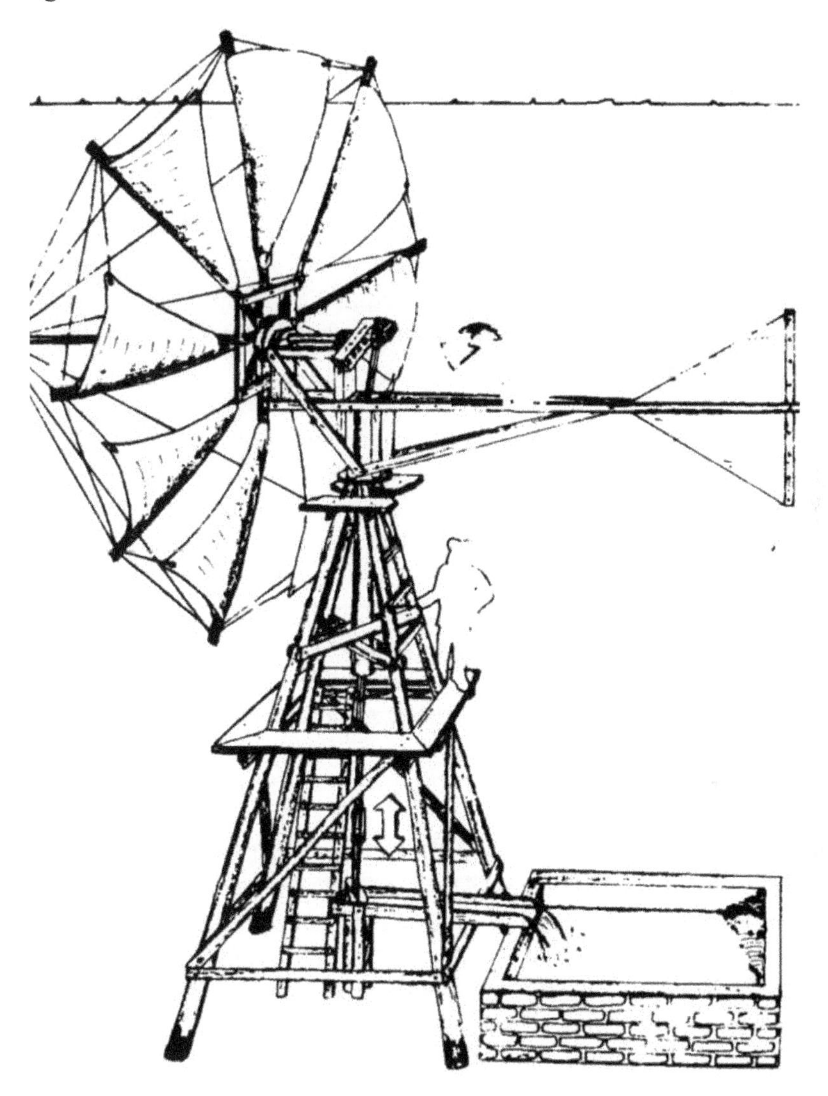

or shielded (literally), and conceptually opaque or incomprehensible (figuratively). Moreover, just those properties that Borgmann attributes to devices can be attributed equally well to 'black boxes'. It is not *possible* to get inside them, since they are both sealed and opaque. It is not *necessary* to get inside them either, since in principle it is always possible to replace the three-termed function that includes input, 'black box' and output, with a two-termed function that links input to output directly. All we really care about is that manipulation of the former alters the latter. (Brittan, 2001: 175, emphasis in original)

Acknowledging the centrality of black boxing (for an update on the literal and figurative use of the 'black box' concept, see Shindell, 2020), Brittan elaborated on a follow-up text of his:

> Now I want to make a very controversial claim. Wind turbines are for most of us not things but merely devices. There is therefore no way to go beyond their conventionally uncomfortable appearance to the discovery of a latent mechanical beauty. Thinking for a moment reveals that except for the blades, virtually everything is shielded (Including the towers of many turbines), hidden from view behind the same sort of stainless steel that contains many electronic devices. Moreover, the machinery is distant from anyone save the mechanic. The lack of disclosure goes together with the fact that wind turbines are merely producers of a commodity, electrical energy, and interchangeable in this respect with any other technology that produces the same commodity at least as cheaply and effectively. The only important differences between wind turbines and other energy-generating technologies are not intrinsic to what might be called their design philosophies. In other words, although they differ with respect to their inputs (i.e. fuels) and with respect to their environmental impacts, the same sort of functional description can be given a fossil fuel plant. There is but a single standard on which to evaluate wind turbines. It should not be wondered at that they are, with only small modifications between them, so uniform. (Brittan, 2002: 69)

I follow Brittan in the argument about the functional equivalence between, and the interchangeability of, fossil fuel machines plants and a version of wind turbines/parks, on the grounds of their constitution, both, as black boxes. But, precisely because of this equivalence, I disagree with the statement that the two differ 'with respect to their environmental impacts'. To show what is at stake here, I suggest that we do not stay at what is displayed by wind turbines of the black box kind but also try to look at what is concealed by it, at what the wind-intercepting blades are invisibly connected to.

Turbines of the black box kind, the kind that we find in wind farms/parks, industrial wind turbines, are connected to a long-distance transmission network. It is this network that links them to the consumers of the electricity generated by them. Noticeably, this is the same network that links consumers to electricity generating plants based on fossil fuels. Someone in, for example, Athens, Greece, can consume electricity transmitted through this network that represents a mix of electricity generated at fossil fuels plants and industrial wind parks. The mix can actually include electricity generated at nuclear plants, including ones located at a country different from that of consumption (see Tympas et al, 2013) – incidentally, this explains why industrial wind parks and nuclear energy plants can be part of the same corporate portfolio. If industrial wind turbines and fossil fuel plants are interchangeable as a commodity at the point of their consumption, it is the sharing of the transmission network that makes this interchangeability possible.

Fossil fuel plants cannot be replaced by industrial wind parks because when there is no or little wind they need to be in operation. This means that industrial wind parks are added to the total of fossil fuels plants, they are not replaced by them. Moreover, the available fossil fuel plants have to operate in addition to industrial wind parks because they represent a huge capital investment that cannot remain idle. For as we know very well from the study of the availability of mass production machinery, once available it must operate, with its production inducing consumption (Hounsell, 1985). In the case of mass production of electricity through fossil fuels electricity generators, this is known as the 'load factor' imperative: new consumption must be encouraged so as to match the generating capacity of production machinery (the long-distance transmission network is here of constitutional importance, see Cohn, 2017). This is why national and transnational drives that target the increase of the percentage of electricity generated by wind do not assume that the overall consumption of electricity – not to say the overall consumption of energy – will decrease. Suggestively, while the increase in electricity consumption that was generated by industrial wind plants was impressive in Europe between 2012, when the '2020 target' for a green energy transition was adopted, Europe consumes more energy in 2020 today than in 2012 (Sanchez-Nicolas, 2020).

The black boxing of industrial wind turbines is critical when it comes to isolating the wind-intercepting turbine from the rest of the technological materialities involved in making the wind energy available to a consumer. More specifically, it conceals the long-distance transmission network that starts right behind the wind-intercepting turbine blades of wind farms. This transmission network is connected to (and is, therefore, inseparable from) the turning of the turbine blades by the wind. This connection to the transmission network starts right behind the blades, where there is an apparatus that is supposedly self-regulatory (self-controlled, automatic), and continues throughout the long-distance transmission line, which is also loaded with apparatus of this kind. More precisely, it starts right behind the blades with regulatory mechanisms based on feedbacks between the wind and the transmission line. Differently put, the black boxing allows for the display of all this as a two-termed function, as if there is only the wind, as the source of electricity, and the electricity switch on/off button at the consumption. It conceals from a consumer that this is actually a three-termed function, the invisible function being the one where wind and fossil fuel electricity become mixed in the drive for the dynamic reproduction of a growth of energy consumption. It follows that whatever the environmental impact of this kind of (black boxed) wind farm turbines may be, it is part and parcel of the environmental impact of fossil fuel plant turbines.

As I suggested elsewhere (Τύμπας, 2018, chapter 7; Tympas, 2005), relying on the transmission network and the multitude of feedbacks mechanisms (now falling under what is called 'smart grids'), all the way from the generation of electricity currents and signals (strong, as in electric power, or weak, as in electric communication) to their consumption, is not actually all that automatic. It requires skilful labour to maintain the balance in the transmission network, to keep the network functional. And it is precisely this social labour that is concealed through black boxing. To elaborate on this, I may refer to the functional equivalence of energy and computing black boxes. Brittan has insightfully prepared us for this by writing that the 'wind in – electricity out' function of black boxed turbines, which is the only one visible, 'is roughly the same kind of comprehensibility that is involved when we note the correlation between punching numbers into our pocket calculators and seeing the result as a digital readout' (Brittan, 2002: 70). I have had the opportunity to argue that displaying the digital input and output of numbers at the exterior of a black box, while encasing at its interior the social labour required to produce the output for a certain input based on the production of an appropriate computing analogy (between the computer and the computable), defines all computing technology, from the emergence of capitalism to the present (Tympas, 2017, 2020). Brittan had correctly pointed to the technical similarity between black boxed turbines and pocket calculators. I suggest that the similarity is more general: black boxing wind energy is technically similar to computer digitalization, present-day industrial wind turbines and digital computers are products of the same socio-technical drive.

Black boxing versus repairing: on ecological reparation

Based on everything said so far, in this section I will move on to argue that the black boxing that we find in both industrial wind turbines and digital computers devaluates, through its concealment, the need for social labour. In the case of computing, this is the skilled labour to produce and modify, maintain and repair so as to have a proper analogy between the computer and the computable (Tympas, 2017, 2020). What is it that is concealed in the case of industrial wind turbines? To answer this question, we may rely on a comparison between industrial-closed wind structures and the open ones introduced in the first section.

It is not enough to invite attention to the long-distance transmission network that is concealed by the black boxed version of wind energy technology; we also need to place the emphasis on what is not concealed by the wind energy structures that are not black boxed. Here the Lasithi Plateau case can help us, but not only because of the use of a more inviting materiality at the interception of the wind that captured Brittan's attention, the wheel with reparable cloth sails. In the case of the Lasithi Plateau wind energy structures, I find extremely important the visible low-tech water tank at the basis of each of these structures, which was functionally equivalent to the (also visible) water tank attached to the wind pumps of the American West wind pumps introduced earlier. These water tanks, for local energy storage, represent the direct opposite to the long-distance energy transmission.

If we want a direct comparison, which involves wind electricity, at the opposite end of the long-distance transmission network of black boxed industrial wind turbines I would place the batteries at the basement of the late interwar American farm household, which were completing the circuit of its wind energy technology. The key here is that the batteries and the wind turbine were connected through a local circuit, not a long line, as in the case of industrial wind parks and fossil fuel plants. Research on details is here necessary. I would suggest that it ought to start from the fact that this local wind electricity circuit was operating at an electric tension and a current that was as low as possible, thereby placing the emphasis on flexibility, adaptability, maintenance, repairability. Virtues privileged by the case that we know more about, that of the wind electricity advanced by the community tinkering that we find in the aforementioned history of the UK Scoraig community (Latoufis and Tympas, 2020), and, in the present, in the 'open source' of Latoufis and the like-minded members of his global networks.

The closed versus open (and long-distance transmission versus local storage) wind electricity contrasts are actually coupled by many more ones: large versus small size, highly synthetic versus more natural materials, centralization versus decentralization, business and ownership schemes that favour the big capital/corporations versus privileging the immediate producer, commodity versus need, use versus exchange value, profit versus care. And, last but not least, here agreeing with Brittan, black boxed wind electricity machinery is distant from anyone save the mechanic Brittan, 2001, 175); black boxed technological design by a distant expert is the opposite to technical repair by everybody. A machine designed by the scientifically trained engineer, which can only be abandoned when it stops functioning after a limited time versus something defined by its ongoing susceptibility to local reparation. The news regarding what happens with industrial wind farm turbines after they run the course that they are designed for confirms the centrality of ongoing-local-open reparation, an ecological reparation: from the start to the end, from mining for wind farm turbines (Sonter et al, 2020) to burying them when they die (Martin, 2020), the technology of black boxed industrial wind turbines reproduces the cause of the environmental crisis.

There is by now a sense that there is a problem somewhere with the dominant version of present-day wind energy technology, with relying on it for a transition to renewable energy and environmental sustainability (Rumpala, 2017; Siamanta, 2021; Sovacool et al, 2021). This invites a rethinking/reconceptualization of what such transition may actually require, what renewable energy may actually be. The chapter sought to contribute towards such rethinking. We may now state fully its argument, an argument for pursuing openness as a prerequisite of ecological reparation. One could only pose *in* wind structures like the one in the cover of the Baker book, because, unlike present-day industrial wind turbines, they had no interior and exterior, no inside and outside, no surface and back, no concealed and displayed part. Differently put, through black boxing (typical of the dominant version of present-day technology, from industrial turbines to digital computers), we are presented with an environmental-friendly energy 'source', a 'renewable energy source', the wind, while the destruction of the environment through the machine (the long-distance transmission network) that follows is concealed.

Without the black boxing, there would actually be no reason to talk about the 'environment' in the first place (while we lack a history of the concept 'environment', we now know enough to argue that it emerged in the context of historical capitalism, see, for example, Benson, 2020). What is actually concealed by the black boxing is that there is not just the wind-intercepting blades and the button at the other end. We should not then talk about renewable energy sources, as the wind (and we could argue the same about the sun) by itself does not guarantee renewability. The black boxing makes the reproduction of an environmentally destructive energy machine invisible by allowing a view only to the source of energy. It is on the grounds of the black boxing that we see the source and not the machine, we see the exterior, an environmental source, but not the interior, the machine destruction of the environment. It is this that makes it look as if renewability is possible by a mere digital click (switch on) of a button by an energy consumer.

References

Baker, T.L. (2012) *American Windmills: An Album of Historic Photographs* (reprint edn), Norman, OK: University of Oklahoma Press.

Benson, E.S. (2020) *Surroundings: A History of Environments and Environmentalisms*, Chicago: University of Chicago Press.

Brittan, G.G., Jr. (2001) 'Wind, energy, landscape: Reconciling nature and technology', *Philosophy and Geography*, 4(2): 169–84.

Brittan, G.G., Jr. (2002) 'The Wind in One's Sails: A Philosophy', in M. Pasqualeti and P. Gipe (eds) *Wind Power in View: Energy Landscapes in a Crowded World*, San Diego, CA: Academic Press, pp 59–79.

Cohn, J. (2017) *The Grid: Biography of an American Technology*, Cambridge, MA: MIT Press.

Evangelodimou, E. (2018) 'The "forest" of windmills at UNESCO's doorstep', *Τα Νέα*, 17 July. Available from: www.tanea.gr/2018/07/17/english-edition/the-forest-of-windmills-at-unesco-s-doorstep/

Fraenkel, P.L. (1986) *Water Lifting Devices*. Rome: Food and Agriculture Organization.

Heymann, M. (1998) 'Signs of hubris: The shaping of the wind technology styles in Germany, Denmark and the United States, 1940–1990', *Technology and Culture*, 39(4): 641–70.

Hoogervorst, N. and Land, G.V. (1983) *Why Windmills Applied: Socio-Economic Determinants of Windmill Use for Irrigation on the Lasithi Plateau in Crete*, Wagingen: Wageningen University Press.

Hounsell, D. (1985) *From the American System to Mass Production, 1800–1932: The Development of Manufacturing Technology in the United States*, Baltimore, MD: Johns Hopkins University Press.

Latoufis, K. and Tympas, A. (2020) 'The craft of small wind turbine making: The windmills of Scoraig and the alternative technology movement in the UK', *Digital Culture and Society*, 6(1): 187–91.

Μακριδάκης, Γ. (2006) *Συρματένιοι, Ξεσυρματένοι, Όλοι: Χιώτες Πρόσφυγες και Στρατιώτες στη Μέση Ανατολή, Αφηγήσεις 1941–1946*, Πελιναίο: Χίος.

Mann, R.D. (1979). *How to Build a 'Cretan Sail' Wind-pump: For Use in Low-speed Wind Conditions*, London: Intermediate Technology Development Group.

Martin, C. (2020) 'Wind turbine blades can't be recycled, so they are piling up in landfills', *Blomberg*, 5 February. Available from: www.bloomberg.com/news/features/2020-02-05/wind-turbine-blades-can-t-be-recycled-so-they-re-piling-up-in-landfills

Morgan, R.P and Icerman, L.J. (1981) *Renewable Resource Utilization for Development*, Oxford: Pergamon.

Pasqualetti, M. and Stremke, S. (2018) 'Energy landscapes in a crowded world: A first typology of origins and expressions', *Energy Research and Social Science*, 36: 94–105.

Righter, R.W. (1996) *Wind Energy in America: A History*, Norman, OK: University of Oklahoma Press.

Rumpala, Y. (2017) 'Alternative forms of energy production and political reconfigurations: Exploring alternative energies as potentialities of collective reorganization', *Bulletin of Science, Technology & Society*, 37(2): 85–96.

Sanchez-Nicolas, E. (2020) 'Why is EU off track for 2020 energy efficiency target?' EUobserver, 11 February. Available from: https://euobserver.com/energy/147407

Shindell, M. (2020) 'Outlining the black box: An introduction to four papers', *Science, Technology and Human Values*, 45(4): 567–74.

Siamanta, Z.C. (2021) 'Conceptualizing alternatives to contemporary renewable energy development: Community Renewable Energy Ecologies (CREE)', *Journal of Political Ecology*, 28(1): 247–69.

Sonter, L., Dade, M.C., Watson, J.E.M. and Valenta, R.K. (2020) 'Renewable energy production will exacerbate mining threats to biodiversity', *Nature Communication*, 11: 1–6.

Sovacool, B.K., Hess, D.J. and Cantoni, R. (2021) 'Energy transitions from the cradle to the grave: A meta-theoretical framework integrating responsible innovation, social practices, and energy justice', *Energy Research and Social Science*, 75: 1–16.

Τύμπας, Τ. (2018), *Αναλογική Εργασία, Ψηφιακό Κεφάλαιο: Ιστορία των Τεχνολογιών Υπολογισμού και Αυτοματισμού στην Ενέργεια και την Επικοινωνία*, Αθήνα: Angelus Novus.

Tympas, A. (2005) 'Between telecommunications efficiency and stability: Towards an historical approach', in G. Kouzelis, M. Pournari, M. Stoppler and V. Tselfes (eds) *In-Use Knowledge*, Bern: Peter Lang, pp 197–212.

Tympas, A. (2017) *Calculation and Computation in the Pre-electronic Era: The Mechanical and Electrical Ages*, London: Springer.

Tympas, A. (2020) 'From the display of a digital-masculine machine to the concealed analog-feminine labour: The passage from the history of technology to labour and gender history', *Historein*, 19(1).

Tympas, A. Arapostathis, S. Vlantoni, K. and Garyfallos, Y. (2013) 'Border-crossing electrons: Critical energy flows to and from Greece'. In P. Högselius, A. Hommels, A., Kaijser and E. Vleuten (eds) *The Making of Europe's Critical Infrastructures: Common Connections and Shared Vulnerabilities*, London: Palgrave/Macmillan, pp 157–83.

Van de yen, N. (1977) *Construction Manual for a Cretan Windmill*, Amersfoort: Steering Committee for Wind Energy in Developing Countries, Pub. # SWD 77–4.

PART VIII

Growth<>Flourishing

24

Algorithmic Food Justice

Lara Houston, Sara Heitlinger, Ruth Catlow and Alex Taylor

Ecological reparation

In recent years, practices of reparation (broadly including maintenance, repair, remediation and regeneration) have become lively sites of scholarly interest in the social sciences, particularly in relation to environmental concerns. Some scholars have critiqued the idea of repair as a framing for ecological reparation, seeing repair as too invested in a return to normative relations that are already broken (Middleton, 2018). Indeed, some have argued that repair and maintenance work can keep orders running long past their end dates, masking and displacing the opportunity for system change and alternative approaches (Ribes, 2017).

For others, the idea of repair still has something to offer. Jackson writes about repair as the kind of hope that we need, as we think from – and respond to – 'broken worlds' (Jackson 2014). Henke and Sims disambiguate between forms of repair action: 'repair as maintenance' works within existing orders whereas 'repair as transformation' brings new configurations into being (2020). Henke and Sims suggest that both are required for repair within the context of the Anthropocene, a recursive idea of 'repairing repair' (2020: 122).

What is clear to all interested in reparation, is that when moments of breakdown come under study, they are often revealing: decentring subjects and showing up relations in their complexity, in their fragility and in their ever-changing temporality (Houston, 2017). To study repair is to think about the possibility of action in this moment; to examine decisions that are taken about what endures and what is let go. On the one hand, that might mean thinking about building new orders in the world's aftermaths: working with the conditions at hand and without the expectation of solutions (Tsing, 2017). On the other, reparation might also mean un-making damaging systems through design for decline (Tonkinwise, 2019; Lindström and Ståhl, 2020).

Algorithmic Food Justice

The politics of ecological reparation is increasingly played out in spaces of anticipatory governance. Here, repairing problems involves the conceptualization of new visions of future systems, in which algorithms increasingly loom large as reparative agents. In the case of smart cities, for example, the responsibility for emissions reduction is delegated to networked infrastructures and Big Data, which are intended to produce carbon efficiencies through real-time data gathering, analysis and control (Gabrys, 2014). This is a 'repair as maintenance' vision (Henke and Sims, 2020), where computational efficiency is given priority so that existing logics of valuing and exchange can go on (relatively) uninterrupted.

In the last decade, 'smart' algorithmic technologies have also taken on a much larger role in agricultural production. For example, the measurement, mapping and sensing of soil and crops has been used to maintain and maximize yields, and algorithms have become central to logistics, helping to reduce crop losses and optimize the supply chains that distribute food. Computational systems have also become central figures in future visions for urban food, where automated vertical farms have attracted considerable hype.

In the *Algorithmic Food Justice* project we have investigated the ways that algorithmic imaginations are being developed within urban food ecologies. In prior work, we observed shortcomings in 'ecocidal' and 'broken' smart city visions (Heitlinger et al, 2018; Houston et al, 2019). In this chapter, we work with urban agricultural communities, technologists, artists and researchers to propose our own, reparative future visions. We critically explore the potential that blockchain might have for creating algorithmic networks that enact transformative repair. We build on existing work that uses blockchain to reconfigure relations between human and more-than-human life forms, using alternative systems of valuing and exchange such as commoning (Teli et al, 2018).

We take up these ideas around more-than-human algorithmic agency and entanglement using justice as a framing. As a form of reparation, we take justice to mean grappling with complex accountabilities towards some sort of recuperative action, even where this may be difficult, or even impossible. Indeed, given that the idea of justice is a foundationally human concept of governance, our idea of extending this to include more-than-human actors is ultimately destined to fail. However, this chapter will demonstrate how trying and failing at more-than-human design for digital infrastructures, foregrounds alternative value systems and modes of exchange that create playful and difficult dissonances. Designing future systems with food justice in mind helps us to encounter the many trade-offs and synergies in the (co)production and sharing of food, towards the aim of multispecies flourishing.

In this chapter, we will describe two (out of a series of three) workshops that took place in late 2019, where we co-designed algorithmic futures for a thriving multispecies urban food commons along with urban growers, community organizers, technologists, artists and researchers. We will reflect on the way they created conditions of possibility for more-than-human relations, including how they surfaced speculative futures in which the algorithmic afforded imagining alternative, more-than-human value systems. Through partial and improvised efforts to mimic and speak for the more-than-human, workshop participants attuned themselves to reparative relations.

Understanding blockchain

In order to contextualize the workshop descriptions that follow, we briefly summarize several important aspects of blockchain technology. In short, blockchain is a distributed ledger technology: rather than having a centralized ledger that records and stores the transaction of assets, blockchain ledgers are maintained in multiple places at the same time. Therefore, blockchains offer significant tools for decentralized organizing. Experimental blockchain communities have formed decentralized autonomous organizations (DAOs). A key idea is that these social-digital architectures can open up different types and granularities of exchange, helping to facilitate distributed and self-governing networks that are able to recognize and reward often excluded forms of labour – for example the work of care (DisCO.coop et al, 2019).

Two affordances of DAOs are particularly significant: first, the ability to issue tokens to each member and, second, the ability to automate organizational rules through the use of 'smart contracts' – software-based protocols that can execute contract-like forms without human intermediaries when certain conditions are met (Nissen et al, 2018). The issuing of tokens and the creation of smart contracts can partly automate the processes of distributed rule-making (for example relating to sharing and voting), and, depending on the design of the system, they may register the contributions made by actors. For example, the blockchain-based co-operation platforms Backfeed (Pazaitis et al, 2017) and Commonfare (Teli et al, 2018) use the 'reputation' token as a way for community members to collectively address issues of contribution and reward for participation in an emergent, consensus-based fashion. Our work attempts to add to these discussions, albeit through the production of conceptual prototypes rather than computational platforms.

More-than-human blockchains

For many scholars, the possibility of ecological reparation lies in crafting alternative relations with more-than-human others, which is as much an ontological as a practical project. In her work on the Planthropocene, Myers calls for 'art, experiment and radical disruption to learn other ways to see, feel and know' (2020, np) that can form the basis for different modes of relating. Our work has been inspired by two critical art projects that use blockchain to prototype new modes of algorithmic interrelation with other species: terra0 and Zoöp. In a 2016 white paper, the artists Paul Seidler, Paul Kolling and Max Hampshire speculate on how to create a DAO from a forest – by buying a piece of forest, mapping its ecosystem relations using networked sensors, and encoding a corresponding model on a blockchain using smart contracts. When sensors report that trees are grown, the terra0 DAO can then sell its own logging rights for cryptocurrency which it spends on managing itself, eventually buying itself back from the artists.

A Zoöp is a piece of land that is managed for the benefit of its more-than-human inhabitants. In this co-operative legal structure, they are represented by a Zoönomic Foundation of humans who use analogue and digital sensors to understand their needs. This project emerged from a speculative arts-led workshop and the initial vision includes a role for blockchain in mapping and managing exchanges (Kuitenbrouwer and Rui, 2019). In these projects blockchain technologies become a test site for different forms of algorithmic relations, where a turn to automation is not about creating efficiencies, but about crafting some form of autonomy for land and forests. In both cases the moral and ethical questions are not simply delegated to blockchain as an agent of repair, but are set into wider social, legal and economic infrastructures that aim to free more-than-humans from (at least some) aspects of human-centred value extraction.

These projects also grapple with the vexed question of how more-than-human interests are represented in human-led governance processes – a site of interest for many scholars in the social sciences more broadly (Bastian et al, 2017). Despite scaffolding more-than-human autonomy through algorithmic automation, all of this inevitably takes place within a human frame, created by human designers and developers. These projects demonstrate that the project of designing for more-than-humans will always be partial and incomplete. However, even attempting to do so prompts us to reach beyond the value systems that underpin many of our societies' extractivist imaginations. In the next section we will build on our previous work about the project (Heitlinger et al, 2021) to explore where we succeeded and failed at more-than-human or 'posthuman' design (Forlano, 2017; Clarke et al, 2019).

Algorithmic Food Justice workshops

The Algorithmic Food Justice project involved three workshops that took place in the autumn and winter of 2019. They were held at Spitalfields City Farm – a 1.3-acre urban farm housed on the site of a former railway goods depot near London's financial centre, which has been running since the late 1970s. The site hosts growing areas and community facilities such as a shared outdoor kitchen, and is home to farm animals such as donkeys, pigs and ducks. One of the authors has undertaken participatory research there for the last decade. For brevity, the following sections will expand on the second workshop and show how these strands of thinking carried on into the third workshop.

Workshop 2, entitled 'Now London is a City Farm', was attended by approximately 20 participants, including community growers, artists, researchers and technologists. The workshop used an arts-based research approach developed by one of the authors, known as LAARRPing (Live Art Action Research Role Playing) (Catlow, 2017). In essence, LAARRPs are expansive exercises in collaborative world-building, where each participant roleplays a character, and the group collectively adds detail and complexity around scaffolded scenarios. They differ from other forms of roleplaying in that they have been adapted to address a research question. Scholar Natasha Myers argues: 'if you had to consult the plants to ask permission to use them, industrial agriculture, strip mining, clear-cutting and the expanding concrete of urban sprawl would be inconceivable' (2020, n.p.). She writes: 'just ask, they will tell you'. We designed our LAARRP as an imaginative method to do this 'asking'.

Following existing trajectories in design and futures research, our aim was to use this highly embodied technique to create provocative and critical prototypes that could start discussions about both the development of blockchain and the ethics of more-than-human relations, design and futures research (Dunne and Raby, 2013; Muidermann et al, 2020). Rather than playing with probable or plausible futures, we chose a speculative approach in which participants were transported to the year 2025, five years after the fictional Great Food Emergency of 2020, when all available land in London was being reclaimed for food production. We LAARRPed for just under two hours and afterwards held a debriefing discussion. Later, we transcribed the recorded audio. When analysing these we looked for three specific themes: how participants had figured future modes of algorithmic governance; how they had constructed more-than-human relations; and how they had conceptualized justice within their improvisations (the latter two were often deeply intertwined).

On the morning of the workshop, we reminded participants of the outline scenario of the LAARRP and split them into two groups: (1) the

Greater London City Farm Assembly (GLCFA), the governing body of the whole of London; and (2) the E1 City Farm Assembly (E1CFA), the governing body of the local City Farm. Participants were invited to choose a governance portfolio to represent during the upcoming Assembly meeting (which formed the setting of the roleplay game). Participants could choose to represent the following portfolios: Coordination, Health, Agriculture, Security, Culture, Justice, Resources and Waste Management, Education, Energy, Infrastructure, and Assembly Liaison. In order to represent more-than-human entities, we asked participants to formally adopt a 'companion species' (Haraway, 2003), by choosing a badge showing birds, insects, farm animals, honeybees, soil, plants, trees, sensors, water, air, or weather. Participants wore this throughout, to signal their connection.

Participants then fleshed out their own character, by filling in an official-looking Identity Certificate that we had designed. We scattered character traits and potential plot points across these materials: possible roles included Land Conversion Co-ordinator, Extinction Rebel, Live-in Guardian Community Police Officer, Human Hand Pollinator, Single Parent and Mental Health Service User. Each was accompanied by a short description, for example Human Hand Pollinators 'pollinate crops by hand, a job that used to be performed by insects before their demise. We use our craft to build crop diversity in the hope that pollinators will return. Our job is under threat by robots'.

Once each group had taken their seats at the two separate Assembly meeting tables, we performed a ritual of communion, imagining our fictional characters in our left hands and our companion species in our right hands, rubbing them together until we felt a bonding heat. From that moment on we had to speak as our new characters, making no reference to the artifice of the game. The gameplay involved two of the authors acting as chairpersons – leading each of the groups through a pre-designed Meeting Agenda which presented a series of propositions that participants had to improvise around. We worked with questions around membership, fair shares and conflict resolution that are also being experimentally negotiated in blockchain initiatives (Pazaitis et al, 2018; Teli et al, 2018; DisCO.coop et al, 2019).

The Chair used a technique called 'hot seating' to kick off discussion. The first Agenda item, for example concerned a false claim on some bags of salad:

> Someone had claimed that they were entitled to five bags of salad … luckily, there was someone from the Coordination Committee who discovered that they weren't a member of this city farm … could you just remind us what happened? (EC1FA Chair)

Here, the Chair invited the representative from the Coordination portfolio to invent something about the governance of group membership in the

EC1 Assembly (hereafter referred to as the Coordination Rep). Subsequent participants had to build on, or gently modify this contribution – in the process collaboratively creating a semi-coherent scenario.

Algorithmic imaginaries and the more-than-human

The local E1CFA and London-wide GLCFA differed significantly in the algorithmic imaginaries that they put forward. In response to the bags of salad problem here outlined, the E1CFA group improvised an extensive digital infrastructure of control. According to the E1CFA Coordination Rep, the person who had claimed the salad had been 'tending a lot of the sensors that are in the park outside Bethnal Green, so they built up a lot of credits within their own farm membership – and they assumed that our farm was part of their network'. In the following five minutes, other players expanded on this to include: an automated digital credits system, a food tracking app linked to a distribution centre, identity scanning on the farm gates and some (contentious) proposals to build profiles of farm members using personal data and aggregations of their social media feeds. This sequence of play represents a wholesale intensification of today's algorithmic logics and, interestingly, it was derailed by the EC1FA Culture Rep, raising the question of how these systems would apply to more-than-humans. The infrastructure failed at this point because no accommodation had been made for the materiality or meaning of more-than-human others.

Conversely, for the London-wide GLCFA in the post-Emergency period, digital infrastructures were 'starting to disaggregate … the problem [is] that we're now running different systems that aren't entirely compatible … people have gone from sort of grids and much more independent systems that don't quite overlap' (GLCFA Infrastructure Rep). The GLCFA helps communities to set up energy and water microgrids in London localities using scrap materials, as the GLCFA Energy Rep outlines: 'we can … give microgrid options to local communities and then they can rely on each other for energy so that that community, if they can define what that community is we can supply them with enough energy from anything around them'. This is a vision of community-led and managed infrastructures that (unlike their neighbours in EC1FA) don't use digital technologies to monitor and track human and more-than-human entities.

Both groups proposed using algorithms to facilitate multispecies communication, with varying levels of detail. In the E1CFA group, algorithmic relations were later proposed as a way to facilitate interspecies communication across an 'urban biome', as the EC1FA Culture Rep explained:

> 'one of the ideas … was to set up a whole very dense network of sensors across the city, which measure the health of the sort of urban biome

> … through that, we could sort of gauge the happiness of the various species … you know [the] relationship with our species.'

Here, tracking is not about the maximization of productivity and the awarding of fair shares, but about understanding the wellness of other species.

Interestingly, when the problem of interspecies communication got too great both groups turned to the seemingly all-encompassing possibilities of AI. In the E1CFA, the Culture Rep quips: 'we need an AI, like to mediate between all the species – that's a solution'. In a conversation about interspecies sharing, the GLCFA the Energy Rep also turns to an all-knowing AI: 'This is where the algorithms will be coming in, logging and running these processes would be much safer to leave to an algorithm, or an AI possibly – that speaks both insects and microbial languages and human'. This turn to technological solutions is heard, but not taken further by either of the groups.

Many of the Agenda prompts included references to multispecies entanglements, and therefore it was no surprise that the roleplay prompted powerful testimonies from companion species. Speaking as a bee, GCLFA Agriculture Rep gave an emotional account of the ongoing bees' strike:

> '[You're] expecting us to pollinate all your crops so that the humans can have their food. And yet [you] make it hard for us to have a decent life and to be able to have a diversity of different flowers that we make nice honey for baby bees. So we've had enough … We're only pollinating wildflowers. We're not pollinating any of your crops anymore until this is resolved.' (Bee, companion of GCLFA Agriculture Rep)

The bees' testimony holds the other human members to account for their part in the conflict. Moments like these produced an electric strand of irony that ran through the gameplay, where the participants are playing together in a fictional scenario that they know is a proxy for interspecies conflict unfolding in the world right now.

Overall, both groups came up with remarkably similar strategies for multispecies communication, most notably by continuing to hold spaces for cross-species empathy and appreciation. The E1CFA group set up 'multispecies assemblies' and 'one-to-one working groups' so that species would get to know each other's contributions better (this notably required a serious amount of 'unlearning' for the humans). Both proposed 'cultural pieces' – for the E1CFA this was 'Bacteria Week', a festival that promoted and celebrated the valuable contributions of under-acknowledged bacterial agencies, while the GLCFA wanted to create a multispecies musical. Likewise, both groups developed new rituals to thank other species for their contributions. Finally, appreciation was shown by the giving of material rewards – the E1CFA group proposed moving bacterial communities to a new spot for a little holiday, plus giving them a little nutrient agar.

Many of these visions were humorous, but at the same time involved genuinely seeking new modes of multispecies understanding. The absurdity often arose in the recognition of human exceptionalism, as this exchange in the GLCFA aptly demonstrates:

Agriculture Rep:	Do all the species like musicals? …
Infrastructure Rep:	I've heard that trees don't like a musical …
Chair as Companion Species Tree:	The trees have their own great musicals that the humans just don't even hear.

In fact, the breaching of the human-political frame of the Assembly meeting, turned out to be a significant dynamic of the game, where more than-human members repeatedly pointed out that deliberative democracy was simply not sufficient to recognize the complexity of their exchanges. These sequences of play produced clear disputations of anthropocentricism, and also began to suggest other types of relation.

One of the richest examples took place in the GLCFA group, where the Liaison Rep was concerned by the security of the farm, and the stealing of pumpkins (a real-world problem for this Spitalfields Farm grower). After cycling through a discussion of how to educate, punish or rehabilitate (human) offenders, the group moved on to consider the limits of the concepts of justice in a more-than-human world. The Resources Rep asks: "Is justice an anthropocentric concept though? You couldn't expect a carnivorous animal to be just to its prey – how would you mediate between a cat which has to eat rodents and things, between its things it's feeding on?"

For the Coordination Rep the ecosystem is more important than relations predicated on the actions of individual subjects, and they explore soil as a model for human and more-than-human relations:

> 'It's more like building a system that's robust enough to be to hold tensions or people … taking more. Like if you look at soil as a model. Mostly if you create the right conditions and the structure is right then most organisms will exchange fairly and that usually works. And there are a few that will take and not give back, but still overall, the system works. So, I think it's about thinking about how to create a wider system that creates the right conditions for things to work, but not always and that's ok.' (GLCFA Coordination Rep)

Here we get to the relations of interdependence more commonly found in more-than-human ontologies, where processes rather than subjectivities are foregrounded. Though interestingly for the Energy Rep there are some hard limits: 'Driving other species into extinction is a no-no'. What constitutes more-than-human *injustice* is easier to figure than a shared concept of justice.

The conversation also engaged an interesting discussion of the agencies entailed in 'growing' the pumpkins in the first place. The Infrastructure Rep argued that growing is 'still being regarded in quite a possessive [way] like "we grew the pumpkins"; no, the pumpkins grew and lots of other things contributed.' Again, here the assumed agencies are being turned upside down. The pumpkins are not owned, because they were not grown, but grew themselves. For the Infrastructure Rep this shift to relations undermines the concept of justice entirely because 'it's very difficult then to come down with the justice system to see exactly where does justice sit?' in a shift from entities to relations.

Trying and failing at posthuman design

The LAARRP concluded with group reflections. The scenario developed by the E1CFA group was very positive for humans: in 2025 food is being grown in homes, parks, rooftops and all the abandoned buildings of the city. The allocation of food for humans is based on a needs assessment for those living in the local community. However, needs and contributions are tracked by an extensive algorithmic ID system with few privacy constraints across a network of Farm Nodes. Interspecies relations are primarily worked out in multispecies assemblies and one-to-one working groups while the happiness of other species is also monitored through all kinds of sensing of the urban biome. There is enough food for the humans, though London's population have needed time to adapt to the new diet.

The GLCFA vision was less optimistic. In 2025, some humans are still going hungry, although the situation has improved in recent years. Water conflicts and animal rustling were once commonplace, but community relations have recently stabilized. There is a thriving circular economy, where organic waste materials are carefully recycled into food for other species and where obsolete technology is being reclaimed to set up localized infrastructures that are managed by communities to meet their own needs. Re-wilding is happening across human political borders which are becoming less and less important.

As part of this discussion, participants raised the limitations of the committee format and the 'companion species' model that we had chosen in addressing interspecies relations, as outlined in the following exchange:

E1CFA Justice Rep:	I think at today's meeting it became clear that most of the time it's humans talking again still … there's strong concentration of the power there. Despite the fact that we all had non human representatives. Like air said, nothing.
Infrastructure Rep:	Yeah water was pretty minimal as well.
Energy Rep:	Yeah, it's all hot air.

Our decision to include more-than-humans as 'companion species' (Haraway, 2003) was intended to evoke a hybrid politics, however the act of improvising during the gameplay was cognitively taxing and therefore human participants only sporadically inhabited their more-than-human characters.

In our project, we took the established approach of inviting 'human intermediaries' (Bastian, 2017: 21) such as urban growers practicing forms of regenerative agriculture, to speak about the more-than-human others that they had an intimate knowledge of, hoping that this would be less 'representation by human ventriloquists' (Pitt, 2017: 92) and more a generative facilitation. In retrospect, our decision to set up the gameplay using a format from representational politics (the committee) cemented the dynamics of one species speaking for another within a human-centric frame. We risked that even playing with humanist concepts like justice materialize more-than-humans as a knowledge object within an already given structure, and therefore collapse the transformative possibilities generated by new more-than-human ontologies as they are pulled into a familiar frame, that is oriented towards democratic, political, human subjects (a 'humanist posthumanism'; Wolfe, 2009).

The LAARRP was ultimately an exercise in failing at more-than-human design, however the process of failing opened up very lively spaces to creatively deconstruct human governance structures on behalf of the more-than-humans that they have come to know through community growing. An important quality of the LAARRP was its liveliness, in creating a collective world and a shared emotional experience connected to the improvisational qualities of the gameplay. The shared recognition of irony in playing anthropomorphic relations that were clearly impossible (for example, that bees could go on strike) yet were also clearly reasonable (most people supported the bees' demands). For Haraway:

> Irony is about contradictions that do not resolve into larger wholes, even dialectically, about the tension of holding incompatible things together because both or all are necessary and true. Irony is about humour and serious play. It is also a rhetorical strategy and a political method. (Haraway, 1991: 149)

In this sense, if ontological transformation is the cornerstone of interdependent ways of thinking (Myers, 2020), then the LAARRP has demonstrated the potential of the movement between (human) ways of knowing the world as human subjects and algorithmic objects, and the always un-knowable ways of knowing brought to bear by more-than-humans.

Workshop 3: Blockchain prototypes

In the third workshop we built on the LAARRP gameplay, by selecting eight ideas from the game to use as discussion prompts for participants about how blockchain could potentially be used as part of new infrastructure of more-than-human relating. To help participants think through their ideas we also produced a set of syntax cards with terms from conditional programming ('IF', 'THEN' and 'ELSE') plus other words such as TOKENS, ACTOR, ASSET, DURATION, LOCATION and EVENT, drawing on Nissen et al's work on the GeoCoin project (2018). During the morning session, a smaller group of technologists, artists and community organizers were given the task of designing blockchain-based DAOs along with their governance rules, which were then discussed by a larger group (of previous participants) in the afternoon. We briefly describe each of the three conceptual prototypes here.

Group 1: The fellowshit of dark matter

The first group worked with the following scenario from the LAARRP:

Education Rep:	Can I speak as soil. With my soil hat on. The pile of waste that you're talking about. Is it just human excrement?
Resources Rep:	It's a mix of farming manure like cattle and sheep.
Education Rep:	So everything organic. So for me, that's food. So I wouldn't call that waste, that's wonderful.

This group sought to capture the relational and emergent qualities of valuing in the design of socio-technical system that nurtured multispecies relations. The system utilized an app, where humans could post waste materials and make them freely available to others within the community, but there was also weekly ritual, 'where storytellers would gather … to tell the story of how this waste is being transmutated into useful things from all sorts of perspectives … from cows' perspectives' (Participant KG).

Since multispecies entities would have trouble using the app directly, the ceremony provides a way for human dramatists to inhabit their perspectives;

spotting opportunities for waste materials to be used for their own species' ends, and also celebrating the transformations of waste that had already been completed. During the ceremony, Transmute tokens would be awarded to those 'transmutations'. The DAO promotes a multispecies 'circular economy' with an expansive sense of value that is not only focused on material utility but the meaning of unfolding interspecies relationships as they exchange materials within an ecosystem.

Group 2: DAO-n to earth

The second group worked with the issue of the mis-claimed salad bags described previously:

> 'That individual … tends to a lot of the sensors in the park outside Bethnal Green. They built up a lot of credits within their own farm membership. And they incorrectly assumed that our farm was part of their network of food sharing. … And because they weren't part of the same shared group or food node as we are – that was where the confusion came in.' (E1CFA Coordination Rep)

The DAO-n to Earth group devised a DAO that helps to coordinate the exchange of tokens (currencies) between all the farms in the London Food Network. In short, the value of each farm's currency is tied to the soil health data that is acquired automatically from sensor arrays. The weight of each community's 'reputation' token changes according to whether its soil improves, stays the same or degrades over a given period of time. The system also includes "sensor sentinels" – sensor arrays connected to an Artificial Intelligence algorithm – that monitor soil health in real time. If a rapid decline is detected it tells all actors on the soil exchange, triggering a market crash of that farm's tokens, but also a social solidarity mechanism where other farmers are asked to step in and assist.

Group 3: The Corn Council

The third group worked on the lack of acknowledgement experienced by plants:

> 'There hasn't been overall enough acknowledgment of what sacrifices we [plants] make. Because principally we are generating most of the air and most of the food … we did originally discuss having rituals that actually thanked the plants and I don't know what's happened to that. I just know that the plants are carrying a load that no-one is acknowledging.' (GLCFA Infrastructure Rep)

Participant ET suggests that alienation from the conditions of food production is a core condition for many humans, and proposes that 'rituals, for example, or ceremonies can bring back the value and having a connection to what you consume'. This idea forms the basis for the final proposal of the Corn Council, a system for repairing the disconnect between humans and other species. The group draw on the logics of the Brave browser's Basic Attention Token, which is a blockchain initiative intended to more visibly and fairly redistribute a proportion of the advertising monies received by website owners, directly to website users. However, in their prototype, the Corn Council DAO will reward humans with tokens simply for spending time with plants in a non-instrumental way (though tokens will also be available for care-taking work such as pruning and watering). Accruing more reputation tokens by spending more time with plants allows DAO members a greater say in votes about the DAO's management.

Carrying the LAARRP scenarios over into the blockchain prototyping workshop was a generative way to continue the discussion around multispecies justice and questions of algorithmic governance. Rather than working together to propose solutions for problems (as in a 'hackathon' style design event), the prototyping process began from speculative formations where the entanglements of humans and more-than-humans had already been made visible as relationships, processes and wider ecosystems. In addition, the LAARRP excerpts drew attention to interesting social formations such as ceremonies and rituals that might not have been explored otherwise in this type of co-design setting. At the same time, this short workshop only began to surface the complexity of the issues at stake, but the prototypes themselves remained highly partial and provisional.

Attuning to reparative relations

We want, in closing, to turn back to reparative relations. Both the LAARRP and prototyping did not give rise to an extended or coherent narrative in the ways we'd hoped for. Seen through a prism of ecological reparation, we might see this as a failure on the part of the workshops. Collectively, we failed to imagine fully-fledged futures, interweaving algorithmic systems and blockchains with multispecies relations, and distributing agencies across heterogeneous actors that might afford alternative value systems. The participants' repeated experimentation – from treating more-than-humans as subjects engaging in active communication and exchange (as in the bees' dispute), to considering soil as a model for future relations of justice – revealed, in a sense, the impossibility of speaking for or standing in for more than human species.

At the same time, however, we learned from the ways participants played up to and beyond the limits that the structures of the workshops imposed. Thus, what we found revealing was the larger groups' fluid movement between different ontological formations, for, if you will, momentary, just plausible and only partial ways of improvising more-than-human relations. We saw this, for example, with the GLCFA who went from seeing the pumpkin as stolen property, to an agent in its own right, who grew itself in relation to humans with the help of air and water. What we might say, then, is that transformation and the possibilities for change came not from wholesale visions, but from the chance for fleeting and partial co-minglings, for the right conditions to ask 'what if ...?': 'what if we might speak for soil?'

The chance to play, to tinker and to mimic (no matter how partial and sometimes ludicrous) offered, then, the possibility of imagining more-than-human relations, the possibility for making reparations (for example increased soil health, fairer interchanges in multispecies accounting, transmutation of waste materials) in the aftermath of human exceptionalism. The LAARRPs and prototypes created 'experimental propositions' (Hustak and Myers, 2012: 78), the chance to be moved by more-than-human relations in uneasy and unfamiliar ways, akin to Hustak and Myers' *involutionary momentum*, a bodily *leaning in* that 'helps us to get a feel for [the] affective push and pull among bodies, including the affinities, ruptures, enmeshments, and repulsions among organisms constantly inventing new ways to live with and alongside one another' (Hustak and Myers, 2012: 97). From Hustak and Myers, we discover such momentary improvisations become the basis for attunements with affectively charged ecologies; in other words, for building alliances, affinities and connections we might care for and nurture in new ways.

How can ideas of repair and reparation help us to think about the huge challenge of moving at least partially towards such new sets of relations that take the lives of more-than-humans seriously? The LAARRP certainly produced a new kind of experience and new challenges for creating a just multispecies food system. Empathizing, improvising, laughing, experiencing irony and being moved by soliloquies about plant sacrifice provided new ways to understand the liveliness of the stakes at play here, beyond policy debates. The ability of species to call out other species was a moment where the politics of reparation came sharply into play: as the bees, for example, called out the damage that the humans were doing into starkest and most personal terms. The game brought problems down to local and individual scales: not addressing the conditions for *all* bees, but exploring conditions for *this* bee, sat across the table.

Rather than a return to normative relations, we have come to understand that even the most material of repairs (the replacement of one machine part with another) involves crafting a new set of relations: a new capacity and a new temporal unfolding (Houston, 2017). Repair, like care, involves 'tinkering' (Mol et al, 2010): altering relations to create new conditions and then understanding how practices or performances also transform;

and treating the overall process as a tentative and iterative endeavour. If the ecologies on this damaged planet are to be recuperated then we will need tools like LAARRPing to think about systems and performances from a multispecies perspective (however partially) – bringing new registers of feeling to those encounters, which are rich with empathy and exchange. Repairing ecologies will mean transitioning from ontologies that enable extraction, destruction and extermination into something different – opening up a space for speculative futures that can enfold these kinds of transitions.

Acknowledgements

We wish to thank all workshop participants and Spitalfields City Farm for showing us what a thriving multispecies food commons looks like. This project was funded through an EPSRC NotEqual Network+ grant.

References

Bastian, M. (2017) 'Towards a more-than-human participatory research', in M. Bastian, O. Jones, N. Moore, and E. Roe (eds) *Participatory Research in More-than-Human Worlds*, Abingdon: Routledge, pp 19–37.

Bastian, M., Jones, O. Moore, N. and Roe, E. (eds) (2017) Participatory Research in More-than-Human Worlds, Abingdon: Routledge.

Catlow, R. (2017) 'Artists re:thinking the blockchain introduction', in R. Catlow, M. Garrett, N. Jones and S. Skinner (eds) *Artists Re:thinking the Blockchain*, Oxford: Torque Editions, pp 21–37.

Clarke, R., Heitlinger, S., Light, A., Forlano, L., Foth, M. and DiSalvo, C. (2019) 'More-than-human participation: Design for sustainable smart city futures', *Interactions*, 26(3): 60–63.

DisCO.coop, The Transnational Institute and Guerrilla Media Collective (2019) 'If I Only Had a Heart: A DisCO Manifesto: Value Sovereignty, Care Work, Commons and Distributed Cooperative Organizations'. Available from: https://disco.coop/wp-content/uploads/2019/11/DisCO_Manifesto-v1-1.pdf [Last accessed 4 October 2022]

Dunne A. and Raby, F. (2013) *Speculative Everything: Design, Fiction, and Social Dreaming*, Cambridge, MA: MIT Press.

Forlano, L. (2017) 'Posthumanism and design' *She Ji: The Journal of Design, Economics, and Innovation*, 3(1): 16–29.

Gabrys J. (2014) 'Programming environments: Environmentality and citizen sensing in the smart city', *Environment and Planning D: Society and Space* Volume, 32(1): 30–48.

Haraway, D. (1991) *Simians, Cyborgs and Women: The Reinvention of Nature*, New York: Routledge.

Haraway, D. (2003) *The Companion Species Manifesto: Dogs, People and Significant Otherness*, Chicago, IL: Prickly Paradigm Press.

Heitlinger, S., Foth, M., Clarke, R., DiSalvo, C., Light, A., and Forlano, L. (2018) 'Avoiding ecocidal smart cities: Participatory design for more-than-human futures', Proceedings of the 15th Participatory Design Conference: Short Papers, Situated Actions, Workshops and Tutorial, 2(51): 1–3.

Heitlinger, S., Houston, L., Taylor, A. and Catlow, R. (2021) 'Algorithmic food justice: Co-designing more-than-human blockchain futures for the food commons', *Proceedings of the 2021 CHI Conference on Human Factors in Computing Systems*, 305: 1–17

Henke, C.R. and Sims, B. (2020) *Repairing Infrastructures: The Maintenance of Materiality and Power*, Cambridge, MA: MIT Press.

Houston, L. (2017) 'The timeliness of repair', *Continent*, 6(1): 51–55.

Houston, L., Gabrys, J., and Pritchard, H. (2019) 'Breakdown in the smart city: Exploring workarounds with urban-sensing practices and technologies', *Science, Technology, & Human Values*, 44(5), 843–870.

Hustak, C. and Myers, N. (2012) 'Involutionary momentum: Affective ecologies and the sciences of plant/insect encounters', *differences*, 23(3), 74–118.

Jackson, S. (2014) 'Rethinking repair', in T. Gillespie, P. J. Boczkowski, and K. A. Foot (eds) *Media Technologies: Essays on Communication, Materiality, and Society*, Cambridge, MA: MIT Press, pp 221–40.

Kuitenbrouwer, K. and Rui, A. (2019) Report Zoöp Workshop I: Legal Representation for (Collective Bodies of) Non-humans, Het Nieuwe Instituut. Available from: https://neuhaus.hetnieuweinstituut.nl/en/report-zoop-workshop-i-legal-representation-collective-bodies-nonhumans [Last accessed 3 January 2021].

Lindström, K. and Ståhl, A. (2020) 'Un/Making in the aftermath of design', *Proceedings of the 16th Participatory Design Conference: Participation(s) Otherwise*, 1: 12–21.

Middleton, J. (2018) Mending the sensible: Ontoexperiments for a politics of matter, PhD thesis. https://doi.org/10.17635/lancaster/thesis/642

Mol, A., Moser, I. and Pols, J. (eds) (2010). *Care in Practice: On Tinkering in Clinics, Homes and Farms*, Berlin: Verlag.

Muiderman, K., Gupta, A., Vervoort, J. and Biermann, F. (2020) 'Four approaches to anticipatory climate governance', *WIREs Climate Change*, 11(6): 1–20.

Myers, N. (2020) 'How to grow liveable worlds: Ten (not-so-easy) steps for life in the Planthroposcene', *ABC Religion and Ethics*, 28 January. Available from: www.abc.net.au/religion/natasha-myers-how-to-grow-liveable-worlds:-ten-not-so-easy-step/11906548 [Last accessed 3 January 2021].

Nissen, B., Pschetz, L., Murray-Rust, D., Mehrpouya, H., Oosthuizen, S. and Speed, C. (2018) 'GeoCoin: Supporting ideation and collaborative design with smart contracts', *Proceedings of the 2018 CHI Conference on Human Factors in Computing Systems*, 163: 1–10.

Pazaitis, A., De Filippi, P. and Kostakis, V. (2017) 'Blockchain and value systems in the sharing economy: The illustrative case of Backfeed', *Technological Forecasting and Social Change*, 125: 105–115.

Pitt, H. (2017) 'An apprenticeship in plant thinking', in M. Bastian, O. Jones, N. Moore and E. Roe (eds) *Participatory Research in More-than-Human Worlds*, Abingdon: Routledge, pp 92–106.

Ribes, D. (2017) 'The rub and chafe of maintenance and repair', 6(1): 71–76.

Rozas, D., Tenorio-Fornés, A., D'iaz-Molina, S. and Hassan, S. (2018) 'When Ostrom meets blockchain: Exploring the potentials of blockchain for commons governance', *SSRN*, 30 July. Available at: https://ssrn.com/abstract=3272329 or http://dx.doi.org/10.2139/ssrn.3272329 [Last accessed 3 January 2021].

Seidler, P., Kolling, P. and Hampshire, M. (2017) 'Terra 0 – can an augmented forest own and utilize itself?', in R. Catlow, M. Garrett, N. Jones and S. Skinner (eds) *Artists Re:thinking the Blockchain*, Oxford: Torque Editions, pp 63–72.

Stark, L. (2016) 'Larp as technology', *Technosphere Magazine*, 15 November. Available at: https://technosphere-magazine.hkw.de/p/Larp-as-Technology-rEFzwyH3u6VuVo9zCGXz6 [Last accessed 3 January 2021].

Teli, M., Lyle, P. and Sciannamblo, M. (2018) 'Institutioning the common: The case of Commonfare', *Proceedings of the 15th Participatory Design Conference: Full Papers*, 1(6): 1–11.

Tonkinwise, C. (2019) '"I Prefer No to" Anti-Progressive Designing', in G. Coombs, A. McNamara and G. Sade (eds) *Undesign: A Critical Practice at the Intersection of Art and Design*, Abingdon: Routledge.

Tsing, A.L. (2017) *The Mushroom at the End of the World: On the Possibility of Life in Capitalist Ruins*, Princeton, NJ: Princeton University Press.

Wolfe, C. (2009) *What Is Posthumanism?*, Minneapolis, MN: University of Minnesota Press.

25

Being Affected by *Páramo*: Maps, Landscape Drawings and a Risky Science

Alejandra Osejo and Santiago Martínez-Medina

Introduction

We work at Instituto Alexander von Humboldt, a Colombian research institute in charge of biodiversity studies. Colombia is one of the most biodiverse countries in the world, and sadly, it is also one of the countries where the loss of biodiversity is recorded in real-time. Urgency and relevance are familiar feelings pushed by working in this context. For that reason, science, and the authority it confides, is essential in the kind of responsibilities that the institute needs to cover: to inform policy with the best scientific information possible within the framework of funding cuts and a turbulent national political climate, which does not always prioritize environmental issues.

We are both anthropologists, working in a small social scientists' team amidst many natural scientists. Our role is to analyse the multiple relationships between people and biodiversity in Colombia, integrating our results with the natural sciences on biodiversity. Multidisciplinary work is not always easy, and our position can be described as uncomfortable. In thematic terms, our academic work is related to *páramo*, defined by natural sciences as a tropical high mountain ecosystem. This chapter articulates diverse stories, experiences and conversations around páramo and the natural and social sciences. We are particularly interested in overcoming the aspiration of integration in interdisciplinary work and look for ways to relate diverse epistemologies, methods and practices, which are only partially coherent, what Jensen and Morita call *sophisticated conjunctions* (Jensen and Morita, 2020). This chapter is a step further in experimenting with what form these sophisticated conjunctions can take in the context of our quotidian job, especially when this conjunction needs to take the form of a relationship between methods, their materials and the affections that methods and materials convey. Our intuition is that the conservation and restoration of Colombian páramos also require repairing relationships beyond the páramo as 'nature' or 'páramo without people'. In order to participate in this task, the biodiversity sciences would benefit from these sophisticated conjunctions, which we imagine as ecologies of affections that feed sciences that risk novel articulations.

In this piece, our ethnographic strategy takes the form of a series of conversations. First, between us, this chapter is the result of a long process of thinking together about field experiences that we share only in our dialogue. Most of the field notes used in this chapter belong to Alejandra, but we say here that *we* lived said ethnographic moments together because both of us return to the experiences thanks to Alejandra's field notes. Santiago's role was to be a conceptual partner and writer, a fellow traveller returning to the páramo through Alejandra's experiences. Inspired by Strathern (1999), we can say that we live páramo as a fieldsite differently if we return to it together in writing. Central to this possibility have been Santiago's own experiences in páramo, even if he was not with Alejandra, and did not visit the same places that she did. Our ethnography is made in duet, mixing materials and concepts related to páramo. Second, this chapter is also informed by debates and reflections held over several years with colleagues from the biological sciences and geography and fellow social scientists about their practices concerning the páramo. We consider the institute's role in producing knowledge about páramos as ecosystems, so we use our own experience in this multidisciplinary environment. We are interested in thinking about the differences in which páramo emerges as an increasingly debated assemblage in our country and our institute. Finally, our dialogue is not only with fellow researchers and páramo *campesinos*.[1]. Besides people, we talk with scientific artifacts: representational devices made by natural and social scientists. Building in a relationship between artifacts and affect, we reflect on the possibilities of a crossing inspiration among sciences to inspire alternative forms of ecological repair.

Delimitation and scale

From a natural sciences perspective, it can be said that páramo is a particular tropical Andean high mountain ecosystem. In Colombia, the páramo is currently at the centre of public attention. It is considered a strategic ecosystem for capturing, storing, and providing water to millions of people. Campesinos also inhabit many páramos. Since the beginning of the twentieth century, most of them settled in these places, fleeing the repetitive cycles of political and economic violence plaguing the country. Despite the extreme temperatures, the altitude, and the remoteness, campesinos made themselves people of the páramo. They made many of these páramos agricultural landscapes with potato and onion crops, with cattle that can endure the climate and high altitude, hunting and using forests and grasslands for their survival.

Now, páramo is facing the threats that extractivism looms over it, mainly through mega-mining. This situation is in line with trends in Latin America. Here, what in the Global North is called the 'Anthropocene' acquires a radical configuration that transforms landscapes overnight, a force of occupation, dispossession, displacement and death for a wide variety of human and non-human beings (De la Cadena, 2019). In 2009, a Canadian mining company requested an exploitation license that affects a wide area of the Santurbán páramo, from which hundreds of thousands of people obtain water, many of them in the city of Bucaramanga, in the Department of Santander. The páramo thus began an unprecedented process of communalization, in which a great variety of interests, including those of the inhabitants of Bucaramanga, artisanal miners, campesinos and environmentalists, undertook a wide variety of collective actions around a 'common interest that is not the same interest' (Osejo and Ungar, 2017; De la Cadena, 2019). Although it has not been cancelled, the mining project was temporarily halted. However, the controversy surrounding how to use, conserve and repair these places to preserve water put the páramos at the centre of the country's environmental concerns, with unexpected effects. One of them was the National Development Plan's issuance by the Colombian Congress in 2011 through Law 1450. This regulation prohibited mining and agricultural practices in these ecosystems, articulating scientific arguments about their biodiversity (Sarmiento and Zapata, 2016) and determining the elaboration of maps of the páramo boundaries to make prescriptions possible in their interior.

This Law elevated regional environmental policies that attempted to regulate productive activities in the páramos at the national level. In these regional regulations, expressed in the declaration of regional protected areas and control exercised by environmental authorities, agriculture was already presented as a set of 'pressures' that 'degrade' this ecosystem. The most problematic issue in this Law is that it does not differentiate between the great variety of productive activities developed in these ecosystems, including small family economies and large commercial producers with a great capacity to invest resources and agrochemical technology in crops. Furthermore, this national Law does not recognize the multiple assemblages between people, páramo and crops. This inability produced a radical version, where the idea of 'degradation' resents all human presence, understands it as harmful and puts into action a perspective of nature conservation, which is based on excluding the human being (Ingold, 2005), without recognizing that the páramo is also the result of human practices. In this arithmetic, the "páramo without people" must be tended to guarantee water for cities and ensure that these places continue to fulfil these functions in the future. In this way, since 2011, the 'paramo without people' idea has been consolidated as the primary determinant of the national policy around these ecosystems. Thus, the Law reorganized diverse and distant places that have permanently been inhabited, adhering to criteria of truth in whose constitution the sciences of biodiversity played an important role. This articulation between natural sciences and law could contribute to colonial logic that defines areas as natural, empty and pristine and therefore subject to appropriation. In this way, conserving and restoring the páramo became part of the problem by separating people from their inhabited places and dissociating social and ecological justice.

Making maps of the páramos' limits at the 1:25.000 scale was the main strategy established in the Law to enforce the prohibitions as mentioned. This task was entrusted to the Humboldt Institute. It turned out to be more difficult than provided. As a technical activity, the delimitation showed that it is not the same to answer the question about what a páramo is to respond to the order to demarcate its limit. The challenge for our Humboldt colleagues involved with the said maps was to identify the site where the high Andean Forest ends and where the páramo begins: the latter's lower limit. Their instruments then focused on the *forest-páramo transition zone*, the contact zone between the two ecosystems. These scientists developed complicated methodological protocols to analyse flora and fauna and parameters of their potential distribution along the altitudinal gradient. The plants and animals were summoned in the name of the páramos in the experts' models. The result is inevitably subject to satellites' resolution capabilities, computer programs, equations and geographic information systems in the form of maps. Thus, the transition between the páramos and the forest was modelled, and at its lower limit, a line that separates them was drawn in maps. In this way, a version of the páramo defined by mathematical models and satellite photos emerged. The environmental authorities used this line shortly after to delimit the actual páramos in Colombia, with all its diversity, making it feasible to comply with the Law. A line that returned fixed a transition that scientists knew was mobile, subject to variations in their algorithms' probability and all kinds of environmental changes.

Thanks to this line, the páramo became a demarcated space on the map, an ontological condition that is not the same as that of its existence as an ecosystem. An ecosystem is an assemblage of entities fluid and indeterminate. Between the forest and the páramo, there is a slow transition, felt when climbing the mountain: a transformation that does not draw a limit in the sense of line, but rather in the sense of the power of spreading into each other that both forest and páramo have (Deleuze, 1981). Thus, climbing the mountain, hikers know when they are in the forest because thicket covers them, and they know when they are in the páramo because they are immersed in an 'open' landscape, where the tallest plants do not exceed their height. There is only a gradual change between the two experiences, a subtle variation, never a line. While the line is drawn, the relationship between the páramo and the forests changes. Operating what Ingold calls 'logic of inversion' (2011: 145), even while it is traced, we stop seeing the line to suppose the content, in this case, the páramo, which is no longer a world of relationships between all kinds of heterogeneous entities, including humans, to become spots on a map made up of polygons. Inversion of the line to the demarcated form: between the páramo as an ecosystem, that is, as a habitat for living entities, to the páramo as space.

> How have we arrived at such an abstract and rarified concept to describe the world in which we live? My contention is that it results from the operation of what I have called the logic of inversion. … In a nutshell, inversion turns the pathways along which life is lived into boundaries within which it is enclosed. Life, according to this logic, is reduced to an internal property of things that occupy the world but do not, strictly speaking, inhabit it. A world that is occupied but not inhabited, that is filled with existing things rather than woven from the strands of their coming-into-being, is a world of space. (Ingold, 2011: 145)

These models and the maps they produced have 'violently simplified landscapes and human and non-human forms of life' (Viveiros de Castro, 2019). For a long time, the campesinos of the high mountains had an intimate relationship with the cold, the fertility of the earth, the wind and solar radiation, the potato, the bears, the onions and the cows, a relationship in which the main issue was survival. Thus, their task was to inhabit, make habitable, become with, tune in, affect and be affected in a corporeal way. These practices clearly imply transformations in the páramo, some of them deleterious to humans and non-humans, such as the removal of native vegetation, the use of chemicals in crops, livestock grazing that compacts the soil, water pollution and the use of heavy machinery. Therefore, the radical separation from the human produced by the Law is neither the best nor the only alternative. On the contrary, the effect produced by this particular way to conserve cut the possible futures of the inhabitants of these places. This kind of conservation and restauration has also constituted from above a particular version of páramo, one in which the question is no longer to become with páramo, to tune in to it, but to know whether you are inside it. As a result, once published, between 2014 and 2018, and in the hands of environmental authorities, this new version of páramo has transformed people's historical relationships with the páramo. During the following years, the campesinos we worked with told us that, due to the delimitation, water concessions for farmers were denied, credits for agricultural production were limited and construction licenses for new housing were rejected, arguing that 'they were in the páramo'. Once the map was published, campesinos' question will also be whether his property is inside or outside a map with the power to restrict their activity. The new version of páramo made the challenge of survival even more problematic for the campesinos, an unexpected result of the 'logic of inversion' that organizes the map and the delimitation, thus creating a series of conflicts that require repair.

The question of the possibilities of many campesinos' existence is imposed, although in an irresolute way, by the map. Despite the rigour and scientific work that produced a map at a scale of 1: 25,000, its scale makes it incapable of saying anything about small properties. The most everyday detail, the most intimate question, that of the campesinos, their cows and crops, is invisible to the map in many of its points. In many cases, the map cannot resolve the question of whether one is inside or outside, precisely the one posed by the map, in a country with significant doubts about its cadastral property records. For us, this is no minor issue and has been the subject of discussion with some colleagues in charge of making the maps. When questioning one of them about this issue, he replied: 'You and I have different problems. My problem is the páramo; your problem is the property. I cannot see the property from my maps and my models. I am not able to see properties smaller than 250 m. You are not able to see the páramo that I am seeing.'

The discussion is still open today, and the maps that generated the boundaries are the subject of public debate. Various stakeholders have generated demands and reactions, arguing the lack of participation of those affected by these decisions. As a result, the inhabited páramos began to be visible in public debates. Different courts and judicial instances have ruled, becoming fundamental players, with a national scope, in constructing other versions of the páramo. Human and non-human relationships, unseen in the delimitation maps, began to appear in new laws and judicial sentences. Thus, some delimitations must be reviewed, and páramo inhabitants must be included in the decision-making processes regarding these ecosystems. Besides, in 2018, due to the exacerbated conflicts resulting from these decisions, the paramos Law became more flexible (Osejo et al, 2020), allowing productive activities under some conditions. Still, they must be regulated and tend towards sustainability. Thus, the current legislation requires differentiating agricultural activities according to the ecosystem's impact and developing new maps based on biological and hydrological information. These maps determine in which places within the páramo these activities can be developed. Thus, maps make new maps necessary.

Together, these two types of maps – maps that delimit the páramo and maps that delimit productive and conservation areas within the páramo – will continue to affect human and non-human life. In this way, it is essential to problematize these maps and their production processes. How is it possible that maps do not consider and include the relationships they affect, being extremely determinant of human and non-human relations in the páramo? What role have the social sciences had, or can they have in the act of mapping the páramos and making these relationships visible? What capacity do the social sciences have to affect maps and, in this way, project sophisticated conjunctions around the complex act of mapping? These questions will be explored from the experience of mapping 'social life in the páramo' as social scientists in interdisciplinary teams at the Humboldt Institute, the one in charge of delimitation maps.

Páramo landscapes: inaccurate materials

Relationships between the sciences at the institute we work in are not horizontal. In the interdisciplinary practice, the campesinos' narratives, the ethnographic observations, the ways of understanding and establishing relationships find a very restricted place in documents constructed with maps, plans, graphs and tables. In the broad field of biodiversity sciences, social sciences must make a place for themselves by exploring with methodologies that allow results to be shown, trying to avoid the usual accusation of little objectivity or mere anecdote. Thus, while delimitation maps were elaborated using biological models, in another team, we discussed the consequences of such delimitation and anticipated that it was necessary to make visible both the people living in the páramos and their relations with the ecosystem.[2] We wanted to show that the restriction of all productive activities in the páramo was not a good way to repair past damage and anticipate the future of the páramos since, in essence, it is based on the exclusion of the human from these places. In this context, we worked with páramo communities to understand their relationships with these places and, above all, communicate these results to our biologist colleagues. Therefore, we assembled various social science methods in the field. Strategically, we decided to use maps to communicate our results to natural scientist: we used social mapping techniques, making maps a fundamental means of interaction with the inhabitants of the páramo. In this way, as we will see, other maps were produced. The purpose of this social cartography was to show what the delimitation maps did not allow us to see. That is, to scale other relationships in which the human dimension had a proper place. Consequently, multiple and diverse páramos emerged.

In this section, we want to analyse two of them: stacked maps and painted landscapes. The first one is called stacked because it results from layering maps made by campesinos. In each layer, people solve a question, locating what we want to visualize on a municipality's sketch that we prepared previously – water sources, forests, schools and roads, houses, meeting places and animals. Stacked maps manage to show relationships between the people and the páramo from a top-down perspective made with campesinos. When a map that collects people's 'social' lives is stacked on a map of 'natural' areas, socio-ecology emerges as patterns of occupation, represented by spots within the sketch.

These maps are often difficult to make. It is not easy to locate oneself within the sketch. It does not necessarily coincide with the way people imagine the municipality. It is difficult to resolve the issue of distances. For example, I perceive a distance as a journey on a known road between my house and the neighbour's house, but this distance must become one between points on the map. Over time, we refined the method. The sketch is not enough, so it is better to start the work with a sketch crossed by rivers and roads, lines also present in the territories through which life passes, and which, therefore, facilitate locating on the map (Ingold, 2007). Once the attendees get to work, the map becomes populated with elements within the exercise's proposed plan. Most of the time, it is enough to follow a path and locate a point, perhaps a communal room or the house of someone you know. Once those summoned agree on these starting points, the map is quickly resolved. In the end, the map shows everyone's and no one´s work. The stacked charts seem to collect the stark information of subjects who, thanks to the map, make themselves occupants of the sketch. Whatever each has drawn ceases to be theirs. to become a piece of information that can be enlarged or made smaller, resists change in scale without being transformed.

Painted landscapes answer another question. It is about understanding the experiences which people live as páramo. The indication was to paint the páramo 'as it is for each one', 'as each one feels it'. The starting point is a blank sheet of paper. The result is colourful landscapes full of heterogeneous elements: bears and *frailejones*,[3] houses, cows, dogs and horses. Each landscape is a point of view: the point of view of the artist who has their own particular way of inhabiting the páramo, and at the same time, each landscape is an invitation to see 'from the point of view' of whoever paints. 'Wherever there is a point of view, there is a subject position', recalls Viveiros de Castro (2004a), just before noting: 'which is activated by the point of view will be a subject' (2004a: 467). In these drawings, the páramo itself is a relationship, emerging as an inhabited landscape. Always from a place, from a body, the landscape has a horizon, has an up and down, left and right. It is made from a standing hominid bipedal body. The result is an image in which the páramo is the whole drawing, not a part of it. An image in which the páramo is 'everywhere' instead of 'out there' (Ingold, 2011: 111). There is no 'out there' in the páramo inhabited by the body. Looking at several of these drawings, it becomes clear that the elements do not necessarily repeat themselves. In some landscapes, frailejones are gigantic and are mixed with potatoes. In others, streams divide the drawing as the roads do in others. If there are fences in some of them, they do not separate what is human from what is not: even the bear that walks in these drawings shares the stage with cows. If in stacked maps the socio-ecological is the effect of overlapping layers, separation cannot be made in these painted landscapes. Human elements are ecologically intertwined, and non-humans are part of a sociology of complex relationships.

Insofar as the participant was asked to paint the páramo as they experience it, it can be assumed that what is painted on each sheet resonates in that 'surplus value' of landscape affordances (Rose and Wylie, 2006). Painted landscapes are also enumerations of elements that make the páramo a páramo from the point of view of the páramo inhabitant. A selection of features to which the campesino responds through drawing. Without wishing to be exhaustive, the painted landscapes are framed within the material possibility provided by the sheet of paper. Still, that edge is not a limit for what is presented in the drawing. For example, mountains are cut, but both who draw and who observe can imagine mountain lines continuing. Unlike maps whose sketches separate what is not, the edge of the sheet of paper also hints at the experiential continuity of what is drawn. Each landscape has a beyond, as it has a depth, even though the drawing area ends. No one painting a landscape in this way can say that they have painted the entire páramo.

Once completed, the maps are stacked, and the painted landscapes are viewed in sequence, side by side. A lot can be said about these landscapes; they concentrate many of the affective relationships typical of inhabiting; they show relationships, insist on problems, place humans and more than humans in the same plane, and always offer a point of view. However, they cannot be added together. Scientific maps, on the contrary, put objectivity into action as a loss of a point of view. Stacked, it seems that they reached all the voices and the elements. At the institute, by conversing with geographers, ecologists and biologists, the stacked maps were more easily able to enter the reports that refer to how people 'occupy', 'use' and 'transform' the páramo. Some of these maps could even be scanned, turned into polygons and pixels, and albeit inaccurately, stacked on top of satellite images and cadastral cartography. "It would be ideal," said a fellow geographer who has dealt many times with that curious translation from social cartography to satellite cartography, "when doing the exercise, people had a GPS so they could locate the points exactly." Then there wouldn't be any problems with the 'subjective distances' that separate the elements on the map since the GPS makes the 'occupant' of the sketch a human sensor (Martínez-Medina, 2020). The painted landscapes are not, for their part, susceptible to any improvement in these terms. What they show is not mappable on a plane.

Figure 25.1: Staked maps made by campesino inhabitants of the páramo in the project 'Páramos y Sistemas de Vida'

Figure 25.2: Painted landscape made by campesino inhabitants of the páramo in the project 'Páramos y Sistemas de Vida'

The relationships they summon are not scalable. In the 'popular objectivist epistemology' of our tradition, Viveiros de Castro continues:

> the category of the object supplies the telos: to know is to objectify – that is, to be able to distinguish what is inherent to the object from what belongs to the knowing subject and has been unduly (or inevitably) projected into the object. To know, then, is to desubjectify, to make explicit the subject's partial presence in the object so as to reduce it to an ideal minimum. In objectivist epistemology, subjects as much as objects are seen as the result of a process of objectification. The subject constitutes/recognizes itself in the objects it produces, and the subject knows itself objectively when it comes to see itself from the outside as an 'it.' Objectification is the name of our game; what is not objectified remains unreal and abstract. The form of the other is the thing. (2004a: 468)

Too quickly described as 'anecdotal' by our natural sciences colleagues, the painted landscapes propose another way of knowing where knowledge is only possible by occupying the point of view presented by the artist. That means an invitation to become other, transform oneself and become adept at the affections that cross the body in another arrangement of relationships: an invitation to become with another body. Whether this is possible or not, looking at these drawings is already receiving another point of view as an offer. Natural scientists, for whom the painted landscapes are 'inaccurate and anecdotal', abort that invitation and disregard that claim too quickly.

Body maps and body affect

The mapping of the relationships between campesinos and páramo continued beyond the stacked maps and painted landscapes, as well as the interest in dialoguing with the natural sciences. An ecologist from our team proposed to make visible what the campesinos know and how they use the plant biodiversity of the páramo. To take this knowledge into account, she built huge databases with plant uses reported by campesinos. In addition, she had to find and standardize the common names of the plants with their scientific names. For this, she collected plants with their coordinates, described their botanical characteristics, and took photos. In this way, these campesino-biodiversity relationships could be scaled, allowing botanists, experts in páramo vegetation, to analyse and compare them with specimens from biological collections or scientific publications. However, this database remained closer to biologists and moved away from campesinos and other relationships that we wanted to make visible.

For that reason, we decided to draw on a map that would allow connecting páramo plants with those men and women inhabiting the páramo. To this end, we drew sketches of male and female bodies, printed plants photos and asked them to locate the body organs and connect with coloured wool, plants, and organs, linking body and plant through the ailments they cure. With this map we saw how specific forms of knowledge about and with the páramo emerged in practical ways, hardly scalable through the databases created by our colleague and the maps we had previously made. With this web of relationships, we saw different ways of being with the páramo related to how campesinos live and feel with other-than-human beings. Protection, help, solidarity, gratitude and care were words associated with the body map. In this way, our map tried to show those ignored relationships.

We used the body sketch as a surface that made the connection between the bodies and the páramo visible. For us, noticing this link was revealing. The exercise functioned as a reverse figure to the stacked maps. In these maps, people located themselves in a plan of their municipalities, and for continuity, in the páramo as a mappable place. In the body sketches, it was the páramo that appeared *within* the campesinos. Together, maps and body sketches showed a double mutual inclusion, in which neither the people nor the páramo could be reduced one into the other (Law, 2004). Mutual excess and inclusion, the reverse figure spoke of connection. A map that showed people as occupants of the páramo was joined by a sketch made of other inhabitants of the páramo, the plants, part of campesino bodies. The

Figure 25.3: Body maps made by campesino inhabitants of the páramo in the project 'Páramos y Sistemas de Vida'

campesinos began to look like inhabitants, in the sense of beings that come to be with páramo traversed by the páramo. At the same time, páramo is no longer an 'environment', but another part of their bodies.

We could not have responded to what the sketch showed us if we had not given ourselves the task of 'taking seriously' what the exercise produced. Before our eyes and thoughts, campesinos' bodies began to look different from ours because the páramo did not pass through us, at least not in the same way. Walking through the páramo, we cannot recognize dozens of plants that campesino women used to relieve their bodies, even passing by. Faced with a headache or an upset stomach, we could not have made any of those páramo plants help us. Amid these musings, Mrs Paulina, one of the campesino women in the workshop, pushed us to think again and more slowly. When her turn came, in front of the female body sketch, Paulina did not know where to paste the *pimpinela* image, a plant from *her* páramo. Without eagerness, the woman was wondering where to place the photo. We knew from the database of our colleague that this plant was used for a wide range of ailments, irritations and inflammations. However, Paulina could not bring herself to put the photograph somewhere on the body. Finally, and with the image still in hand, after much thought, the woman explained: "To cure sadness, drink lemon balm and pimpinela water. It must be *stone water*, that which is accumulated by the rain of the páramo.

It is kept in a bottle with pieces of coral and blessed by a priest." Later she told us: "My daughter suffers from sadness. Doctors call it depression. But for me, it is sadness. I keep a bottle of páramo water for her. That relieves her a lot." Paulina did know how to use the pimpinela. Her doubt was the product of the sketch, of knowing that the sadness cannot merely be put here or there in the generic human body. In the same way, the photograph of the plant was insufficient to account for this remedy. What cures sadness is not only the fresh pimpinela collected in the páramo, but its mixture with lemon balm from her garden, with stone rainwater, coral, and the priestly blessing. Neither the body sketch nor the image could summon the meeting of diverse body elements and the remedy in this place in the Andes and Mrs Paulina's care practice.

Today we understand that Paulina's remedy is a composition of heterogeneous elements dissolved in páramo water. Lemon balm becomes a remedy with wild pimpinela, water, the blessing and coral. People become with the gathering of elements that is páramo, in a process of mutual transformation. 'Ontologically heterogeneous partners become who and what they are in relational material-semiotic worlding. Natures, cultures, subjects, and objects do not preexist their intertwined worldings' (Haraway, 2016: 12–13). Maps that put people into action as occupants of the páramo carry the risk of preventing us from imagining other kinds of stories, such as those in which sad bodies are relieved by Paulina's remedies. 'Becoming-with, not becoming, is the name of the game; becoming-with is how partners are, in Vinciane Despret's terms, rendered capable' (Haraway, 2016: 12–13). In this case, becoming capable is gaining the ability to respond to the remedy. Pimpinela, páramo, water, sadness, linked by the 'miracle of attunement' (Despret, 2004: 125), posing a challenge to the scientific understanding according to which depression – supposedly depression, says Paulina – the human body and the páramo are before and beyond the relationships that constitute them.

Being affected by páramo

Articulation is the term Latour (2004) uses to refer to how bodies are tuned thanks to devices, artifacts, standards, objects, methods and ways of doing. Latour uses an illustrative case to elaborate on this concept: apprentice perfumers and a box of scents. The apprentices learn to make the kind of discrimination between different smells using the substances in the box. The scent box makes a series of odours available to these students' bodies which learn to be *affected*. These apprentices only become *noses* – master perfumers, experts – since they acquire a richer, more complex and therefore better articulated odorous world. Jensen and Morita (2020) propose that concepts also dispose us to be affected by certain qualities of the world so that those worlds are always possible in relation to those concepts. Our conceptual/methodological/sensorial devices articulate us in different ways. In this sense, we can say that how we know, experience and live the páramo – our scent boxes – not only allow us to be affected in a particular way but also allow the emergence of divergent affections that mobilize us.

Being affected by páramo is itself an expression of multiplicity. The páramo can establish very diverse articulations, excite different affections and mobilize people of very different interests. It is common to witness moments when the páramo diverges. As part of the project we were working on, a group of biologists, ecologists, agronomists and anthropologists visited the farm of Mrs Claudina for the first time. A renowned botanist, an expert in the vegetation of the páramo, took the lead during the walk, explaining how plants participate in water regulation giving the liquid to the cities. Plants and soils are responsible for conserving water in times of drought. During summers, they allow water retained in these high areas to flow to the lowlands. Together with other processes, these soil and vegetation capacities produce what some scientists and policymakers call 'ecosystem services', a utilitarian way to imagine the relationship between people and nature (Hermerlingmeier and Nicholas, 2017). For this reason, the professor explained: conserving the páramo is a necessity for those of us who consume water in cities, a large part of the Colombian population.

Mrs Claudina listened attentively and in silence to the professor's lessons and looked curiously at the group of students and researchers who visited her. She led the way, pointing out the fences the group had to cross as they traversed their potato crops and pastures where she and her neighbours keep livestock. The tour's purpose was to reach the source of a river located on her farm from which at least 25 families collect water. Through swamps and wetlands, the teacher showed how the ecology of the páramo works. We felt the humidity on our feet, heard the water begin to run and shivered from the mountain's cold.

When arriving at the water source, Mrs Claudina explained how the fences that everyone saw protected it from the cows. Barriers were part of an agreement between all the neighbours to care for the site. Undaunted by the professor, the same one who had called each plant with strange names in Latin and who insisted on the 'strategic' nature of the páramo as an ecosystem service supplier, she corrected him right there: "This páramo gives us water," she said, "But only to us. Not to all the cities of Colombia. Not to everyone." There was a profound silence. The comment did not go unnoticed, but there was no reply. The professor did not feel challenged, although the explanation was directed to him. They were both talking about versions of páramo that are not the same, despite occupying the same space simultaneously. During the walk, the professor showed how páramo emerges in his practice: a specific biophysical relationship between altitude, plants, soils, and so on. For him, the group walked on a páramo, a network of relationships with the quality of capturing, storing and providing water. On the other hand, Mrs Claudina was walking through her páramo, through the place her family made habitable. She walked through her farm that is not in any páramo but is part of the páramo that her feet tread. For the professor, we were at the materialization of a series of relationships, abstracted from the study of many páramos. An example of what a páramo is like (indeed also 'degraded' by cows, pastures, potatoes and Claudina): for the woman, we were at the place she inhabits; the landscape in which she is a campesino woman, from where she has a point of view from which it can be said that this is her páramo. *Equivocation* is the term that Viveiros de Castro coined to refer to this type of situation in which two speakers, saying the same thing, refer to different things (2004b). Ontologically divergent deployments from practices that are not the same. Páramo is a set of partially connected articulations of bodies, practices and affections.

Through the distributions of plant and animal species, algorithms tried to establish the border between the forest and the páramo (see the first section of this chapter), responding to a version of páramo historically constituted by scientists like the professor. In these maps, an ecology that, by definition, excludes humans becomes the measure that determines people's practices. In Mrs Claudina's páramo, the potato and the cow are inhabitants because their ecology is not separated from the ties of life and death on which survival depends. The matter is further complicated when the map, no matter how exact, puts the páramo into action as an area from which Mrs Claudina, her parents and her cows must be excluded or regulated. Once again, articulation is the key to understanding this situation. The difference between the professor and Mrs Claudina is that the first is part of the practices included among the experts that make the maps promoted by the Law. Therefore, the páramo that affects the professor can be mapped. In many ways, the delimitation maps are maps of the páramo that affect scientists as the professor. His expertise is articulated by the map with environmental authorities and state practices. As noted by Blaser, ontological conflicts are also conflicts between worldings with different capacities to act in relation to the State (Blaser, 2013).

In this case, as in so many others, the sciences that map the limits of the páramos work without listening to the claims of those affected by the páramo in a way that exceeds how these sciences, and the state is affected. Inaudibility is not a problem of simple willpower. On the contrary, we suggest that the scientists' concepts, methods, devices and standards make them incapable of affections that exceed their knowledge and practical tools (for instance, the blindness of the properties smaller than 250 m). Social sciences travel the opposite way, considering the relationships between humans and non-humans in these places. However, even if the stacked maps, the painted landscapes and the body map partially show these relationships, they cannot speak for the whole páramo. Trying to become articulated with maps, they lose the ability to show what is needed in the first place. Mrs Claudina's embodied relationships, the history of those cows and those fences, the depth of her way of inhabiting the site cannot be mapped. She is a point of view entangled with the inhabited and habitable páramo, a point of view which, as we have already learned, the 'popular objectivist epistemology' of the sciences cannot reach using its usual tools. However, the conjunction between stacked maps, body maps and painted landscapes has allowed us to imagine ways of being affected that differ from how we can be affected, a *conjunction* without the possibility to become an *integration*. There is no sum. The arithmetic of affect and multiplicity does not count wholes and does not reduce to one, in this case, to a single páramo that a line can delimit.

Delimitation scientists respond to the affections that mobilize them, articulated with their methods. There is a mountain that they can 'touch', one that emerges as a result of their practices of care, as Paula Ungar points out (2018). The articulation between science and the state is complex, full of encounters and equivocations. So, it is not a matter of abandoning the tools that make the science of ecosystem delimitation what it is. Nor is the task one of inclusion one which exhaustively seeks to understand what the páramo is in its many manifestations entirely. Instead, it is about allowing oneself to be affected by 'inaccurate materials', such as the painted landscapes, such as the experiences of Mrs Claudina and such as the remedies of Mrs Paulina, to affect their concepts and their ways of doing things. The task is to allow the difficulties that the campesino women propose to the professor, a map that somehow recognizes that it cannot map everything. A profusion of materials whose relationships are complicated and cannot merely be put one on top of the other to proceed as a sum, but that say something about their differences. This consideration applies, of course, also to the social sciences and for their affections.

The páramo's delimitation is not a finished story. When the delimitation maps were produced in 2015, the state did not immediately accept them, and now, five years later, the Law does not completely exclude the campesinos. Páramo's delimitation conflict between science and campesinos is not an inevitable story either. Following Stengers and Despret (Latour, 2004; Stengers, 2010; Despret, 2016), we think science is more interesting, more scientific, and riskier to the extent that it is more richly articulated. In this sense, developing an openness to other affections can provide us with a more complex approach to the Colombian páramos. The sophisticated conjunctions between knowledge, practices, devices, and sciences imply articulations between affective worlds that are not easy to generate. They also imply getting our bodies (physical bodies, bodies of knowledge, bodies of care) attuned to more possibilities. We intuit that only by risking being affected by those other affective worlds, in other words, nurturing sophisticated affective conjunctions, we will be able to tend towards an ecological reparation that also means reparation for excluded humans.

Notes

[1] Peasant is the most literal (and frequently used) translation of the Spanish word 'Campesino'. However, this translation does not convey all the meanings involved in the Campesino category, such as different aspects associated with their ways of life, land ownership, characteristics of family work, values and cultural elements related to reciprocity and the redistribution of production in community relations, the conditions of marginalization and exploitation, among others. For this reason, some authors recommend using the term in the original language (Koopman, 2007, Wood, 2012).

[2] This team worked on the project 'Páramos y Sistemas de Vida' funded by the European Union and developed by the Humboldt Institute between 2011 and 2015.

[3] *Espeletia,* commonly known as *frailejones,* 'big monks', is a genus of plants with a close relationship with páramo.

Acknowledgements

This chapter was written thanks to funding from the Instituto de Investigación de Recursos Biológicos Alexander von Humboldt, Bogotá. Previous versions of this chapter were presented at the panel Feminist Science Studies in the Global South: On Natures, Naturals, and the Fate of the Earth in the 2019 AAA/CASCA Annual Meeting and in the Ethnographic Transdisciplinary Engagements Roundtable and Webinar, 2018. We appreciate the contributions of the organizers and commentators at these events: Kristina Lyons, Rachel A. Cypher, Lesly Green and Carolina Angel. We also appreciate the comments of Ana Maria Garrido, Paula Ungar, Natalia Kazalinsk and the editors of this book.

References

Blaser, M. (2013) 'Ontological conflicts and the stories of peoples in spite of Europe: Toward a conversation on political ontology', *Current Anthropology*, 54(5): 547–68.

De la Cadena, M. (2019) 'Uncommoning nature: Stories from the Anthropo-not-seen', in P. Harvey, C. Krohn-Hansen and K.G. Nustad (eds) *Anthropos and the Material: Anthropological Reflections on Emerging Political Formations*, Durham: Duke University Press, pp 35–58.

Deleuze, G. (1981) 'Sur Spinoza', Cours Vincennes, Saint Denis, 17 February. Available from: www.webdeleuze.com/textes/38

Despret, V. (2004) 'The body we care for: Figures of Anthropo-zoo-genesis',. *Body & Society*, 10(2–3): 111–34.

Despret, V. (2016) *What Would Animals Say if We Asked the Right Questions?*, Minneapolis, MN: University of Minnesota Press.

Haraway, D. (2016) *Staying with the Trouble: Making Kin in the Chthulucene*. Durham, NC: Duke University Press.

Hermelingmeier, V. and Nicholas, K.A. (2017) 'Identifying five different perspectives on the ecosystem services concept using Q methodology', *Ecological Economics* 136 (June): 255–65.

Ingold, T. (2005) 'Epilogue: Towards a politics of dwelling', *Conservation and Society* 3(2): 501.

Ingold, T. (2007) *Lines: A Brief History*, New York: Routledge.

Ingold, T. (2011) *Being Alive: Essays on Movement, Knowledge and Description*, New York: Routledge.

Jensen, C.B. and Morita, A. (2020) 'Deltas in crisis: From systems to sophisticated conjunctions', *Sustainability*, 12(4): 1322.

Koopman, S. (2007) 'Campesino. Spanish for social change'. Available from: www.spanishforsocialchange.com/search?q=campesino&max-results=20&by-date=true (Last accessed December 2020).

Latour, B. (2004) 'How to talk about the body? The normative dimension of science studies', *Body & Society*, 10(2–3): 205–29.

Law, J. (2004) *After Method: Mess in Social Science Research*, Abingdon: Routledge.

Martínez-Medina, S. (2020) 'Lo que pliega la colecta: conocimientos, científicos y especímenes para otras ciencias posibles', *Antípoda. Revista de Antropología y Arqueología*, 41: 31–56.

Osejo, A. and Ungar, P. (2017) '¿Agua sí, oro no? Anclajes del extractivismo y el ambientalismo en el páramo de Santurbán', *Universitas Humanistica*, 84: 143–66.

Osejo Varona, A., Ungar, P. Escobar, D., Mendez, M.C., Pachón, F. and Valencia, L. (2020) 'Desafios y posibilidades de la actual política de páramos: Diálogos en torno a Guerrero y Sumapaz'. *Biodiversidad en la Práctica*, 5(1): e740.

Rose, M. and Wylie, J. (2006) 'Animating landscape', *Environment and Planning D: Society and Space*, 24: 475–9.

Sarmiento, C. and Zapata, J. (2016) 'Instrumentos jurídicos para la protección de los páramos', in M.F. Gómez, L. A. Moreno, G. I. Andrade and C. Rueda (eds) *Biodiversidad 2015*, Bogotá: Instituto Humboldt.

Stengers, I. (2010) *Cosmopolitics I*, Minneapolis, MN: University of Minnesota Press.

Strathern, M. (1999) 'The ethnographic effect I', in M. Strathern, *Property, Substance and Effect. Anthropological Essays on Persons and Things*, London: Athlone Press.

Ungar, P. (2018) 'The mountains we touched – The role of care in delimiting ecosystems', *Weber – the Contemporary West*, 35(1): 4–18.

Viveiros de Castro, E. (2004a) 'Exchanging perspectives: The transformation of objects into subjects in Amerindian ontologies', *Common Knowledge*, 10 (3): 463–84.

Viveiros de Castro, E. (2004b) 'Perspectival anthropology and the method of controlled equivocation', *Tipití: Journal of the Society for the Anthropology of Lowland South America*, 2(1): 3–22.

Viveiros de Castro, E. (2019) 'On models and examples: Engineers and bricoleurs in the Anthropocene', *Current Anthropology*, 60(20): S296–S308.

Wood, J. (2012) *A Word about the Word Campesino*, Heifer International. Available from: www.heifer.org/blog/a-word-about-the-word-campesino.html (Last accessed December 2020).

26

Ordinary Hope

Steven J. Jackson

The opening lines of *The Human Condition*, Hannah Arendt's (1958) sweeping 'reconsideration of the human condition from the vantage point of our newest experiences and our most recent fears' (5) describe the Sputnik launch in 1957.

> In 1957, an earth-born object made by man was launched into the universe, where for some weeks it circled the earth according to the same laws of gravitation that swing and keep in motion the celestial bodies – the sun, the moon, and the stars. To be sure, the man-made satellite was no moon or star, no heavenly body which could follow its circling path for a time span that to us mortals, bound by earthly time, lasts from eternity to eternity. Yet, for a time it managed to stay in the skies; it dwelt and moved in the proximity of the heavenly bodies as thought it had been admitted tentatively to their sublime company.
>
> This event, second in importance to no other, not even the splitting of the atom, would have been greeted with unmitigated joy if it had not been for the uncomfortable military and political circumstances attending it. But, curiously enough, this joy was not triumphal; it was not pride or awe at the tremendousness of human power and mastery which filled the hearts of men, who now, when they looked up from the earth toward the skies, could behold there a thing of their own making. The immediate reaction, expressed on the spur of the moment, was relief about the first 'step toward escape from men's imprisonment to the earth'. And this strange statement, far from being the accidental slip of some American reporter, unwittingly echoed the extraordinary line which, more than twenty years ago, had been carved on the funeral obelisk for one of Russia's great scientists: 'Mankind will not remain bound to the earth forever.'

Image 26.1

Image 26.2

These lines also name the principal sentiment against which the alternative viewpoint that follows is offered. Like Arendt, I am unsettled by the dream of escape captured in this scene and countless others played out in the subsequent enthusiasms of sci fi writers, space aficionados, and billionaires – a dream that seems only to have grown as our sense of planetary fragility has intensified. If I imagine myself in this scene, Arendt's or Musk's, I find myself looking not up and out, but *down*: back along the tail of the rocket to consider the world falling away below. What has been broken, in the world and our relationship to it, that makes escape seem the best and only option? Has our homesickness really gone this far? And what work, here and now, does the dream of escape leave undone?

In the time since Arendt's book, the fantasy of escape has only become more intense. The dream of space itself has changed, from the collective high-modernism of Cold War competition to private plaything of billionaires. So has our collective understanding and lived experience of planetary precariousness, as the looming reality and disparate effects of climate change begin to bite and transform the world.[1] The forests are burning, plains turn to deserts, the ocean is rising, and strange storms abound. There is much to flee, and much to atone for. But for these very reasons, there is much reason and need for hope.

Like the Sputnik fantasy articulated by Arendt, the alternate view offered here recognizes and builds from the reality of a broken world. But it responds, as do other chapters in this volume, not with escape but with *care* – and the ongoing work of repair by which this and other crucial relationships are sustained, honoured, made whole, and extended through time (which is not to say held constant and eternal), even under the debased conditions of the present. For me, however, there is another puzzling term in this equation, and one which figures alongside the problems and possibilities of ecological reparation explored so poignantly by others here.

For the ethos of escape and salvation seems to me very different than the concerns with repair, remediation and resurgence which frame this volume. This difference has *always* I believe lain behind the commitment to care and repair versus other more productivist or escapist visions of transformation, and opens for me another puzzle in the long history of our entanglements with the world (and our much shorter history of academic writing about them). If one cares about broken worlds, and not in a dystopian or apocalyptic way after the fetish of despair and destruction that animates fascist movements old and new, or the pathology of 'damage' (Tuck, 2009) that threatens to lock individuals and communities in place, one is drawn – inevitably, I believe – to the question of hope.

Hope and social theory

Hope has had a long and complex history within the annals of modern social theory and its precursors, and has always been associated with the dissolution and reinvention of worlds. For Thomas More (1967 [1516]) and other early modern theorists, the yearning for utopia was closely associated with the disorders of the day, most notably those attached to the breakdown of medieval religious and scholastic authority and the disruptions of a rising capitalist and imperial system. Later thinkers like Campanella (2009 [1602]) and Bacon (2008 [1626]) would add technocratic content to this vision, offering the first in a long series of secular updates to earlier religious ideals – most notably those of Augustine of Hippo (whose 'city of God' added an otherworldly anchor and exemplar to classical conceptions of political order); and Thomas Aquinas, who distinguished between a 'drunken' and a 'rational' hope, and noted how the latter, understood as a kind of divine gift and will towards God, could motivate and inspire worldly human activity. Subsequent enlightenment thinkers from Descartes (1985 [1649]) and Hobbes (1998 [1651]) to Hume (2007 [1738]) extended and secularized this understanding, offering an individualized and quasi-psychological account of hope as grounded in the contemplation of uncertain futures whose effects may produce, per Hume, either joy (hope) or pain (fear). In the *Critique of Pure Reason*, Immanuel Kant elevated the question 'For what may I hope?' to one of three central questions of philosophy, alongside his more famous takes on epistemology ('What can I know?') and ethics ('What should I do?'). Dealing variously with aspirations towards happiness, individual moral improvement, and the improvement of humankind as a whole ('hope for a better world'), Kant argued for the practical and moral character of hope as a rational alternative to other worldly attitudes – most notably those of despair, resignation or a narrow and destructive self-interestedness.

Modern social theory of a critical bent has questioned and for the most part rejected this enlightenment legacy. Work in the Marxian tradition – and other instances of what Paul Ricoeur (1970) has termed the 'hermeneutics of suspicion' – has tended to approach hope, whether in religious or secularized form, as a form of idealism or mystification that obscures recognition of the essential structural conditions of the world, including those upholding unequal social orders. In this line (but with exceptions returned to shortly), where hope appears it is more likely to be framed as a kind of deception or false consciousness (Lukacs, 1972 [1920]), propagated and sustained through the emergence of the modern culture industries (Horkheimer and Adorno, 1947), the actions of ideological state apparatuses (Althusser, 1971), or other powerful instruments of culture and ideology. Under this false hope view, the obfuscating effects of hope under liberal and neo-liberal orders tend to be politically quietist, forestalling necessary forms of collective action and perpetuating unequal social orders through the always-deferred and frustrated promise of transcendence. In inspiring dreams and desires that can never be fully delivered, such effects can also be, as Lauren Berlant (2011) has more recently argued, 'cruel'.

This brief chapter departs from both these traditions to suggest a different track. It offers a speculative argument for the centrality of hope and its intimate connection to the projects of ecological care and repair explored elsewhere in this volume. Drawing on disparate work in theology and pragmatist philosophy, heterodox traditions of critical theory, and emerging work in black, queer, and indigenous scholarship, it asserts five basic propositions:

- That hope is not predictive (and can therefore be disappointed but never falsified or disproved, whether by history or critique).
- That the measure of hope is not accuracy, but efficacy: its ability to hold and sustain more meaningful forms of action and relationality in the world.
- That hope may be expressed in orientations towards change and transformation, but also in forms of modest patience and enduring (which are not fully reducible to passivity, resignation, or 'mere' waiting).
- That hope is above all a property of ordinary work – a characteristic of our ongoing ongoingness in and with the world – from whence comes it depth and power.
- And that hope is a collective accomplishment – something we do in and with the world, in concert with others and with things.

Along the way, it will explore several key relationships that are essential I believe to the nature and presence of hope in the work of ecological repair, remediation and resurgence: the relationship between hope and worldmaking (including, crucially, the future); between hope, loss and struggle; and between hope, method and critique (including as practiced in the interpretive social sciences today). It concludes with a return to Arendt, her concept of *amor mundi*, and an argument for a more hopeful practice of care for the world.

Hope, vertical and horizontal

An important part of the stickiness and resonance of hope is grounded in its spiritual and religious roots – a point poorly recognized under the secularist tendencies of the (post-?) Enlightenment university. It is therefore no surprise that hope's most interesting intersections with critical thought come from those working at the margins of critical theory and various forms of religious or spiritual thinking, often to the chagrin of their more orthodox interlocutors. In his remarkable 2015 book, *Redemptive Hope*, religious studies scholar Akiba Lerner traces the influence of Jewish and Christian messianic traditions in the work of Walter Benjamin and Ernst Bloch, the twentieth-century critical theorists in whose thought hope plays its most prominent role, along with a parallel series of developments in the traditions of American pragmatism and neo-pragmatism.

For Benjamin, hope is a fundamentally recuperative act, produced from the gleanings of memory and culture (thus Benjamin's famous and unfinished

Arcades project (1999)) that point, in fragments, towards an as-yet unknown future of redemption. Despair constitutes an essential moment within this dialectic, and it is destruction that gives rise to the messianic force of hope (a point Benjamin often chose to express through paradox: for example, his often cited claims that 'there is an infinite amount of hope, but not for us'; or 'hope is given to us only for the sake of those without hope' (1969)). For Bloch (1995), continuing this thread after Benjamin's death in 1943, the 'principle of hope' in messianic traditions expressed the power of the 'not-yet': a restless and innate yearning and projection into the future that could provide a more potent basis for justice and social change than the scientific pretensions of the Marxism of his day (with its tendency to foreclose the future under the presumed structural certainties of historical development).

The theological roots which shape this way of thinking were, in Lerner's striking characterization, *vertical* in orientation, projecting hope 'upwards' in space (for example, towards heaven) and 'forwards' in time (towards remote or indeed eschatological futures that would end time altogether). This verticality constituted the power and, for Benjamin and Bloch, the revolutionary potential of the messianic tradition, and thus the force that each and others in the subsequent 'School of Hope' in philosophy and theology sought to harness. But verticality also carried dangers that were indeed central to the nineteenth-century critique of idealism and religion (and remain central to the ecological concerns of this volume): that its projections 'up' tended to devalue the importance and care for present conditions; that the passing nature of the material world made it less worthy of care, respect and an autonomous (extra-human) realm of existence; and that its promise of eventual redemption amounted to a politics of quietist waiting and infinite postponement.

Compare this with the orientation of Benjamin and Bloch's near contemporaries William James and John Dewey and subsequent theorists in the American pragmatist tradition. As Lerner and others (Westbrook, 2005; Koopman, 2009) have noted, hope plays a central role in pragmatist philosophy, but it follows a distinctly *horizontal* track: pragmatist hope expresses 'the desire to be a part of something larger than oneself, but – instead of directing these desires for great meaning upwards – directs them horizontally by inspiring us to look to other people as sources for hope and fulfilment' (111). Here too the stakes of hope are high, its achievement hard-won, and therefore not to be read in any particularly light or sunny way, as reflected in Martin Luther King Jr.'s reflections on the labour of turning 'the fatigue of despair into the buoyancy of hope' (1981). This hope is also radically uncertain – and in the expansiveness of its vision and as a purely *predictive* matter, is bound to disappoint (while doing powerful work along the way); as Rorty wrote, our hope for democracy rests ultimately 'on our loyalty to other human beings clinging together against the dark, not our hope of getting things right' (1982: 166). Finally, pragmatist hope – Lerner's 'horizontal' tradition – is grounded not in prediction or perception, but above all in a disposition and orientation towards *action*. As Cornel West explains:

> Optimism adopts the role of the spectator who surveys the evidence in order to infer that things are going to get better. Yet we know that the evidence does not look good ... Hope enacts the stance of the participant who actively struggles against the evidence in order to change the deadly tides of wealth inequality, group xenophobia, and personal despair. Only a new wave of vision, courage, and hope can keep us sane. ... To live is to wrestle with despair yet never to allow despair to have the last word. (1997: xii).

The sections that follow build on these foundations to elaborate three essential relationships that I believe are central to the work of repair, remediation and resurgence foregrounded in this volume: between hope and worldmaking; hope, loss and suffering; and hope, method and critique.

Hope and worldmaking

My first point concerns the relation between hope and what philosopher Nelson Goodman (1978) has addressed under the suggestive language of *worldmaking*. For Goodman, worldmaking calls attention to the practices of rendering – 'all the ways of making and presenting worlds' – by which our basic frameworks and senses of the world (including alternate worlds) are called into expression, negotiated with others, changed and in some instances left divergently intact.

There are important strengths to this conception that fit it well for the concerns of hope and repair tackled here. One is its inherent plurality – Goodman's insistence that 'there are many worlds, if any' (Goodman, 1984: 127). The making and proliferation of worlds (or 'versions') is what we do each day, working within the bounds and resources of *past* such makings. (Goodman characterizes this position as neither realist nor anti-realist, but *irrealist*.) It is also worth noting, despite the language of 'making', the friendliness of Goodman's account to notions addressed elsewhere in this volume as repair. Thus, worldmaking for Goodman is made up of composition and *de*composition; weighting and ordering; deletion and supplementation; and deformation. As Goodman insists, 'worldmaking as we know it always starts from worlds on hand; the making is a remaking' (6).

But to be made in this expansive way, worlds must also be *thought* (and thinkable), and thought takes work – and the willingness or predilection to undertake it. To engage in this work embeds a claim and presupposition that we *can* make worlds, and are not merely and purely on the receiving end of such things. This brings us back to the principles of action and efficacy advanced earlier (to wit: that 'the measure of hope is not accuracy, but efficacy: its ability to hold and sustain more meaningful forms of action and interaction in the world'). For the same reason once again, as Ernst Bloch observed, hope can be disappointed but never disproved.

I would make two additions to Goodman's account. The first concerns a *material* quality of worldmaking that goes underdeveloped in his more meaning and category centred account of the term. This is more than the by-now reflexive reminder to include 'stuff' in our analysis (though the ideational roots of hope and the tendency to regard it as a kind of attitude or 'feeling' may merit special caution in this regard). Rather it is to note that in making worlds we work with *things* as well as meanings, objects as well as categories, and the worlds that result are in fact complex amalgamations of both, with no clear priority or precedence between them.

My second addition to Goodman would be around the *temporality* of worldmaking – and by extension: the timeliness of hope. While not named by Goodman as such, worldmaking is also a deeply temporal phenomenon, and essential to the process by which futures are brought to bear and pass, even where thickly entangled with the recruitment and recuperation of the past. Here we find an essential relationship between hope and the work of *imagining* – whether understood at the level of individual or small group actors, or what recent scholarship in the social sciences (Taylor, 2003; Jasanoff and Kim, 2015) has come to term *social* (or *sociotechnical*) *imaginaries*: understandings and orientations to the world that draw into being wider collectives and orient them towards shared sets of common(ish) and thinkable futures. This assigns hope an inventive or world-disclosing property, and an essential role in the ideation and movement towards alternative futures. But hope's contribution to the future is expressed not *only* in orientations towards change (another common misconception); rather, hope may be expressed with equal force, if less fanfare, through the more modest arts of endurance, resilience and patient tending – what Donna Haraway (2016) has evocatively described as 'staying with the trouble'. Hope can operate without concrete and specific claims to the future – and in this sense, without 'a future' at all. This sense is well captured in feminist science fiction, for example, Octavia Butler's *Parable of the Sower* (1993), or Margaret Atwood's *Maddaddam* trilogy (2002, 2009, 2014). Keeping things going, holding on a little longer, gardening on the edge of disaster – these too are part of the temporality of worldmaking and the 'not-yet-ness' of hope.

This temporal ordering of hope – the weight of the pasts it recuperates and reinvents, and the potentialities of the future it claims or merely holds open – are also central to the revolutionary potential scholars like Bloch and Benjamin saw in hope. In this way, following Frederic Jameson (1972, cited in Munoz, 2009), we may observe how hope "provincializes" the present, making current reality seem narrow, suspect and incomplete, rather than the iron cage limiting and dominating our possibilities it otherwise so often seems to be. In this it points both backwards and forwards, making histories (always plural) a field of fragments and possibilities rather than a sentence, and the future a contested and open-ended terrain. It could have been and might yet be otherwise: such is the worldmaking power of hope.

Hope, loss and struggle

My second point concerns the intimate relation between hope, loss and struggle, and how a deflationary notion of hope may help us to see deeper roots and connections between them. I have written elsewhere (Jackson, 2014) about how repair emerges from and constitutes a kind of 'aftermath', a point numerous other scholars, in this volume and elsewhere, have substantiated in greater depth and clarity (including around the essential role and relationship of care in these contexts). I have since come to favour the concepts of 'the wake' and 'wake work' offered by scholar of the African diaspora Christina Sharpe (2016). Writing out of the afterlives of slavery, Sharpe notes the rich and entangled meanings of the wake – the aftermath of a slave ship making the middle passage; a vigil over the dead; a rising (slow or sudden) to consciousness – and offers 'wake work' as a language for the myriad practices by which Black lives navigate, reclaim, flourish within, and sometimes simply suffer and survive these worlds. Linked in this way, the wake is a space of terror, pain and suffering. It is also (and for this reason) a space of hope, and a source of what Fred Moten (2003) has argued for as a radical and improvisatory political potential.

Related ideas may be drawn from the field of queer theory. Writing against both pragmatic accommodation and anti-idealist and anti-utopian movements in the field, Jose Esteban Munoz's *Cruising Utopia* (2009) draws on Benjamin and Bloch to argue for what he describes as the 'not-yet-ness' of queer futurity. Examining a series of artistic practices and events from before, around and slightly after the Stonewall rebellion of 1969, Munoz calls out what he terms (pace Bloch) the 'anticipatory illumination of art', understood as 'the process of identifying certain properties that can be detected in representational practices helping us to see the not-yet-conscious' (3). Such practices help to constitute what Bloch called *concrete utopias*: alternative futures that arise from the situated struggles of historically placed groups, whose meaning and identity may in fact be called into being by their shared orientation to a common future (however remote and far from present circumstances that future might be). Such acts and imaginings are premised on and portend a 'then and there' that is different and other from an inhospitable 'here and now'. A similar stance is offered in Sedgewick's (2002) argument for 'reparative' (over 'paranoid') reading – a fractal and recuperative mode of interpretation in which "hope, often a fracturing, even a traumatic thing to experience, is among the energies by which the reparatively positioned reader tries to organize the fragments and part-objects she encounters or creates." (24)

A third source for this way of thinking can be found in a growing body of indigenous work around the relationship between hope, endurance and resurgence amidst ongoing projects of dispossession, erasure and cultural genocide under settler colonialism. Kim TallBear (2019) has argued for the central role of hope in 'caretaking relations', offered as an alternative to forms of 'American dreaming' premised on indigenous erasure, elimination and the subordination of other-than-human neighbours. Kyle Whyte (2020) points to the slow and ongoing work of relation building – the restoration of long-abused relations of consent, trust, accountability and reciprocity between indigenous and settler societies that must be tackled before any hope for just and effective climate action can occur. William Lempert (2018) has argued for 'the generativity of hope in the post-apocalyptic present' and shown the essential presence and dynamism of hope among indigenous media communities and creators around the world. Hope is no less central to the forms of 'alterlife' that Murphy (2017) argues will be essential to any possibility of decolonial chemical relations in the Great Lakes region of North America, or to the forms of violence, resilience and endurance captured by indigenous writers from Thomas King (2012) to Tommy Orange (2018) to Cherie Dimaline (2019). This work and insistence offers an essential corrective to monotonic narratives of loss, deficit and 'damage' (Tuck, 2009) that have long marred relations between settler and indigenous societies both within and beyond the academy.

Even in this impossibly brief description, it will be clear that the sensibilities here run counter to the vertical tradition outlined, and also to any particularly upbeat understandings of the nature and operation of hope. There is no model or imagination of transcendence, no simplistic or empty celebration of solution, resistance or escape: there is no Sputnik for racism. There is also no sense in which hope somehow escapes and floats above the devastating experience of pain, loss and suffering, both personal and historical; to the contrary, hope is rooted in these experiences and grows precisely out of them. Nor finally is the investment of hope grounded in a confidence of outcome. To know or believe to know the future reflects a different kind of strength and privilege. There are actors who can do this, because their claim to control or organize the future is stronger. For the rest, the accounts here suggest important addendums to the earlier propositions on hope: that hope is most powerful, and most needed, where hard-won; that radical insecurity and uncertainty, both individual and collective, may be a powerful engine of hope (though no guarantor of its emergence); and that hope comes not from the absence of pain, loss and struggle but in its enduring and formative presence.

Hope, method and critique

My final point turns the camera back to consider the relationship between hope and the forms of scholarship, method and critique practiced in the interpretive social sciences today. It asks what role hope plays in scholarly practice itself (always a form of wake work or repair); and the related but separate question of how we might think towards and practice an empirical programme of hope, understood as a question of method, topic and our own ethical placement and anchoring in the world.

This question is central to the leading anthropological work on the topic, Hirokazu Miyazaki's *The Method of Hope* (2004). Drawing on Benjamin and Bloch, Miyazaki's book follows the case of the Suvavou people in present-day Fiji and their long-frustrated effort to secure reparations and acknowledgement for the loss of their ancestral lands. It shows how community relations are extended and held through this indefinite and indefinitely postponed claim on the future, including the seemingly inextinguishable belief that 'somewhere in the government archives, there is a document that will validate their claim once and for all' (45).

This of course can be told as a story of the relationship between hope, loss and struggle, like those just explored. But it is also a story of method, and an answer to the question of how an account of method conditioned on hope might overcome the limits of representation in anthropology vis-à-vis the live and open-ended nature of the world; or as a friendly reviewer summarizes the problem: 'representation fails because it is always too late to capture a past that has already flown and too slow to apprehend a present that is therefore always a past' (Abramson, 2004: 531). This echoes, as Miyazaki argues, a fundamental discrepancy also marked by Bloch: namely, between the contemplative and backwards-looking quality of philosophy, and the projective and forwards-looking quality of hope, which constitutes a fundamentally different basis of relation with the world.

Scholarly work itself – no more nor less than other ordinary forms of work – also partakes in or refuses a world of hope. Understood as a method of discovery and understanding, hope carries key properties in this regard. In its tendency to disrupt and open up new possibilities, hope is intimately linked to the essential methodological virtue of surprise, or what Bloch wrote about as the 'astonishment' of hope. In its mixture of ambition and modesty towards the future – its orientation to efficacy and action, but refusal of authoritative or over-confident predictive claims – hope also establishes an important ethical grounding and balance, moving from accounts centred on deficit, absence and loss to abundance, resurgence and engagement (to name just three terms figuring centrally in this volume). In this, scholarship merely borrows the method and plenitude of the world itself, positioning itself *within* rather than above or outside the messiness of worldly flows and churns. This virtue is I believe essential to critique, even forms of critique that are themselves sceptical or dismissive of hopeful claims. Thus for example immanent critique (of the sort specified under the negative dialectics of Adorno, say) proceeds on the basis not just of contradictions but also hopes that are *also* immanent in the social (we might now say socio-ecological) worlds around us. Feminist and other standpoint epistemologies rely for their force and sustenance on the same kind of situated solidarities and commitments. In the absence of this grounding, critique runs the risk of cynicism, despair and its own particular brand of withdrawal or takeoff, becoming unmoored from efficacy or relationality and providing few reasons or resources for entrance. This establishes a style and standard of care sometimes neglected within contemporary research and teaching communities. From this standpoint, like the Suvavou people studied by Miyazaki, it may be the patient nurturing and mutual transmission of hope, rather than the always disappointed search for revolutionary transformation or a historical agent, that forms the central task and challenge of critical scholarship today.

Conclusion

But what does this have to do with the projects of repair, remediation and resurgence that form the shared leitmotif of this volume? I believe all three are central to our ever-unfolding programmes of endurance and flourishing in the world today – what Anna Tsing and collaborators have evocatively called the 'arts of living on a damaged planet' (Tsing et al, 2017). I believe they are also antithetical to the dream of transcendence and escape with which this piece opened – and that such dreams are, as the cliché puts it, part of the problem, not the solution. A new class of settlers practicing escape pod salvationism is not what is needed now (has it ever been?). What is needed are more creative and committed forms of wake work, more and better ways of staying with the trouble, and futurisms that are oriented to staying rather than going. If any of this is to be found, it will be in a richer and more inclusive horizontalism that draws strength from the *aroundness* rather than the *aboveness* of hope. And the multiple forms of care and repair that a worldly hope ultimately powers and subtends.

Which brings us back to *The Human Condition*, and Arendt's urgent plea to 'think what we are doing'. As argued by Ella Myers (2013), Arendt's one-time working title for the book, *Amor Mundi*, or 'Care for the World' reveals a strain or emphasis not widely pursued by subsequent generations of Arendt interpreters, but which offers an intriguing (if speculative) bridge between the political and earthly aspects of Arendt's thought. In Myer's treatment, 'care for the world' pluralizes and concretizes both the 'subject' and 'object' of (democratic) worldly action. Care is something we do together, as concrete and specific 'we's', but in combinations that are larger than an 'I', and also vis-à-vis worlds which are themselves *plural* and *specific*, and thus endowed with a specific ethical standing that exceeds a purely humanist frame (without giving up on Arendt's specific interest in and attention to the human). This makes the kinds of actions associated with worldly care both expansive and political in nature (including, as Dipesh Chakrabarty (2021) has recently argued, in their planetary and intergenerational effects). While this perspective shares much in common with subsequent feminist work, the specifically Arendtian flavour called out by Myers emphasizes the *commonality* of the world, both in the sense of being a 'shared home for all' and a go-between (or 'mediating power') between humans seeking to build and sustain this common world, which provides foundation enough for the more-than-human (but not truly 'post-human') ethics that Myers is interested to build.

As the work in this volume makes clear, inventing new (and old) forms of justice, solidarity and care for the world is essential if our communities, and our planet, are to remain habitable and hospitable places. This in turn will rest on new (and old) forms of hope that must also be cultivated and extended, including in the face of the most dire predictions of planetary peril and change, which otherwise seem as likely to inspire cynicism and despair as engaged and committed forms of action. This hope will be *thick* rather than *thin*: connected to the living of lives and to patient entanglements with the things, people and places around us. If thin hopes are external, reliant on gifts or grants from elsewhere and the actions of distant or disconnected others, thick hope keeps agency at home – though with no easy or predictive confidence of outcome. If thin hopes need heroes or saviours, thick hope needs neither, apart from the patient and ordinary forms of heroism entailed in navigating with care a broken and dynamic world. If thin hopes continue their search for a saving power, thick hope is content with what it can build from the fragments and pieces around it. No need for fantasies of escape or transcendence, when there's plenty of work here at home. No need for sputniks launching into the sky, when there are wakes and wake work all around. No need for the vertical, when the horizontal is in such rich and plentiful supply. No need for The Future, when futures are what we build and live each day.

Hope of this ordinary character is needed now more than ever. The version puzzled towards here would give up on hope's long anchoring in the sky, and the long-deferred dream of transcendence associated with escape pod salvationism. It would also multiply the cast of characters with whom and with which we can learn from and build, making our imagination and practice of horizontality a good deal more earth-y, and perhaps more humanly just as well. This perspective has much to offer to the projects of ecological reparation described in this volume. It also has much to offer to worlds of technology, where thin hopes and rockets abound.

To give up the sky, I believe, is not to give up hope. Nor is it to give up on action, or the crucial anchoring to action that hope provides. It is instead to insist on the reality and simple *efficacy* of hope – even, perhaps especially, after the rockets have left. This hope is humble, patient, even boring. By the standards of religion or critical theory's long frustrated search for transcendence, it seems hardly worth noticing. But it has the advantage of being anchored in the ordinary ways we navigate, survive and occasionally flourish in the world, including with the people and things around us. It is also widely distributed, and integral to the forms and relations of care we can choose to exercise towards each other. It supports and draws energy from a politics of material participation (Marres, 2012), rather than one of renunciation, subordination or escape. It engages with broken worlds (and all our worlds are broken) from the standpoint of their strengths, their resilience and their potentialities, and not from the standpoint of their lacks, absences or deficits. It is essential to and fuelled by caretaking relations, and suitably nurtured and supported, can lead towards modes of engagement that are modest, creative and as multiple as the world itself – including in the face of problems and changes whose immensity may otherwise overwhelm our capacity to think and act.

As this last point and the chapters in this volume make clear, it is also a hope expressed and enacted through work, including in and with what some used to erroneously separate out as the 'natural world' around us: a feet-in-the-mud, dirt-under-fingernails hope, and not one expressed in the plaintive or beseeching gaze towards heaven. A hope that looks *down* and *around*, rather than *up*, to imagine the prospective multiplicity and plenitude of the worlds around us: if only we could think and care to build them.

If there were a credo to this way of thinking – if I could sneak onto the launchpad and scrawl something on the skin of the rocket – it would be this:

> *The world is always breaking, carrying much of what we care about into loss and oblivion. There is nothing and no one to save us. We are always and everywhere alone,* but for the profuse and teeming worlds around us.
>
> *Now let us get to work.*

Note

[1] Note though that the 'we', here and elsewhere, hides a lot: some experiences of precariousness go back centuries; for some, the apocalypse has already happened; both the consequences and culpability for these changes are distributed unevenly; and some, per Simmel's (1972) 'tragedy of whomever is lowest', are more on the receiving end of environmental loss and planetary change than others. The 'we' here is also meant to be modest and conversational, and not a massified we bundled or claimed for authority and power. As Tyson Yunkaporta (2020) has recently observed, English lacks a good word for this modest and invitational we. The version here should be read like his, as 'us-two': we who may be gathered and thinking together right now.

References

Abramson, A. (2004) 'Distilling Hope (Miyazaki's The Method of Hope)', *Current Anthropology*, 49(3): 530–532.

Althusser, L. (1971) *Lenin and Other Essays*, New York: Monthly Review Press.

Arendt, H. (1958) *The Human Condition*, Chicago: University of Chicago Press.

Atwood, M. (2002) *Oryx and Crake*, Toronto: McLelland and Stewart.

Atwood, M. (2009) *The Year of the Flood*, Toronto: McLelland and Stewart.

Atwood, M. (2014) *Maddaddam*, Toronto: McLelland and Stewart.

Bacon, F. (2008 [1626]) 'The New Atlantis', in B. Vickers (ed) *Francis Bacon: The Major Works*, Oxford: Oxford University Press.

Benjamin, W. (1969) *Illuminations*. Trans. H. Zohn and ed. H. Arendt, New York: Schocken Books.

Benjamin, W. (1999) *The Arcades Project*. Trans. H. Eiland and K. McLaughlin, Cambridge: Harvard University Press.

Berlant, L. (2011) *Cruel Optimism*, Durham, NC: Duke University Press.

Bloch, E. (1995) *The Principle of Hope*. Trans. S. Plaice, P. Knight and N. Plaice, , Cambridge: MIT Press.

Butler, O. (1993) *Parable of the Sower*, New York: Four Walls Eight Windows.

Campanella, T. (2009 [1602]) *The City of the Sun*, Project Gutenberg. Available from: www.gutenberg.org/ebooks/2816

Chakrabarty, D. (2021) *The Climate of History in a Planetary Age*, Chicago: University of Chicago Press.

Descartes, R. (1985 [1649]) 'The Passions of the Soul', in *The Philosophical Writings of Descartes, Vol I*. Trans. J. Cottingham, R. Stoothoff, D. Murdoch, Cambridge: Cambridge University Press.

Dimaline, Cherie. (2019) *Empire of Wild*, Toronto: Random House Canada.

Goodman, N. (1978) *Ways of Worldmaking*, Indianapolis: Hackett Publishing Company.

Goodman, N. (1984) *Of Mind and Other Matters*, Cambridge: Harvard University Press.

Haraway, D.J. (2016) *Staying with the Trouble: Making Kin in the Chthulucene*, Durham, NC: Duke University Press.

Hobbes, T. (1998 [1651]) *Leviathan*, J.C.A. Gaskin (ed) Oxford: Oxford University Press.

Horkheimer, M. and Adorno, T.W. (1969 [org 1947]), *Dialectic of Enlightenment*, New York: Continuum.

Hume, D. (2007 [1738]) *A Treatise on Human Nature*, D.F. Norton and M.J. Morton (eds) Oxford: Oxford University Press.

Jackson, S.J. (2014) 'Rethinking Repair', in T. Gillespie, P. Boczkowski and K. Foote (eds) *Media Technologies: Essays on Communication, Materiality and Society*, Cambridge: MIT Press, pp 221–39.

Jameson, F. (1972) *Marxism and Form: Twentieth-Century Dialectical Theories of Literature*, Princeton: Princeton University Press.

Jasanoff, S. and Kim, S.H. (2015) *Dreamscapes of Modernity: Sociotechnical Imaginaries and the Fabrication of Power*, Chicago: University of Chicago Press.

King, M.L., Jr. (1981) 'Pilgrimage to Nonviolence', in *Strength to Love*, Minneapolis, MN: Fortress Press.

King, T. (2012) *The Inconvenient Indian: A Curious Account of Native People in North America*. Toronto: Doubleday Canada.

Koopman, C. (2009) *Pragmatism as Transition: Historicity and Hope in James, Dewey, and Rorty*, New York: Columbia University Press.

Lempert, W. (2018) 'Generative Hope in the Postapocalyptic Present', *Cultural Anthropology*, 33(2): 202–12.

Lerner, A.J. (2015) *Redemptive Hope: From the Age of Enlightenment to the Age of Obama*, New York: Fordham University Press.

Lukacs, G. (1972 [1920]) *History and Class Consciousness*, Cambridge: MIT Press.

Marres, N. (2012) *Material Participation: Technology, the Environment, and Everyday Publics*, London: Palgrave Macmillan.

Miyazaki, H. (2004) *The Method of Hope: Anthropology, Philosophy, and Fijian Knowledge*, Stanford: Stanford University Press.

More, T. (1967 [1516]) 'Utopia'. Trans. J.P. Dolan, in J.J. Greene and J.P. Dolan (eds) *The Essential Thomas More*, New York: New American Library.

Moten, F. (2003) *In the Break: The Aesthetics of the Black Radical Tradition*, Minneapolis, MN: University of Minnesota Press.

Munoz, J.E. (2009) *Cruising Utopia: The Then and There of Queer Futurity*, New York: New York University Press.

Murphy, M. (2017) 'Alterlife and decolonial chemical relations', *Cultural Anthropology*, 32(4): 494–503.

Myers, E. (2013) *Worldly Ethics: Democratic Politics and Care for the World*, Durham, NC: Duke University Press.

Orange, T. (2018) *There There*, New York: Alfred A. Knopf.

Puig de la Bellacasa, M. (2017) *Matters of Care: Speculative Ethics in More Than Human Worlds*, Minneapolis, MN: University of Minnesota Press.

Ricoeur, P. (1970) *Freud and Philosophy: An Essay on Interpretation*, translated by D. Savage, New Haven, CT: Yale University Press.

Rorty, R. (1982) *Consequences of Pragmatism: Essays, 1972–1980*, Minneapolis, MN: University of Minnesota Press.

Sedgewick, E.K. (2002) 'Paranoid reading and reparative reading; or you're so paranoid you probably think this introduction is about you', in E.K. Sedgewick, M.A. Barale, J. Goldberg and M. Moon (eds) *Touching Feeling*, Durham NC: Duke University Press.

Sharpe, C. (2016) *In the Wake: On Blackness and Being*, Durham, NC: Duke University Press.

Simmel, G. (1972) *On Individual and Social Forms*, ed. D. Levine, Chicago: University of Chicago Press.

TallBear, K. (2019) 'Caretaking relations, not American dreaming', *Kalfou*, 6(1): 24–41.

Taylor, C. (2003) *Modern Social Imaginaries*, Durham, NC: Duke University Press.

Tsing, A.L., Swanson, H.A., Gan, E., Bubandt, N., eds (2017) *Arts of Living on a Damaged Planet: Ghosts and Monsters of the Anthropocene*, Minneapolis, MN: University of Minnesota Press.

Tuck, E. (2009) 'Suspending Damage: A Letter to Communities', *Harvard Educational Review*, 79(3): 409–27.

West, C. (1997) *Restoring Hope: Conversations on the Future of Black America,* Boston: Beacon Press.

Westbrook, R.B. (2005) *Democratic Hope: Pragmatism and the Politics of Truth*, Ithaca: Cornell University Press.

Whyte, K. (2020) 'Too late for indigenous climate justice: Ecological and relational tipping points', *WIREs Clim Change*, 11: e603.

Yunkaporta, T. (2020) *Sand Talk: How Indigenous Thinking Can Save the World*, New York: HarperOne.

Index

References to figures appear in *italic* type.
References to endnotes show both the page number and the note number (220n8).